AF358812

Electromembrane Processes

Electromembrane Processes: Experiments and Modelling

Editors

Luigi Gurreri
Alessandro Tamburini
Giorgio Micale

MDPI • Basel • Beijing • Wuhan • Barcelona • Belgrade • Manchester • Tokyo • Cluj • Tianjin

Editors

Luigi Gurreri
Department of Engineering
University of Palermo
Palermo
Italy

Alessandro Tamburini
Department of Engineering
University of Palermo
Palermo
Italy

Giorgio Micale
Department of Engineering
University of Palermo
Palermo
Italy

Editorial Office
MDPI
St. Alban-Anlage 66
4052 Basel, Switzerland

This is a reprint of articles from the Special Issue published online in the open access journal *Membranes* (ISSN 2077-0375) (available at: www.mdpi.com/journal/membranes/special_issues/ electromembrane_processes).

For citation purposes, cite each article independently as indicated on the article page online and as indicated below:

LastName, A.A.; LastName, B.B.; LastName, C.C. Article Title. *Journal Name* **Year**, *Volume Number*, Page Range.

ISBN 978-3-0365-1530-4 (Hbk)
ISBN 978-3-0365-1529-8 (PDF)

Contents

About the Editors

Luigi Gurreri

Luigi Gurreri is a research fellow in chemical engineering at the University of Palermo. His research activities concern modelling and experimental characterization of membrane processes for various applications, including desalination and salinity gradient power, with focuses on fluid dynamics, transport phenomena, and process simulation. He has participated in many EU-funded projects in this area.

Alessandro Tamburini

Alessandro Tamburini is an assistant professor of chemical process control at the University of Palermo. He has carried out research activities focused on the experimental and numerical analysis of complex systems including multiphase stirred tanks and membrane-based units. He has participated in many EU-funded and national projects on water desalination and renewable energy technologies as well as salinity gradient power processes.

Giorgio Micale

Giorgio Micale is a full professor of chemical process and plant design at the University of Palermo. His core research topics are the study of conventional and renewable energy desalination processes, salinity gradient power processes, computational fluid dynamics, mixing and multiphase flows, and computer-aided process engineering. He leads his research group and is responsible for several EU projects.

Preface to "Electromembrane Processes: Experiments and Modelling"

The increasing demand for water and energy poses technological challenges to the implementation of efficient concepts for a sustainable development. In this perspective, electromembrane processes (EMPs) can play a crucial role in green chemistry schemes oriented towards circular economy approaches and renewable energy systems. EMPs are based on the use of ion-exchange membranes under the action of an electric field. Versatility, selectivity, high recovery, and chemical-free operations are their main strengths.

Experimental campaigns and modelling tools are prompting the improvement of consolidated processes and the development of novel concepts. Several application fields have been proposed (in chemical, food, pharmaceutical industries, and others) including desalination, water and wastewater treatment, recovery of valuable products, concentration and purification operations, chemical production, and energy production and storage.

This book is a collection of the scientific contributions in the Special Issue Processes: Experiments and Modelling from the journal *Membranes*. It is focused on recent advancements in EMPs and their applications based on the development of cutting-edge engineered systems via experiments and/or models.

Luigi Gurreri, Alessandro Tamburini, Giorgio Micale
Editors

membranes

Editorial

Electromembrane Processes: Experiments and Modelling

Luigi Gurreri , **Alessandro Tamburini** * and **Giorgio Micale**

Dipartimento di Ingegneria, Università degli Studi di Palermo, viale delle Scienze Ed. 6, 90128 Palermo, Italy; luigi.gurreri@unipa.it (L.G.); giorgiod.maria.micale@unipa.it (G.M.)
* Correspondence: alessandro.tamburini@unipa.it; Tel.: +39-09123863767

check for updates

Citation: Gurreri, L.; Tamburini, A.; Micale, G. Electromembrane Processes: Experiments and Modelling. *Membranes* **2021**, *11*, 149. https://doi.org/10.3390/membranes11020149

Received: 15 February 2021
Accepted: 19 February 2021
Published: 20 February 2021

Publisher's Note: MDPI stays neutral with regard to jurisdictional claims in published maps and institutional affiliations.

This Special Issue of *Membranes* journal focuses on electromembrane processes and is motivated by the increasing interest of the scientific community towards their characterization by experiments and modelling for several applications.

The increasing demand of water and energy poses technological challenges for the implementation of efficient and sustainable concepts. In this perspective, electromembrane processes can play a crucial role in green chemistry schemes oriented towards circular economy approaches and renewable energy systems. Electromembrane processes are based on the use of ion-exchange membranes under the action of an electric field. Versatility, selectivity, high recovery, and chemical-free operations are their main advantages. Experimental campaigns and modelling tools are prompting the improvement of consolidated processes and the development of novel concepts. Moreover, hybrid and integrated systems offer synergistic benefits for efficiency enhancement. Several application fields have been proposed (in chemical, food, pharmaceutical industries and others) including desalination, water and wastewater treatment, recovery of valuable products, concentration and purification operations, chemical production, energy conversion and storage. Significant advancements have been achieved in recent years, and, currently, research is very active in this field.

In this Special Issue, we have collected contributions on recent advancements of electromembrane processes and their applications, with a focus on the development of cutting-edge engineered systems by experiments and/or models. The Special Issue contains eight articles. One review and three research articles regard electrodialysis, two research articles pertain to reverse electrodialysis, one research article focuses on an integrated membrane process including electrochemical intercalation–deintercalation, and a perspective article regards acid-base flow battery.

Electrodialysis (ED) is a mature technology that has been studied for a variety of applications. Among them, the treatment of wastewater has, in the last few years, attracted very broad attention.

In this Special Issue, Gurreri et al. [1] present the first comprehensive review of studies on ED applications in wastewater treatment for environmental protection and resources recovery, outlining the current status and the future prospect. About 400 relevant higher-quality scientific papers are reviewed and discussed. ED treatments of effluents from various industrial processes, municipal wastewater or saltwater treatment plants, and animal farms, are considered. The review shows that ED and unconventional configurations of ED, i.e., bipolar membrane electrodialysis (BMED), selectrodialysis (SED), electrodialysis metathesis (EDM) and electrodeionization (EDI), have great potential in desalination and valorization strategies of wastewater for a broad range of applications to recover water and/or other valuable components. The main ones are metals, salts, acids and bases, nutrients, and organics. Energy recovery via reverse electrodialysis (RED) is another possibility. The large variety of uses of conventional ED and similar technologies is discussed by analyzing experimental results, process performance, strengths and drawbacks, and techno-economic competitiveness. Therefore, conclusions and outlook are provided, highlighting the main technical challenges, the current status of the process scale in the various applications, and the key points for future R&D. Recent advances and emerging applications are reviewed, along with examples among the few well-established implementations in real environments. The review shows that recent research efforts have led to the development of several enhanced or novel systems, with the possible implementation of techno-economically affordable and competitive (near) zero liquid discharge approaches. Few real plants have been installed due to techno-economic challenges that are still present. Overall, research is opening new routes for the large-scale use of ED techniques in a plethora of treatment processes.

Sheng et al. [2] prepare ZSM-5 zeolite/PVA (polyvinyl alcohol) mixed matrix cation-exchange membranes with high monovalent permselectivity and test them for recovery of either acid or Li^+. Zeolites are crystalline aluminosilicates with pores and cavities of molecular dimensions and have excellent ion-exchange capacity. Commercial low-price ZSM-5 zeolite with 0.5–0.6 nm pore size is mixed at various fractions with PVA for the casting procedure. The prepared membranes are tested for H^+/Zn^{2+} and Li^+/Mg^{2+} selective separation for possible application of ED for acid recovery and saltwater desalination with scaling prevention, respectively. The optimal content of ZSM-5 zeolite of 50 wt% leads to higher permselectivity values compared to other commercial and non-commercial membranes, i.e., $P_{Zn^{2+}}^{H^+} = 34.4$ and $P_{Mg^{2+}}^{Li^+} = 3.7$. This result is achieved thanks to the transport facilitated for protons and small ions by the ZSM-5 sites and to the transport hindered for larger hydrated cations by the microporous crystalline morphology. The SEM analysis shows a uniform dispersion and the XRD patterns exhibits identical characteristic peaks as the pristine ZSM-5 powder. As the amount of ZSM-5 is increased, water uptake and area swelling decrease. Finally, the prepared membranes exhibit high limiting current density and low electrical resistance, thus being very attractive for practical applications.

Uzdenova and Urtenov [3] perform 2-D Direct Numerical Simulations to compare current–voltage curves and parameters of the electroconvective vortex layer in the potentiodynamic and galvanodynamic regimes for an ED dilute channel with forced flow in contact with homogeneous membranes [3]. In this case, electroconvection arises due to uneven distribution of concentration along the channel, which generates a tangential component of electric field. The Nernst–Planck–Poisson and Navier–Stokes equations are solved. The computational domain consists of half channel in contact with the CEM surface. The channel thickness, the channel length, the inlet concentration, and the mean velocity are of 500 μm, 1 mm, 0.1 mol m^{-3} and 3.8 mm s^{-1}, respectively. The model is implemented by the finite element COMSOL Multiphysics$^{®}$ commercial platform. Realistic *i–V* curves are predicted, exhibiting four current regimes, i.e., underlimiting, limiting, overlimiting, and chaotic overlimiting. The potentiodynamic and galvanodynamic modes have similar behavior, apart from the oscillations of current and voltage, respectively, in the chaotic overlimiting region. The limiting–overlimiting and overlimiting–chaotic overlimiting transitions are characterized by hysteresis. Vortices rotating in the same direction are developed in the overlimiting region. Instead, large complexes of several counter-rotating vortices are formed in the chaotic overlimiting regime. The vortex layer has length and thickness increasing as the applied current or voltage increase, and the expansion rate is greater in the range close to the limiting current. This is the range in which the vortex density per unit length decreases up to a minimum, but it then increases with the development of large vortex complexes.

The ion transport at overlimiting currents is also studied via mathematical modelling by Urtenov et al. [4] in order to investigate the formation and properties of the local maximum (or minimum) space charge. Two systems are considered: the depleted solution at a CEM, and a desalination channel between a CEM and an AEM. A 1-D dynamic modelling based on the Nernst–Planck–Poisson equations is performed. At overlimiting currents, the counter-ion concentration decreases in the boundary layer and in the extended space charge region up to a minimum value before the electric double layer at the membrane interface. At this point, the space charge has a non-zero local minimum, whose value depends on the applied voltage. Moreover, the space charge has a local maximum at a certain point between the electroneutral solution and the point of the local minimum. In the simulations, as the imposed cation concentration at the CEM increases, the local maximum of space charge moves towards the membrane until it disappears together with the local minimum. This result shows that the local values of charge density exist due to the limited ion-exchange capacity of the membranes. In a transient system with increasing applied voltage, the local maximum moves similarly to a single soliton-like wave into the solution but changes slowly its size and shape. In the desalination channel, a different behavior is observed. In the case of a KCl solution, two waves of opposite charge move towards each other and interact up to breakdown. As the voltage increases, new waves are not formed, since the solution concentration is practically zero. With an NaCl solution, the difference was that the negative peak is generated later or even does not arise, depending on the applied linear growth rate of voltage.

Reverse electrodialysis (RED), which is the opposite process with respect to ED, converts the mixing free energy of two streams with different salt concentration into electrical energy. RED systems have been widely studied in the last decade. However, its development is still limited to prototypal installations, due to low power density and low membrane performance/durability with natural solutions, as well as high membrane cost. Therefore, several studies are currently performed to improve the process performance regarding different aspects.

Merino-Garcia et al. [5] perform a monolayer surface modification to functionalize heterogeneous Ralex Anion-Exchange Membranes with poly(acrylic) acid (PAA) yielding an increased monovalent permselectivity. After a contact of 24 h with the PAA aqueous so-lution, the modified AEMs are characterized by measuring contact angle, water uptake, ion exchange capacity, fixed charge density, and swelling degree. The electrochemical charac-terization is performed by cyclic voltammetry and electrochemical impedance spectroscopy. Permselectivity and fouling behavior are evaluated via mass transport experiments. The modified membranes show a significantly enhanced monovalent permselectivity and hy-drophilicity, by increasing sulfate rejection by 36–54% and decreasing the contact angle by 15–31%. Higher current responses are also observed, while the electrical resistance exhibits only a small increase. However, in the presence of humic acid, used as model organic foulant, the overall ion transport and the monovalent selectivity of the modified membranes are reduced. Therefore, the technical feasibility of the proposed modification, which may improve the RED process efficiency, is demonstrated, but fouling mechanisms and appropriate pre-treatments and cleaning strategies should be studied.

Jalili et al. [6] present a Computational Fluid Dynamics (CFD) study on reverse elec-trodialysis. Single channels, 12.6 mm long with staggered hemi-circular or triangular non-conductive spacers, are simulated in two dimensions. Numerical simulations are based on the Navier–Stokes and Nernst–Planck governing equations under electroneutral-ity assumption, computing the pressure, velocity, concentrations, and electric potential fields. The open circuit voltage, the cell pair resistance and the peak power density are calculated. By including the pumping power losses, the net power density is computed. The freeware open source OpenFOAM platform (finite-volume method) is used. A factorial design parametrical study is performed with a total of 2×2^4 test cases, given by two values for each of the four quantities, i.e., velocity, temperature, corrugation density and height for each of the two corrugation shapes. Results show that the temperature is the most influential parameter, leading to an increase of 43% in the net peak power density when passing from 25 to 55 °C. In descending order, lower effects are provided by inlet velocity, corrugation density, and corrugation height. The corrugation shape does not affect significantly the producible maximum peak power density. The efficiency of using low-grade waste heat to increase the feed temperature should be evaluated in future works.

The integration of electromembrane processes with pressure driven membrane pro-cesses can offer several solutions in water treatment for different aims.

Xu et al. [7] propose an integrated treatment process of brine with a high Mg^{2+}/Li^+ mass ratio for preparing Li_2CO_3, which is an important raw material. The process included electrochemical intercalation–deintercalation, nanofiltration, reverse osmosis, evaporation, and precipitation. The electrochemical intercalation–deintercalation (EID) method is used to maximize the separation between magnesium and lithium. EID is performed by a two-chamber electrolytic cell separated by an AEM, with $LiFePO_4$ anode and $FePO_4$ cathode (17×20 cm^2). The anode chamber is filled with a 5 g L^{-1} NaCl solution as supporting electrolyte, the cathode chamber with brine (2.05 g L^{-1} Li^+, 120.56 g L^{-1} Mg^{2+}, 360.9 g L^{-1} Cl^- etc.). Lithium of the $LiFePO_4$ anode intercalates to the supporting electrolyte, and lithium of the brine intercalates into the $FePO_4$ cathode, while chloride migrates through the membrane from the catholyte to the anolyte. The EID method can thus extract lithium by subsequent cycles. The EID experiments show excellent separation performance, by reducing the Mg^{2+}/Li^+ mass ratio from 58.5 (brine) to 0.93 in the anolyte with 90.6% recovery of lithium after the second cycle, and low concentration of impurities. Nanofiltration and reverse osmosis are used for purification (removal of residual magnesium and other multivalent ions) and concentration (lithium enrichment) of the EID anolyte. After evaporation (electric furnace) and further removal of magnesium (precipitation with NaOH), industrial-grade Li_2CO_3 is prepared via chemical precipitation. The direct recovery of lithium from the brine to produce Li_2CO_3 is ~70%, but a higher recovery would be feasible in a system with recirculation of the solutions.

Systems based on reversible electromembrane processes may be used for energy storage, which is crucial for a greater penetration of renewable energies. In principle, a single unit may be employed and operated as ED process (charging mode) to separate a saline solution into a concentrate solution (when an energy surplus is available) and operated as RED (discharge mode) to recover energy from the controlled mixing of the two solutions (when energy demand is high). Unfortunately, although the significantly low environmental impact, the resulting battery named concentration gradient flow battery has a low energy density. In this respect, the coupling of bipolar membrane ED and RED may offer a sustainable solution due to acceptable values of energy/power density (about 10 kWh m^{-3}) achievable. This technology, named acid-base flow battery (AB-FB), is based on the reversible water dissociation via bipolar membranes, and stores electricity in the form of the chemical energy of a pH gradient. It has received little attention so far; however, it has an interesting potential.

Pärnamäe et al. [8] present a perspective article with the state-of-the-art and latest developments of the AB-FB. The technology has already been demonstrated at the laboratory scale with maximum energy density and power density in discharge of ~10 kWh m^{-3} and ~17 W m^{-2} membrane, respectively. The discharge power is still limited by delamination issues that may damage irreversibly the bipolar membrane at high discharge currents. Multi-stage scenarios are simulated by a process model, highlighting that, to reach AB-FB applications on the kW–MW scale, further studies should focus on (i) the optimization of the plant design and (ii) the development of improved membranes. These advancements are crucial to reduce the membrane area and the solutions' volume and to increase the round-trip efficiency. Experimental testing of the first 1 kW/7 kWh pilot plant is currently ongoing. A techno-economic analysis is performed for the pilot plant and for a first-of-a-kind commercial plant of 100 kW/700 kWh. With an increase in power density up to 30–40 W m^{-2} and with a projected cost of membranes of 100 € m^{-2} per triplet, the commercial unit would attain a unitary cost of 470 € kWh^{-1}. The development of improved and cheaper components (especially bipolar membranes) and of optimized designs can lead the AB-FB technology to play a role in future systems of energy storage.

In conclusion, the papers in this Special Issue illustrate how experimental activities and modelling approaches can be used at several levels providing a significant contribution to the progress of electromembrane processes through the development of high-performance membranes and of improved engineered systems.

Funding: This research received no external funding.

Acknowledgments: The guest editors are grateful to all the authors that contributed to this Special Issue.

Conflicts of Interest: The editors declare no conflict of interest.

References

1. Gurreri, L.; Tamburini, A.; Cipollina, A.; Micale, G. Electrodialysis Applications in Wastewater Treatment for Environmental Protection and Resources Recovery: A Systematic Review on Progress and Perspectives. *Membranes* **2020**, *10*, 146. [CrossRef] [PubMed]
2. Sheng, F.; Afsar, N.U.; Zhu, Y.; Ge, L.; Xu, T. PVA-Based Mixed Matrix Membranes Comprising ZSM-5 for Cations Separation. *Membranes* **2020**, *10*, 114. [CrossRef] [PubMed]
3. Uzdenova, A.; Urtenov, M. Potentiodynamic and Galvanodynamic Regimes of Mass Transfer in Flow-Through Electrodialysis Membrane Systems: Numerical Simulation of Electroconvection and Current-Voltage Curve. *Membranes* **2020**, *10*, 49. [CrossRef] [PubMed]
4. Urtenov, M.; Chubyr, N.; Gudza, V. Reasons for the Formation and Properties of Soliton-Like Charge Waves in Membrane Systems When Using Overlimiting Current Modes. *Membranes* **2020**, *10*, 189. [CrossRef] [PubMed]
5. Merino-Garcia, I.; Kotoka, F.; Portugal, C.A.M.; Crespo, J.G.; Velizarov, S. Characterization of Poly(Acrylic) Acid-Modified Heterogenous Anion Exchange Membranes with Improved Monovalent Permselectivity for RED. *Membranes* **2020**, *10*, 134. [CrossRef] [PubMed]
6. Jalili, Z.; Burheim, O.S.; Einarsrud, K.E. Computational Fluid Dynamics Modeling of the Resistivity and Power Density in Reverse Electrodialysis: A Parametric Study. *Membranes* **2020**, *10*, 209. [CrossRef] [PubMed]
7. Xu, W.; Liu, D.; He, L.; Zhao, Z. A Comprehensive Membrane Process for Preparing Lithium Carbonate from High Mg/Li Brine. *Membranes* **2020**, *10*, 371. [CrossRef] [PubMed]
8. Pärnamäe, R.; Gurreri, L.; Post, J.; van Egmond, W.J.; Culcasi, A.; Saakes, M.; Cen, J.; Goosen, E.; Tamburini, A.; Vermaas, D.A.; et al. The Acid–Base Flow Battery: Sustainable Energy Storage via Reversible Water Dissociation with Bipolar Membranes. *Membranes* **2020**, *10*, 409. [CrossRef]

 *membranes*

Review

Electrodialysis Applications in Wastewater Treatment for Environmental Protection and Resources Recovery: A Systematic Review on Progress and Perspectives

Luigi Gurreri , **Alessandro Tamburini** * , **Andrea Cipollina** and **Giorgio Micale**

Dipartimento di Ingegneria, Università degli Studi di Palermo, viale delle Scienze Ed. 6, 90128 Palermo, Italy; luigi.gurreri@unipa.it (L.G.); andrea.cipollina@unipa.it (A.C.); giorgiod.maria.micale@unipa.it (G.M.)
* Correspondence: alessandro.tamburini@unipa.it; Tel.: +39-091-2386-3780

Received: 31 May 2020; Accepted: 4 July 2020; Published: 9 July 2020

Abstract: This paper presents a comprehensive review of studies on electrodialysis (ED) applications in wastewater treatment, outlining the current status and the future prospect. ED is a membrane process of separation under the action of an electric field, where ions are selectively transported across ion-exchange membranes. ED of both conventional or unconventional fashion has been tested to treat several waste or spent aqueous solutions, including effluents from various industrial processes, municipal wastewater or salt water treatment plants, and animal farms. Properties such as selectivity, high separation efficiency, and chemical-free treatment make ED methods adequate for desalination and other treatments with significant environmental benefits. ED technologies can be used in operations of concentration, dilution, desalination, regeneration, and valorisation to reclaim wastewater and recover water and/or other products, e.g., heavy metal ions, salts, acids/bases, nutrients, and organics, or electrical energy. Intense research activity has been directed towards developing enhanced or novel systems, showing that zero or minimal liquid discharge approaches can be techno-economically affordable and competitive. Despite few real plants having been installed, recent developments are opening new routes for the large-scale use of ED techniques in a plethora of treatment processes for wastewater.

Keywords: electro-membrane process; electrodialysis reversal; bipolar membrane electrodialysis; selectrodialysis; electrodialysis metathesis; electrodeionisation; reverse electrodialysis; monovalent selective membranes; water reuse; brine valorisation

1. Introduction

The growing water demand in urban, rural and industrial sites poses serious ecological and economic concerns in water management linked to resources depletion and wastes disposal. Several industrial processes use large water volumes, thus producing high quantities of wastewater or spent streams with contaminants and valuable components. In addition, municipal wastewater treatment plants effluents are not directly reusable.

Water recovery offers the possibility of sustainable development. On the other hand, it requires the design and implementation of advanced treatment methods, which represent a techno-economic challenge. In this framework, the zero liquid discharge (ZLD) concept aims at developing strategies to close the material loop, thus minimizing the liquid waste [1–3]. This approach is a specific accomplishment of circular economy [4,5], which proposes "business models based on reducing, alternatively reusing, recycling and recovering materials in production/distribution and consumption processes", by replacing the old perspective of *end of life* [6].

Membrane processes are attracting a great deal of interest, and several studies have led to significant advances [7]. Among them, electro-membrane technologies separate ions by the selective transport through ion-exchange membranes (IEMs) under the influence of an electric field. A large variety of electro-membrane processes has been developed [8–14].

In particular, electrodialysis (ED) produces two streams with different concentrations flowing in alternate compartments separated alternatively by cation-exchange membranes (CEMs) and anion-exchange membranes (AEMs). ED may be cost-effective thanks to properties favourable to the attainment of large selectivity and product recovery, and to the avoided, or limited, need for chemicals [8,9,15–17]. At industrial scale, ED is mainly applied to desalinate brackish water for drinking water production. There have been installed also some ED plants to produce table salt from seawater desalination. However, many studies have focused on application of ED techniques in the (bio)chemistry, food processing, and pharmaceutical industries [8–11,13,17–21], encompassing wastewater treatment, recovery of chemicals or other valuable products, and removal of toxic components [8,9,13,17,20,22,23].

In regard to wastewater treatment via ED, the research exhibits an exponential growth in the last 20 years (Figure 1). More than 75% of the 879 scientific documents published since 1969 up until now fall in this period.

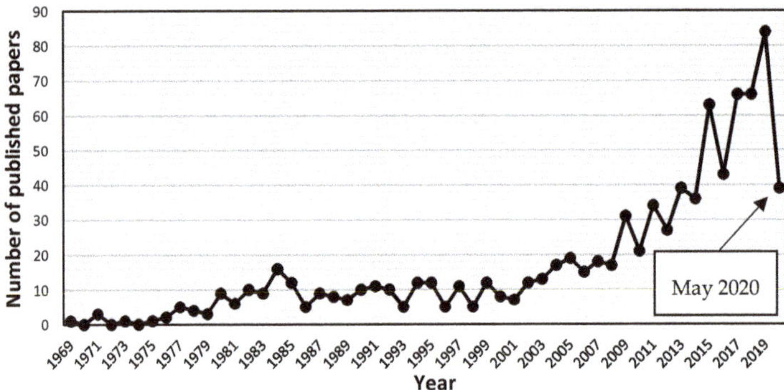

Figure 1. Scientific documents chronology reported on Scopus with "electrodialysis" and "wastewater" search words (article title, abstract or keywords). Source: www.scopus.com, accessed on 7 May 2020.

Many industrial effluents (e.g., from metal finishing, tanning, pulp and paper processing) have a complex composition with contaminants and/or valuable components, e.g., heavy metal ions, acids, organic matter, etc. Similarly, treated effluents from municipal or animal farming sources contain, for example, nutrients, as well as water. Finally, desalination plants' reject brines may provide water and/or salt. Thanks to its ability in separating charged particles, ED methods can effectively recover water and/or other products from these effluents, including electrical energy.

Despite the fact that ED has a long history over more than 120 years, and that intense research activity has been developed, especially in recent years, only a few review articles have been published so far. In 2010, Strathmann [15] described the principle of ED, its operating conditions, and its design features, focusing on water desalination and recent advancements (profiled membranes). ED-related processes were included, by highlighting advantages and limitations. In 2018, our research group published a review paper on ED updated with the most recent developments for water desalination [16]. The main topics were IEMs progress and characterisation, hydrodynamics and transport phenomena, process models, other modelling tools, and ED-related technologies. Recently, Sajjad et al. [17] provided an overview on (waste)water treatment via ED, by discussing the main technological limitations.

The lack of an organic review on ED for wastewater treatment motivated the present work. For the first time, to the best of the authors' knowledge, this paper presents a comprehensive and systematic review of studies on ED applications in wastewater treatment for environmental protection and recovery of resources, outlining the current status and the future prospect. The large variety of uses of conventional ED and similar technologies is discussed by analysing experimental results, process performance, strengths and drawbacks, and techno-economic competitiveness. Recent advances and emerging applications are reviewed, along with examples among the few well-established implementations in real environments.

2. Research Method, Rationale and Structure of the Review

The investigation of the review topic was based on a literature search in the Scopus electronic database. The search words listed in Table 1 were used without limits of date and by excluding only conference papers among the document types. This search found a total number of studies amounting approximatively to 1000, by excluding duplicate results. For the selection process, the eligibility criteria were relevance/consistency for/with the review topic, full text availability and accessibility, number of citations (except for most recent results), and journal metrics. The selected papers were organised in a bibliography on *Mendeley desktop*. The screening by full-text analysis filtered about 400 relevant higher-quality scientific papers to be reviewed and discussed (excluding those used for the fundamentals, Section 3).

Table 1. Search words used in the literature exploration on Scopus.

Search Word		Search Word
Electrodialysis		Wastewater
Bipolar membrane electrodialysis		Effluent
Selective electrodialysis		Spent solution
Selectrodialysis		Recovery
Electrodialysis metathesis	AND	Reclamation
Electrodeionisation		Reuse
Continuous electrodeionisation		Valorisation
Reverse electrodialysis		Regeneration
		Zero liquid discharge

The selected articles were classified into three main categories according to the wastewater origin, and into sub-categories based on the treatment aim, as shown in Figure 2. A third level of sub-classification was used in some cases in order to distinguish among different waste effluents. Sections and sub-sections of the paper will reflect exactly this classification. Therefore, the paper is structured as follows. After a brief description of the process fundamentals (Section 3), Sections 4–6, which are the core of the review, correspond to the three main categories of the classification, and their sub-sections to the sub-categories. Where not specified, the data reported in the review will refer to lab-scale experiments. Otherwise, pilot plants and installations in real environments will be explicitly indicated throughout the paper. Finally, Section 7 provides discussion, conclusions and outlook, highlighting the main technical challenges, the current status of the process scale in the various applications, and the key points for future R&D.

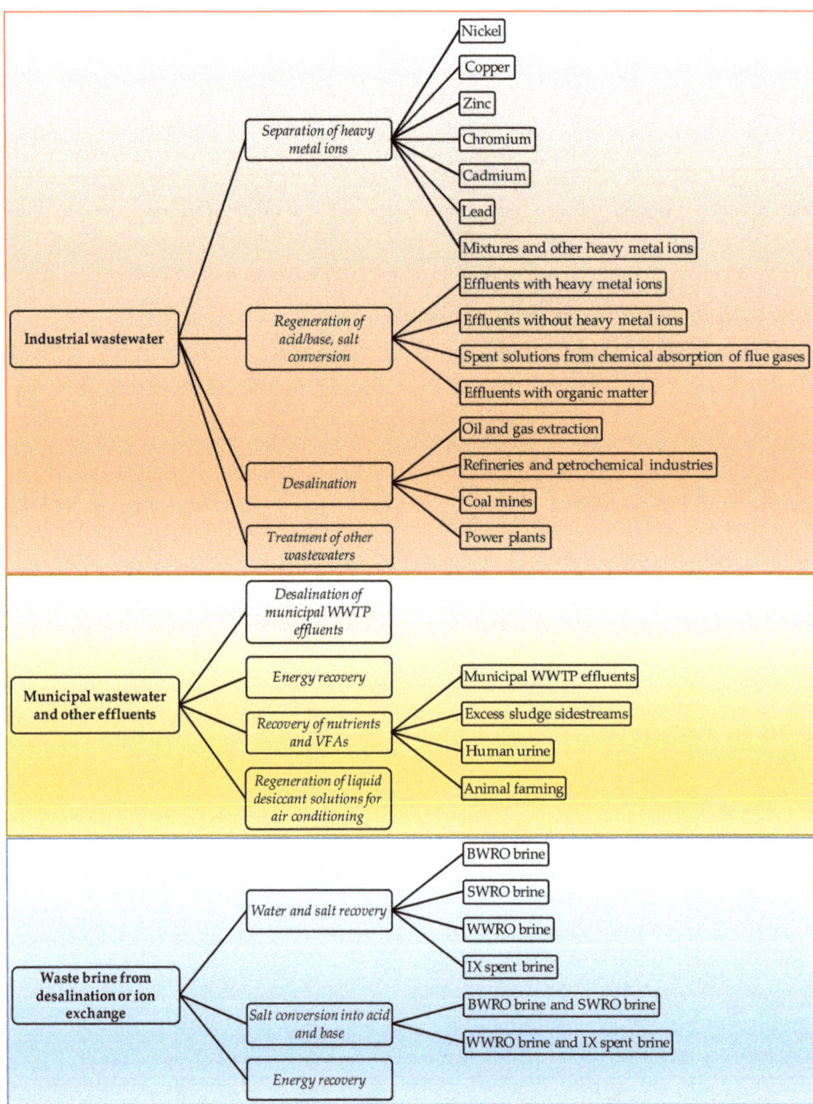

Figure 2. Classification of the selected papers, reflected in the review structure (Sections 4–6 and their sub-sections).

3. Electrodialysis Process Fundamentals

3.1. Working Principle and Design/Operating Features of ED Processes

Figure 3a depicts a scheme of conventional ED. A pile of alternating AEMs and CEMs is arranged with alternating diluate and concentrate channels. At the extreme sides, the ED stack is completed by electrode compartments. Here, the external power supply establishes an electric potential difference, which causes redox reactions. A direct electric current flows through the external circuit as electronic current, and through the stack as ionic current, with cations and anions migrating towards the cathode and the anode, respectively. The co-ion block by the IEMs leads to a selective transport with a resulting salt concentration reduction/increase in the diluate/concentrate channels, respectively. The repetitive unit in a conventional stack, namely the "cell pair", includes an AEM, a diluate, a CEM, and a concentrate. The two inlet feeds may differ each other.

The number of cell pairs in electrodialyzers ranges from few or some tens at bench- or pilot-scale up to hundreds in commercial stacks for real applications. The active area spans roughly from 0.01 to 1 m^2 for the single membrane.

The plate-and-frame configuration is by far the most used. Net spacers (with typical thickness of ~300 μm to ~2 mm [16]) equipped with gaskets are placed between the IEMs in order to create the feed compartments (Figure 3b). The two main designs of the spacer-filled channels are sheet flow and tortuous path [15,16]. In the former pattern, the fluid flows roughly straight along a rectangular channel (Figure 3b). In the latter pattern, the feed moves along a serpentine. U-shaped channels with halfway features are almost common in large units, similarly to tortuous path layouts, while sheet flow channels are more used for small stacks. Membranes with built-in profiles avoid the use of net spacers [15,16], but they have been used only for theoretical or experimental studies [24–28].

The typical range of fluid velocity is 1–10 cm/s. However, along tortuous path layouts, the velocity may be increased to ~50 cm/s to counteract the poorer mixing. In most cases, flow regimes are steady, but turbulence may occur at higher velocities [16].

Batch operations with solution recirculation are typical for lab-scale studies, while continuous processes are basically limited to industrial plants. The "feed and bleed" operation (partial continuous recirculation) is commonly practiced to control water recovery and outlet concentrations [8]. Multi-stage schemes can be devised with several configurations (e.g., multiple hydraulic and/or electrical steps) in order to attain the wanted product features.

IEMs suffer from fouling phenomena less than semi-permeable membranes (e.g., for reverse osmosis). However, depending on the solutions treated, IEMs may experience serious deterioration, resulting in a higher electrical resistance and even in a physical damage. Both suspended and dissolved solids (organic and inorganic) can cause membrane fouling. Organic anions and inorganic compounds can often imply fouling of AEMs and CEMs, respectively [29]. Fouling caused by sparingly soluble salt precipitates is called scaling. Electrodialysis reversal (EDR) is commonly practiced for fouling mitigation. It is applied by periodic switching (cycles of minutes/hours) of electrode polarity (with simultaneous switch of feed solutions). As a result, charged components are removed from the IEM surface by migration in the contrary direction. In addition, feed pre-treatment and stack cleaning-in-place methods (acidic and/or alkaline solutions) can prevent and remove fouling, respectively.

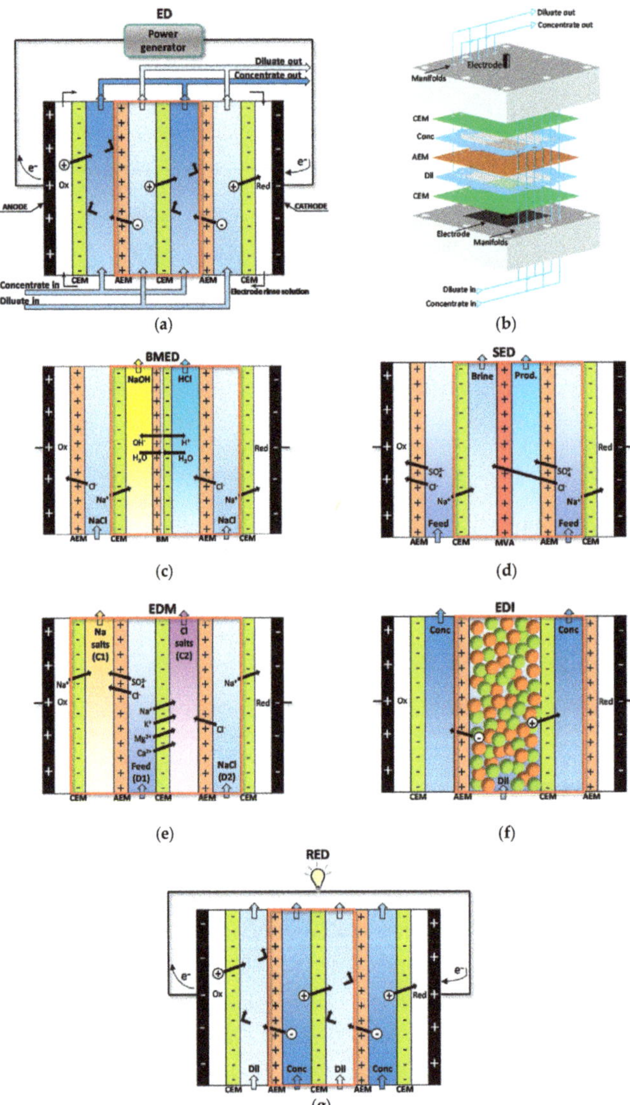

Figure 3. Schematics of ED techniques: (**a**) Conventional electrodialysis (ED); (**b**) Lab-scale ED stack with sheet-flow design (exploded view of one cell pair with an additional CEM); (**c**) Three-compartment bipolar membrane electrodialysis (BMED); (**d**) Selectrodialysis (SED) with MVA (the illustration refers to the fractionation of SO_4^{2-} from Cl^-); (**e**) Electrodialysis metathesis (EDM); (**f**) Electrodeionisation (EDI) with ion-exchange resins filling the diluate; (**g**) Reverse electrodialysis (RED). Red rectangles indicate the repeating units. Panels (**a,b,g**) are reproduced (adapted) with permission from [16], published by Elsevier, 2018.

Bipolar membrane electrodialysis (BMED) uses both monopolar membranes and bipolar membranes (BMs) to generate acid and base by water dissociation (Figure 3c). A BM consists of the overlapping of a cation-exchange layer (CEL) and an anion-exchange layer (AEL), whose inter-layer (thinner than 10 nm [8,30]) promotes water dissociation when a voltage (>0.83 V) is applied, thus releasing H^+ and OH^- [8,10,12,30] at a rate that is six (or more) orders of magnitude larger than in solution [8,12]. This is caused by the catalytic role of the functional groups and/or of the catalyst in the bipolar region, and to the strong electric field (second Wien effect) [31]. The mechanisms of ion transfer and water dissociation are still under study via theoretical approaches and numerical models [31]. Novel preparation techniques based on electrospinning methods can produce high-performing BMs [32,33].

Figure 3c depicts the three-compartment BMED arrangement, which converts salt into acid and base. The repeating cell consists of: AEM, acid compartment, BM, base compartment, CEM, and diluate salt compartment. Protons and hydroxyl ions, generated in the bipolar region by the electric field, cross the CEL and the AEL and migrate to the acid and base channel, respectively, while salt anions and cations (e.g., Cl^- and Na^+) in the salt channel cross the monopolar IEMs and migrate to the acid and base compartment, respectively. Nevertheless, other BMED arrangements have been developed with two-compartment repetitive units, with either BM-AEM or BM-CEM membranes for either acid or base production (and salt feed alkalisation or acidification). These configurations are used when it is possible or desired to obtain only one solution at high purity, with applications in regeneration processes [34–38]. Further BMED configurations include cell triplets with two monopolar IEMs of the same type. The outlet salt stream is sent to the acidic or alkaline channel to attain higher recovery rates [34,35].

Selective ED occurs within electrodialyzers containing monovalent selective membranes (MVMs), which may be anionic (MVAs) and/or cationic (MVCs) and segregate monovalent and multivalent ions. Specifically, the selectrodialysis (SED) process has a three-compartment configuration with an MVM and two conventional IEMs, and fractionates ions by using three different streams [39]. Figure 3d provides a sketch of SED fractionation of SO_4^{2-} from Cl^- contained in a feed solution. The results of the process are the mixture desalination, the product enrichment in divalent anions, and the brine concentration in monovalent ions.

ED stacks may be arranged to perform a metathesis of salts, known also as "double decomposition" (interchange of cations and anions between salts). With a couple of salts, one has:

$$MX + M'X' \rightarrow M'X + MX'. \tag{1}$$

Electrodialysis metathesis (EDM) [40,41] has a four-compartment repeating unit which includes two diluate and two concentrate channels, all being different from each other, divided by two AEMs and two CEMs (Figure 3e). The feed (D1) contains a salt or a salts mixture, while a substitution solution flows along the other diluate channel (D2). From the D1 solution, anions move to the C1 concentrate, and cations move to the C2 concentrate; while from the D2 substitution solution, anions are transferred to the C2 concentrate, and cations are transferred to the C1 concentrate. As a result, the metathesis of salts between feed and substitution solution occurs in the concentrate products. In the example of Figure 3e, Na^+ salts and Cl^- salts are generated inside compartments C1 and C2, respectively.

To boost the ED performance, ion-exchange resins (IXRs) can be inserted inside the channels. The hybridisation of ED and ion exchange (IX) is referred to as electrodeionisation (EDI) or continuous electrodeionisation (CEDI) [8,9,42,43] (Figure 3f). In EDI units, continuously regenerated IXRs beds within the diluate (sometimes also within the concentrate) cause a conductivity increment and a concentration polarisation reduction. Improved ion transport is obtained, thus making it possible to effectively treat very diluted solutions thanks to lower electrical resistances and higher limiting currents. The regeneration in situ of IXRs is carried out by H^+ and/or OH^- from water dissociation occurring at bipolar contacts of

IXRs particles or between IXRs and IEMs [42,43]. EDI is more suitable than its source technologies for producing industrial ultra-pure water [44] and for treating some kinds of wastewater, e.g., wastewater containing metal ions [42]. Moreover, fully regenerated IXRs can ionize and remove weakly ionized species (SiO_2, CO_2, boron, and NH_3) [9,43]. The complex transport mechanism has been argued for several years through various models [42]. It involves the following steps [8,42,43]: ion diffusion through the solution (controlling step), IX, migration across the IXR bed and the IEM, and regeneration of the IXR.

EDI units are arranged with several configurations [8,9,42] by changing the IXRs bed composition and structure (mixed, separate or layered) and the IEMs number, placement and type, also including the employment of BMs as locations for water dissociation. EDI modules may be assembled with several repeating units between the electrodes, similarly to ED stacks. However, other arrangements with few compartments (three or some more) in total, including the electrode ones, were developed. Among them, some units exploit electrode water electrolysis to deliver H^+ ad OH^- for the IXRs electro-regeneration [45–50], thus differing considerably from conventional ED. Limitations in EDI performance may derive from small current efficiencies in operations with high water dissociation [9], and from inhomogeneous flow distributions [42]. The former issue can be solved by identifying optimal values of the applied voltage, thus resulting in the co-existence of water dissociation and electroconvection in the overlimiting regime, which can enhance the process efficiency [51]. The latter issue can be addressed by adopting fixed resin wafers [52].

Reverse electrodialysis (RED) is the opposite process with respect to ED. RED produces electricity by converting the mixing free energy of two streams with different salt concentration (salinity gradient energy, or blue energy or osmotic energy), and is carried out with stacks equivalent to ED units [53–57] (Figure 3g). A high-salinity solution (concentrate, which is actually diluted along the channel) and a low-salinity solution (diluate, which is actually concentrated along the channel) flow through the two compartments of an RED cell pair. The most conventional solutions are seawater and river water, which would provide a maximum theoretical energy density of ~880 kJ/m^3 (equal amounts of both solutions). However, recent studies have assessed the use of waste effluents.

The working principle of RED relies on the electrochemical equilibrium of the co-ion exclusion theorized by Donnan (see Section 3.2), which generates an electrical potential over IEMs immersed between two solutions at different concentration (i.e., different chemical potential). The sum of all membrane potentials of a stack is its electromotive force. It can be measured as the electric potential difference under open circuit conditions (open circuit voltage). When the circuit is closed with an external load, redox reactions at the electrode compartments convert the internal ions flux into the external electrons current. This implies that the voltage over the stack, which corresponds to the voltage over the external load, is reduced when the circuit is closed. Please note that in RED (generator), the cathode and anode are positive and negative, respectively, i.e., with the opposite charge with respect to ED (user). In addition to provide electricity, RED units may produce H_2 via cathode reduction.

The power output depends on electromotive force and stack resistance. Therefore, a trade-off between them, maximizing the power supplied, is due to the effects of the diluate concentration. The RED performance may be significantly affected by the presence of divalent ions, which increase the membrane resistance and reduce the membrane permselectivity [58].

RED stacks are often operated with single pass (once-through), even in lab-scale experiments.

3.2. Ion Exchange Membranes and Mass Transfer

IEMs are dense membranes made by polymeric material with fixed charged groups and movable ions of opposite charge (counter-ions) [12]. IEMs allow counter-ions to pass, while blocking co-ions, which have the same sign of fixed charges. Cation-exchange membranes (CEMs) contain negative fixed groups such as SO_3^-, COO^-, PO_3^{2-}, PO_3H^-, and $C_6H_4O^-$; anion-exchange membranes (AEMs) contain positive fixed groups such as NH_3^+, NRH_2^+, NR_2H^+, NR_3^+, PR_3^+, and SR_2^+ [10]. As mentioned above, bipolar membranes (BMs) consist of an anion-exchange layer (AEL) overlapped to a cation-exchange layer (CEL). Monovalent selective membranes (MVMs) allow monovalent counter-ions to pass, whereas retain multivalent counter-ions [19,28–30]. IEM categories are distinguished on the basis of materials, functional groups, and microstructure [9]. For details on IEM features and methods of preparation, see [8–12,59,60].

The theorisation of co-ion exclusion by IEMs was introduced by Donnan equilibrium [8,61], which implies an electric double layer (EDL) at the membrane–solution boundary [8,9,59,61]. A membrane between two salt solutions with different concentration generates a voltage difference (Teorell-Meyer-Sievers) [9,59]:

$$\Delta\varphi^{IEM} = \frac{RT}{F} \int \sum \frac{t_i^{IEM}}{z_i} d\ln a_i, \tag{2}$$

where $\Delta\varphi^{IEM}$ is the "membrane potential", R is the universal gas constant, T is the absolute temperature, F is Faraday's constant, z_i is the valence, t_i^{IEM} is the transport number in the IEM, a_i is the ion activity. Transport numbers are usually assumed constant [62], and, for a single salt, the following well-known expression is obtained [8,9,59,61]

$$\Delta\varphi^{IEM} = \left(2\, t_{counter}^{IEM} - 1\right)\frac{RT}{z_i F} \ln\frac{a^{SOL,R}}{a^{SOL,L}}, \tag{3}$$

where $a^{SOL,R}$ and $a^{SOL,L}$ are the salt activities at the right and left solution, respectively. Transport phenomena in IEMs and electrolyte solutions are often described through the Nernst–Planck equation, which encompasses the ion flux with three contributions (diffusion, migration and convection):

$$\vec{J}_i = -D_i\vec{\nabla}C_i - \frac{z_i F D_i C_i}{RT}\vec{\nabla}\varphi + C_i\vec{u}, \tag{4}$$

where D_i is the ion diffusivity, C_i is the ion concentration, φ is the electric potential and $\vec{u}$ is the velocity. For strong binary electrolytes, it can become [61]

$$\vec{J}_i = -D_{el}\vec{\nabla}C_i + \frac{t_i\,\vec{i}}{z_i F} + C_i\vec{u}, \tag{5}$$

where D_{el} is the electrolyte diffusivity, t_i is the ion transport number and $\vec{i}$ is the current density. The Nernst–Planck formalism assumes negligible interactions among ions, thus being strictly valid for dilute solutions. Nevertheless, the Maxwell–Stefan and other rigorous, but more complex, approaches are less used [63–65].

From Faraday's law, the electric current is

$$\vec{i} = F\sum_i z_i \vec{J}_i. \tag{6}$$

The transport number of an ionic species is the relative portion of electric current that it carries [8,9,59,60]:

$$t_i = \frac{z_i J_i}{\sum_i z_i J_i}. \tag{7}$$

The IEM permselectivity for a counter-ion is [8,11,60]:

$$P = \frac{t_i^{IEM} - t_i^{SOL}}{1 - t_i^{SOL}}. \tag{8}$$

The IEM permselectivity between two ions (A and B) is [9,59,60,66]:

$$P_B^A = \frac{t_A^{IEM} / t_B^{IEM}}{N_A^{DIL-IEM} / N_B^{DIL-IEM}}, \tag{9}$$

where $N^{DIL-IEM}$ is the equivalent concentration (of A or B) at the membrane–solution interface in the diluted side.

Several methods can evaluate permselectivity and transport numbers. Static techniques measure the membrane potential, while dynamic methods consist of electrodialysis experiments (current efficiency) [11,59,60] or chronopotentiometric measurements [11,60,67,68].

Electrical resistance is another important membrane property. It may be measured by either direct current [11,68–70] or alternating current [11,69,71–75] techniques. The last ones (electrochemical impedance spectroscopy) are more complex, but make it possible to separate EDL and polarisation contributions. At lower solution concentration, membrane resistance exhibits an increasing vertical asymptote [69,72,75]. This behaviour was associated with the IEM morphology by following the micro-heterogeneous model [75,76], which represents the membrane with multi-phase structure [77], also containing the solution from outside. There are various physical models, and also phenomenological models for membrane resistance [78]. Moreover, a diverse behaviour was found by some measurements [74], thus this topic could deserve further investigation.

Salt and water permeabilities are other IEM properties [9,59,79–83] that affect the ED efficiency. The diffusive fluxes of salt and water (from concentrate to diluate and vice versa, respectively) are driven by gradients of concentration and osmotic pressure, respectively. An additional water flux originates from electro-osmosis (water molecules in the salt ions solvation shell).

Concentration polarisation in ED is caused by the difference in the transport numbers between membrane and solution, so that a diffusive transport in solution maintains a constant overall flux [15,16,84]. In particular, salt depletion occurs at the IEM-diluted side, and salt enrichment takes place at the IEM-concentrated side. The Nernst model (film theory) can be used to study transport phenomena in IEM-solution systems [85,86]. As the electric current increases, the salt depletion is possible until reaching a null interface concentration, corresponding to the condition of diffusion-limited current. The theoretical diffusion-limited current density may be related to the mass transfer characteristics in the fluid channel, i.e., to the Sherwood number.

However, current–voltage curves exhibit three regions, including the manifestation of overlimiting currents [87–92]. The first region starts following a linear trend, but more and more pronounced polarisation effects (and larger Ohmic resistances in ED stacks) cause a reduced slope at higher currents. The second region is a low-slope transition step indicating the limiting current achievement (high resistance). In the third region, a secondary current growth occurs. When the change in slope is not evident, Cowan's method (apparent resistance) can identify the limiting current [16,93,94].

The current that is carried by H$^+$ and OH$^-$ generated through water dissociation can partially explain the appearance of overlimiting currents [84,95–99]. Instead, other overlimiting mechanisms involve counter-ions by current-induced convection [67,88,91,100–108]. Electroconvection is the primary process that alters the depleted region for dilute solutions. An extended space charge region develops near the IEM, where the solution is not electroneutral, and inhomogeneous electric fields cause dynamic vortices [87–89,109–116]. The conductive and geometrical heterogeneity in the IEM surface [67,105,111–113,117–121] and other features, e.g., roughness, grade of hydrophobicity, and superficial charge density [102,103,122–125], affect overlimiting mechanisms and current–voltage curve. Another overlimiting mechanism is gravitational convection, which is caused by temperature or concentration gradients [96,111,114,119,126].

Overlimiting regimes may lead to an enhancement of mass transfer [127], and surface modifications can improve IEMs performance [102]. Therefore, ED operations at overlimiting conditions may be considered to enhance the process efficiency. Nevertheless, maintaining ED stacks below the limiting current is a traditional practice [8,103,128] for avoiding dangerous pH values posing risks of fouling and IEM damage [84,95,129,130].

Despite the actual limiting conditions depend significantly on the membrane properties, experimental correlations of the Sherwood number [63,128,131–134] or of the limiting current density [90,131,135–137] are given by similar expressions:

$$\text{Sh} = a\text{Re}^b\text{Sc}^c, \tag{10}$$

$$i_{lim} = dC_i^{bulke}u^b, \tag{11}$$

where Re and Sc are the Reynolds and Schmidt number, respectively, u is the solution velocity, and a–e are coefficients. Many experiments show b values around 0.5, but it can span in a broader range. Actually, power laws fit experimental data well in narrow Reynolds number ranges [128,135], while more complex trends develop at wider ranges (a similar behaviour is exhibited by the friction factor [138–140]). The coefficient c was found to be equal to 1/3 [141,142], but it can assume different values [139,143,144].

Please note that the current–voltage curve of a BM exhibits two limiting zones. At low currents, the first limit is due to salt ions transport. Then, water dissociation occurs. At high currents, the second limit is due to water transport, eventually implying membrane damage.

3.3. Performance Parameters

The performance of ED processes is governed by membrane selectivity and transport properties, non-Ohmic voltage drop given by the membrane potential ("back" electromotive force in most cases, electromotive force in RED), Ohmic voltage drop, and pumping power consumption. The voltage drop over the stack can be computed as [16,145,146]:

$$V_{stack} = V_{ru}N_{ru} + Ir_{el} = \left(\pm r_{Ohm,\,ru}I + V_{non-Ohm,ru}\right)N_{ru} + Ir_{el}, \tag{12}$$

where V_{ru} is the voltage drop over a single repeating unit (e.g., cell pair or triplet), N_{ru} is the number of repeating units, I is the electric current, r_{el} is the resistance of the electrode compartments (negligible in stacks with many repetitive units), $r_{Ohm,ru}$ and $V_{non-Ohm,ru}$ are the Ohmic resistance and non-Ohmic voltage drop in the repeating unit, respectively, the sign "+" applies for ED methods using an electric current provided by an external power supply, i.e., all ED methods except for RED, where the sign "−" applies. The Ohmic resistance encompasses the contributions from all compartments and IEMs (counting also spacer shadow effects [71,147]). $V_{non-Ohm,ru}$ consists of membrane potentials, including polarisation effects. The electric power consumption (or production in RED) can be simply calculated by multiplying the stack voltage drop by the electric current. Actually, the overall power would include the pumping power [16] (it has to be added to the power consumption in most cases, and has to be subtracted to the power production only in RED). The specific energy consumption expresses the energy consumed per product unit volume (e.g., kWh/m^3):

$$E_{spec} = \frac{V_{stack}\,I}{Q_{prod}}, \text{ for continuous operations} \tag{13}$$

$$E_{spec} = \frac{\int_0^\tau V_{stack}\,I\,dt}{v_{prod}}, \text{ for batch operations} \tag{14}$$

where Q_{prod} is the product volume flow rate exiting the module, t is the time, v_{prod} is the product volume at time τ. E_{spec} can be expressed with reference to the transported mass of electrolyte (e.g., kWh/kg):

$$E_{spec} = \frac{V_{stack}\,I}{\left|C_{el,prod,\,in}Q_{prod,\,in} - C_{el,prod}Q_{prod}\right|M_{el}}, \text{ for continuous operations,} \tag{15}$$

$$E_{spec} = \frac{\int_0^\tau V_{stack}\,I\,dt}{\left|C_{el,prod,\,in}v_{prod,\,in} - C_{el,prod}v_{prod}\right|M_{el}}, \text{ for batch operations,} \tag{16}$$

where $C_{el,prod,\,in}$, $Q_{prod,\,in}$ and $v_{prod,\,in}$ are the inlet or initial product concentration, flow rate and volume, respectively, $C_{el,prod}$ is the concentration in the product outgoing from the stack or at time τ, and M_{el} is the molar mass of electrolyte.

The current efficiency quantifies the utilization of the applied current by an ion species:

$$\eta_i = \frac{z_i F\left|C_{i,prod,\,in}Q_{prod,\,in} - C_{i,prod}Q_{prod}\right|}{N_{ru}I}, \text{ for continuous operations,} \tag{17}$$

$$\eta_i = \frac{z_i F\left|C_{i,prod,\,in}v_{prod,\,in} - C_{i,prod}v_{prod}\right|}{N_{ru}\int_0^\tau I\,dt}, \text{ for batch operations ,} \tag{18}$$

where $C_{i,prod,\,in}$ is the inlet or initial concentration of i in the product, $C_{i,prod}$ is the outlet or final (at time τ) concentration of i in the product. The total η for ion mixtures is the sum over all anions or cations. The current efficiency is less than 100% because of unwanted salt and water transport phenomena, water splitting, and current leakage (parasitic or shunt currents through manifolds [146]) [8]. Further performance parameters such as removal efficiency, concentration factor and water recovery, result from easy calculations.

In the special case of RED, an important performance parameter is the electromotive force, given by the open circuit voltage of the stack, which can be estimated as:

$$V_{OC} = N_{ru}(\alpha_{CEM} + \alpha_{AEM})\frac{RT}{zF}ln\frac{a^{CONC}}{a^{DIL}}, \tag{19}$$

where α_{CEM} and α_{AEM} are the apparent permselectivity of CEM and AEM, respectively, and a^{CONC} and a^{DIL} are the salt activity in the concentrate and diluate, respectively. As V_{OC} is usually estimated with the inlet concentrations, it may differ from the actual (local or average) non-Ohmic voltage drop used in Equation (12). When V_{OC} is measured, the average permselectivity can be obtained from Equation (19).

The power density in RED is defined as the power divided by the total membrane area:

$$P_d = \frac{V_{stack}I}{2N_{ru}A}, \tag{20}$$

where A is the active area of one membrane. The voltage over the external load, and thus over the stack, depends on the external resistance ($V_{stack} = I \cdot r_{ext}$). By reducing the external resistance, the stack voltage decreases and the current increases up to short-circuit conditions, where the electromotive force is completely consumed inside the stack. The power $V_{stack} \cdot I$ theoretically follows a parabolic trend as a function of I or V_{stack}, exhibiting a maximum under conditions halfway between open-circuit and short-circuit, in which the external resistance is equal to the stack resistance. The theoretical maximum power density is given by:

$$P_{d,max} = \frac{E^2_{OCV}}{8r_{stack}N_{ru}A},$$ (21)

where r_{stack} is the stack resistance (including cell pairs and electrode chambers). With the conventional couple seawater-river water at ambient temperature, $P_{d,max}$ is in the order of ~1 W/m^2, with the highest measured value being 2.4 W/m^2 [54,148]. As the power density produced by RED units is modest, the pumping power consumption cannot be neglected. Therefore, the net power $P_{d,net}$ (and, more specifically, the net power corresponding to the maximum gross power, $P_{d,max,net}$) is an important performance parameter [149]. As a function of the fluid velocity, the net power (both $P_{d,net}$ and $P_{d,max,net}$) first exhibits an increasing trend, then reaches a maximum, and finally decreases. Similarly to Equation (13), the specific energy production, or energy density, may be calculated for RED operations.

4. Industrial Wastewater

Waste effluents from industry may have different compositions. However, they often contain dissolved ions. Electrodialytic treatments for industrial wastewater can be classified in: separation of heavy metal ions (Section 4.1); regeneration of acid/base, salt conversion (Section 4.2); desalination (Section 4.3). The applications studied for the main types of industrial wastewater are examined first. Then, further studies on other waste effluents are collected in Section 4.4.

4.1. Separation of Heavy Metal Ions

Heavy metals are harmful pollutants characterized by toxicity, carcinogenicity, non-biodegradability, and persistence in the environment and in living beings. Among the treatment processes proposed for wastewater containing heavy metal ions [150], electrodialytic methods have been tested for industrial effluents from several processes (metal finishing, leather industry, etc.) aiming at reuse. For instance, ED can recuperate water and metals from spent baths or rinse waters of plating processes [151], and different EDI configurations offer several alternatives to ED [42,152].

Several studies have focussed on transport phenomena by assessing IEMs properties, pH effect, complexes formation and ions competition, and by developing modified or novel IEMs. Among them, experiments with aqueous solutions of Ni [153–157], Cu [158–162], Zn [157,163,164], Cr [154,156,165–170], Fe [154,156] and Pb [155,157] have been conducted, as well as with mixtures (e.g., Ni, Cu and Pb, or Cu and Zn) [171,172], thus providing important insights on basic phenomena and ED processes. Another crucial aspect is the identification of optimal operative conditions. With this aim, Taguchi's method was adopted for experiments with metal ions present as single salts or salt mixtures [173,174]. It is a powerful method of design of experiments with optimisation of control parameters, and is based on orthogonal arrays that reduce the number of tests. The statistical analysis of the experimental results was performed by analysis of variance, evaluating error variance and relative importance of the various factors. Moreover, validated models can be adopted for sensitivity analyses and optimisation studies [173].

Each of the next sections, from Sections 4.1.1–4.1.6, focuses on main wastewaters with a single heavy metal ion. Finally, Section 4.1.7 focuses on mixtures of metal ions and waste effluents with other metal ions.

4.1.1. Nickel

Nickel is used for the plating processes of metal pieces with a galvanic bath, which is followed by multiple rinse stages with water. ED can be used to treat the first rinse solution, thus recovering a Ni^{2+}

concentrate recycled to the galvanic bath, and a diluate recycled to the rinse stages (Figure 4). In an early study with pilot ED testing, Ni was recovered by 90% (5 g/L rinse wastewater) [175].

Recent experiments with artificial solutions of electroplating Watts' bath (65 g/L $NiCl_2$, 275 g/L $NiSO_4$, 45g/L H_3BO_3, organic additives) recovered ~95–99% of ions and exhibited an acceptable quality of plated pieces [176]. A scale-up was then performed [177], by operating an ED plant introduced within an industrial plating process for 30 days (Figure 4), demonstrating techno-economic feasibility (saving of 3800 US$/y with E_{spec} of 2.8 kWh/m^3 for treating 480 L/day).

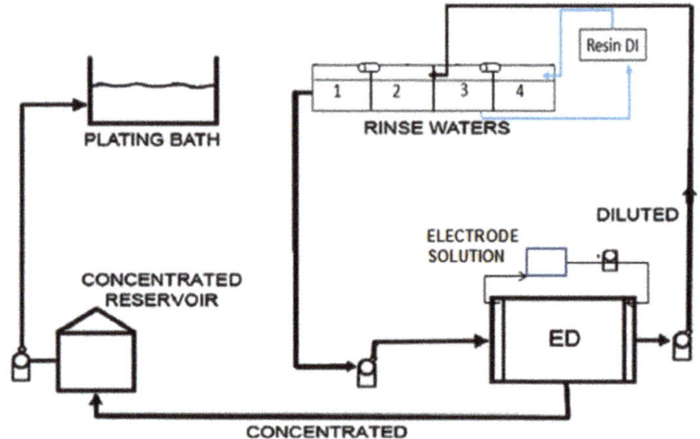

Figure 4. Flow-chart of the ED treatment for Ni electroplating rinse wastewater (the IXR extended the water cycle within the third and fourth tank). Reproduced with permission from [177], published by Elsevier, 2017.

An interesting alternative for ED units with enhanced current efficiencies is given by corrugated membranes [178]. ED was also tested with electroless plating spent solutions, removing harmful ions (HPO_3^{2-}, SO_4^{2-}, and Na^+), and maintaining useful ions (Ni^{2+}, $H_2PO_2^-$, and organic acids) at high concentration [179].

An electrolysis-ED-EDI combined system (13-cell-pair EDI equipped with mixed IXRs in all channels) yielded ~99.8% of Ni^{2+} recovery with 93.9% of purity from a synthetic solution [180]. A simpler two-stage EDI was developed to mitigate the back diffusion [181]. By using a model Ni electroplating rinse solution at 50 mg/L, the first stack diluate effluent (~3 mg/L) was the initial feed of both compartments of the second stack. Mixed beds in the concentrate channels minimized the metal hydroxide precipitation in the 1st stage by limiting the contact probability of OH^- with Ni^{2+} (the lower concentration did not require this measure in the 2nd stage concentrate). Concentration and enhanced purification were accomplished. Ni^{2+} was separated by over 99.8% with E_{spec} of 0.64 kWh/m^3, and the solutions produced were suitable for use in plating and rinsing operations. An economic analysis prospected significant savings compared to chemical precipitation.

Numerous EDI configurations deviating from ED stacks have been tested. They include three-compartment electro-regenerated devices [45,49,50,182], among which there are a hybrid system coupling EDI with capacitive deionisation [183], and a unit without membranes and with electrostatic shielding regions made of graphite powder filling the concentrate compartments [184].

4.1.2. Copper

The ED efficacy in Cu^{2+} separation has been proved. For example, removal percentages of ~97% were attained under optimal conditions [185]. ED can treat and recycle rinsing water from electroless plating [186], and baths and rinse solutions from cyanide electroplating (flowing through concentrate and diluate compartments, respectively) [187].

Non-toxic cyanide-free electroplating baths can be reclaimed by ED. Model rinse waters with 683–1281 mg/L 1-hydroxyethane 1,1-diphosphonic acid and 32–48 mg/L Cu^{2+} (from Cu alkaline strike bath) were used [188]. Despite various ionic complexes formed as the pH was changed, diluate and concentrate products were suitable for recycle in the bath and rinsing processes, respectively (maximum recovery of 99.7% for Cu, 94.4% for organic acid). Further experiments again achieved high recoveries, and pieces of electroplated Zamak alloy exhibited good-quality coatings [189]. However, AEMs properties were worsened (electrical resistance and limiting current), likely because of interactions between organic acids or their chelates and fixed groups. Transport properties were restored only in part by cleaning procedures.

Overlimiting regimes were shown to lead to high values of separation percentage and η [159]. However, copper hydroxide may precipitate causing scaling [190].

An integrated electrochemical process was developed by a pilot system with ED and electrolysis (Figure 5), recovering over 99% of Cu^{2+} and all the water from synthetic solutions (165–504 mg/L), with E_{spec} of ~2 kWh/m^3 in the ED stage [191].

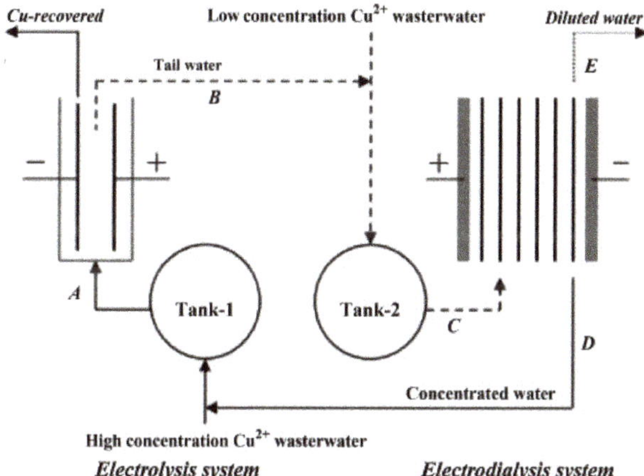

Figure 5. Electrolysis-ED integrated system treating $CuSO_4$ solutions. Reproduced with permission from [191], published by Elsevier, 2011.

Further combined systems aim at treating multiple wastewaters. For example, a process integrating microbial desalination cell, precipitation and ED was developed to treat simultaneously domestic wastewater, Cu wastewater and salt water [192]. In particular, ED finalized the removal of Cu residues and the desalination.

Metal–organic complexes can be separated by ED, which, for example, showed higher efficiencies with Cu–EDTA complexes compared to other electrochemical technologies [193].

EDI processes are suitable for Cu diluted wastewaters, such as plating rinse solutions, as shown, e.g., by a three-compartment device equipped with electro-regenerated layered IXR bed [47].

4.1.3. Zinc

Different plating processes are performed with zinc. Among them, $Zn_3(PO_4)_2$ coating layers are produced via phosphate plating baths with H_3PO_4. Rinse solutions contaminated by different ions (Zn^{2+}, Fe^{2+}, PO_4^{3-}, NO_3^-, etc.) can be treated by ED. For example, 6.5 ppm Zn^{2+} were reduced to ~0.5 ppm, with E_{spec} of ~3.5 kWh/m^3 [186]. ED can also be carried out for Zn cyanide electroplating solutions [187].

4.1.4. Chromium

Hexavalent chromium is another heavy metal commonly used in electroplating. It can exist in the form of several ion species ($Cr_2O_7^{2-}$, $HCr_2O_7^-$, $HCrO_4^-$, CrO_4^{2-}) affected by concentration and pH, and is characterized by high toxicity and carcinogenicity.

In ED experiments with Cr(VI) model solutions, operating conditions (i.e., flow rate, initial concentration, pH, and voltage) were varied, attaining removal percentages of ~79% and ~99% for 50 ppm and 10 ppm as initial concentration, respectively, with $E_{spec} \approx$ 2–4 kWh/m^3 [194]. Therefore, the diluted water can be reclaimed. However, the concentrate has to be cleaned from impurities in order to be recycled.

With this aim, a two-stage selective ED with MVAs was developed [195]. The real wastewater contained Cr(VI) as $HCrO_4^-$ at pH of 2.2, so that it was possible to concentrate it in the first stage along with other ions. By adjusting the pH of the concentrate at 8.5, the most stable form of chromate was the divalent CrO_4^{2-}. MVAs of the second stage retained this in the diluate, while letting monovalent ions to pass (Figure 6). From a 418 mg/L Cr(VI) feed, the first stage achieved a maximum concentration factor of ~1.9, while the second one removed Cl$^-$ by ~45%.

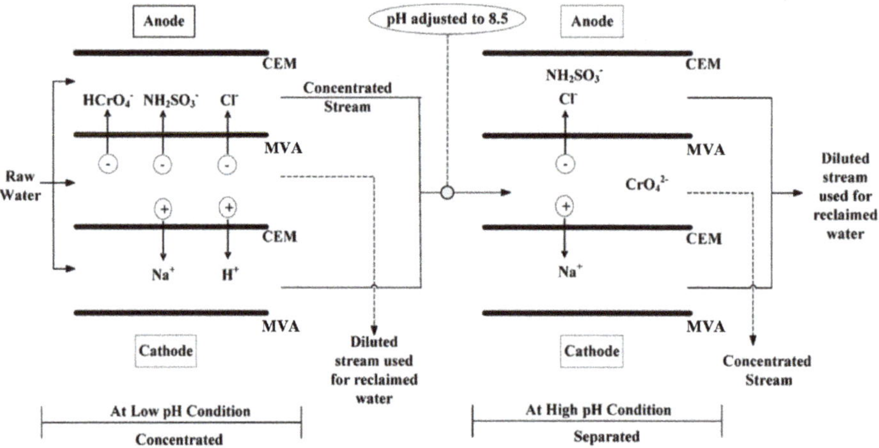

Figure 6. Two-stage selective ED with MVAs for recovering water and concentrate solution from Cr(VI) electroplating wastewater. Reproduced (adapted) with permission from [195], published by Elsevier, 2009.

ED was proposed as a post-concentration step of Cr(VI) diluted solutions after biological treatment [196]. The maximum initial concentration was 100 ppm, simulating the residual concentration of an anaerobic degradation process. A large Cr(VI) retention in the membranes was observed. However, some metal ions were transported to the concentrate channels, leading to a maximum concentration of

570 ppm. The Cr(VI) removal from the feed solution was ~99%, and the volume of concentrate was only ~5.3% of that of the feed. These results suggested the possible reuse of the diluate and concentrate streams.

Cr(VI) separation by EDI processes was studied [197–201]. For example, EDI (four-compartment device, Figure 7a) was combined with IX [198]. After mixed IXRs saturation, electric current was applied by exceeding by 10% the limiting value, removing 98.5% of 100 ppm Cr(VI) with $E_{spec} \approx 0.07$ kWh/m^3.

Cr(III) is less hazardous than Cr(VI). However, it is used in various industrial processes, including plating, and thus can be present in waste effluents. Complexation ED removed simultaneously Cr(III) from a synthetic electroplating wastewater and acetylacetone from a model pharmaceutical waste [202]. The two solutions were mixed, obtaining clathrates formation. These charged chelates were concentrated with removal efficiencies of 99.4–99.5% for Cr and 97.8–99.9% for acetylacetone. The proposed strategy of joint treatment was very promising, but further studies should focus on fouling.

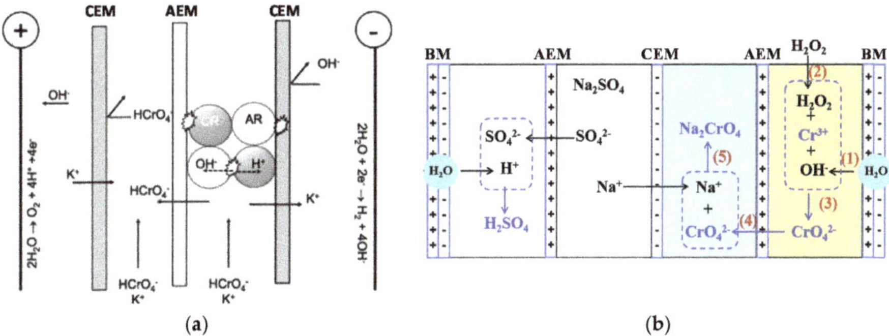

Figure 7. Treatment processes for Cr wastewater: (**a**) EDI with diluate compartment filled by mixed IXRs for separating Cr(VI), water dissociation is highlighted at the bipolar contacts; (**b**) BMED for recovering Cr(III) after oxidation to Cr(VI). Panel (**a**) is reproduced (adapted) with permission from [198], published by Elsevier, 2013. Panel (**b**) is reproduced (adapted) with permission from [203], published by Elsevier, 2020.

Cr(III) was oxidized and recovered as Cr(VI) by BMED [203] (Figure 7b). Cr$_2$(SO$_4$)$_3$ synthetic wastewater (50–1000 mg/L Cr(III)) received OH$^-$ from the BM (1), forming CrO$_2^-$. By adding H$_2$O$_2$ (2), it was oxidized to CrO$_4^{2-}$ (3), which migrated to the recovery chamber (4), where Na$^+$ came from the buffer chamber (5). H$_2$SO$_4$ in the acid chamber was an additional product. Under optimal conditions (e.g., 5.0 g/L Na$_2$SO$_4$ in 100–1000 mg/L Cr(III) solution) ~70% of Cr was recovered. At 500 mg/L Cr(III), η was 67.6% and E_{spec} was 730 kWh/kg Cr in a stack with three repetitive units. By repeating the experiment three times, the recovery increased to ~88%, because of Cr adsorption and release in/by the AEM (removal decreased from ~98% to ~92%).

Cr(VI) and Cr(III) coexisting in wastewater can be removed by electrodialytic techniques. For example, EDI was able to remove both Cr^{3+} and HCrO$_4^-$ from a model solution with 100 ppm of both contaminants, but with higher removal for the latter [204]. The separation from salt mixtures with monovalent or divalent ions is more difficult [205].

Industrial treatments of hides and skins require massive consumption of process water containing chemicals for several manufacturing steps. In particular, salts of Cr(III) are used for tanning processes, which produce wastewater at large volumes and with different contaminants (organics, tannins, and salts). Tanning effluent treatment for recycling purposes has to separate Cr(III) from other ions. ED has been used to recover salt water and Cr(III)-solution by taking advantage of the different selectivity of the membranes towards different ions. For instance, filtered spent tanning effluents flowed through ED diluate, where the membranes retained most of Cr(III), which was present in different ionic and non-ionic forms (total concentration of ~0.27%), while removing other ions (Cl$^-$ by ~91% and SO$_4^{2-}$ by ~51% from concentrations of 3.4% and 3.3% as NaCl and Na$_2$SO$_4$, respectively) [206]. Fouling and chromium leakage were alleviated by dosing EDTA in small quantities and applying EDR. The overall process economics was promising, as the chromium load was lowered by ~33% and the concentrate salt water was usable in pickling operations.

Another strategy for recovering Cr tanning solutions was developed in two steps with (i) monovalent selective ED and (ii) conventional ED [207]. From model tanning wastewaters with concentrations in the order of ~0.1 eq/L, MVCs retained Cr(III) in the diluate and separated NaCl in the concentrate. Then, chromium was concentrated by conventional ED. Optimal pH was around 3, avoiding precipitation and also the competing transfer of H$^+$. Thus, η approached 100% at the second stage. Further experiments on the first step with salt mixtures (Na$_2$SO$_4$, MgCl$_2$ and CaCl$_2$) showed that most of Na$^+$ could be removed ($\eta \approx 70\%$), with a global η of ~97% [208].

Treatment of tannery wastewaters has been based on integrated processes for water recycling where the first operation degraded organics. Photoelectrochemical oxidation [209] or electrocoagulation [210] have been proposed for this purpose, followed by ED or BMED, respectively. The former combined process removed 87.3% of COD and more than 98.5% of ions (Cr was present only in traces in the raw effluent, i.e., with concentration < 0.01 mg/L); the latter combined process removed ~90% of COD and almost all Cr (from 570 mg/L), ammonium and colour, with E_{spec} of 14–30 kWh/m^3.

4.1.5. Cadmium

Cadmium electroplating is a galvanic process performed via alkaline baths with cyanide, and ED represents again an option as recovery process from waste effluents [211]. A simulated wastewater (CdO, NaCN and NaOH at 0.0089, 0.081 and 0.018 mol/L, respectively) flowed through the diluate of a five-compartment ED module. Maximum removals of 86% of CdCN$_4^{2-}$ (which was the predominant complex) and 95% of CN$^-$ were achieved. However, the process efficiency was affected by Cd(OH)$_2$ precipitation on the CEM at the diluate side.

Solutions with similar composition simulating diluted baths (Cd concentration between 1 and 3 g/L) were used in further experiments [212]. No precipitation was observed, with removals of 21.6% and 46.1% for CdCN$_4^{2-}$ and CN$^-$, respectively (η of 13.2% and 59.6%). The process was performed in the same way when feeding the model wastewater in both diluate and concentrate channel, i.e., in a configuration more similar to conventional ED. To recycle the concentrate into electroplating baths, while avoiding efficiency losses and risks of membrane damage, the diluate feed was changed four times. This led to a concentrate concentration increase by 56% for cadmium and 250% for cyanide.

The selective separation between metal ions by EDI, referred to as electropermutation, was achieved in a solution containing Cd^{2+} and Na$^+$ [213]. A cation-exchange resin, modified by natural polyelectrolyte, fixed selectively Cd^{2+} in the central channel of a five-compartment electro-regenerated unit.

4.1.6. Lead

ED has been tested for wastewater containing lead, which originates from industrial processes (regarding, for example, batteries, electronics, printing pigments, explosives, metallurgical processes). With a Pb(NO$_3$)$_2$ model solution, experiments assisted by analysis of variance assessed the effect of concentration (100–1000 ppm), flow rate, voltage and temperature [214]. The separation was affected mostly by the flow rate, and reached ~95% under optimal conditions. Modelling tools validated against experiments showed that the artificial neural network model was more accurate than the simplified model, the former being able to predict the non-linearity of transport phenomena and thus of ED [215].

Another important operating parameter is the pH. Optimal values of 3–5 were found in experiments with a solution at 800 mg/L Pb^{2+} performed with an ED pilot stack [216]. Lower voltages were effective in increasing η (up to ~35%) and maintaining low E_{spec} values (~0.1 kWh/m^3), while higher voltages could regenerate the membranes from the adsorbed ions. The same ED stack was used in a combined treatment process developed with electrolysis and a further ED unit for adsorption in CEMs [217]. The initial concentration had the most significant effects on the ED performance. With optimized parameters (Taguchi method), ED reduced the initial Pb^{2+} concentration of 600 mg/L to ~16 mg/L in the ED diluate. This was further reduced by adsorption to ~1 mg/L, reaching the target required by the Chinese regulation. The electrolysis process recovered ~90% of Pb via cathode deposition from the ED concentrate. Other ED experiments reduced the concentration from 500–1000 mg/L to 1–2 mg/L [218]. Under optimal operating conditions, high values of η were obtained (82.8–72.4%) with E_{spec} of 0.16–0.36 kWh/m^3. Despite several promising results, the feasibility for real Pb-wastewaters has still to be demonstrated.

4.1.7. Mixtures and Other Heavy Metal Ions

Real industrial wastewaters often contain mixtures of metal ions (either ions with similar concentration or impurities), and ED processes can effectively recover water, concentrate ions, and, in some cases, separate different ion species from each other.

Brass (Cu and Zn) electroplating was evaluated by cyanide-free baths with EDTA, by using ED for treating the rinsing water [219]. By adjusting the ED concentrate concentration to that of the original bath, good deposits were obtained. In the overlimiting regime, the recovery of metals and EDTA was more advantageous [220]. The rinse water was prepared with 0.0006 M $CuSO_4$, 0.0014 M $ZnSO_4$, 0.0015 M EDTA, 0.03 M NaOH (conductivity of 5.3 mS/cm), corresponding to 1% of the concentrations in the bath. The diluate solution was replaced once reached the conductivity of ~0.2 mS/cm, and the concentrate solution was not replaced to maximise its concentration. Cu, Zn and EDTA were concentrated more with overlimiting operation (concentration factor of ~3.45 against ~2.94 with underlimiting operation), likely due (i) to water dissociation generating protons that reacted with complexes and insoluble species, and (ii) to electroconvective mass transfer enhancement. Moreover, fouling and scaling were reduced.

High removals were achieved by ED from a real electroplating effluent containing Ni^{2+} and Cu^{2+} (~23 mg/L for both) [221]. A tertiary treatment line was developed for reclaiming a plating wastewater effluent with a mixture of heavy metal ions at low concentration (~1 mg/L) [222]. Microfiltration (MF) and ultrafiltration (UF) removed organics and suspended solids, then ED desalination was conducted, and finally, the concentrate was treated by nanofiltration (NF) or reverse osmosis (RO) to increase water recovery. The ED step removed 97% of Cr^{3+}, Cu^{2+}, and Zn^{2+}, 95% of SO_4^{2-} and Cl^- (from initial concentration of 1000 mg/L), and 85% of COD (300 mg/L initial concentration). In another study, ED was conducted after chemical precipitation of a real Cr(VI) electroplating wastewater (19 mg/L) with minor concentrations of Cu^{2+}, Zn^{2+} and Cd^{2+} [223]. Cr(VI) was reduced to Cr^{3+} by Na_2S and $FeCl_2$, and then precipitated with other metal ions by NaOH dosage (pH = 9). Then, ED diminished the Cr(VI) concentration in the effluent (1–8 mg/L) by up to more than 95%, thus producing a water reusable for rinsing operations. By treating synthetic rinse waters of Cd cyanide electroplating (1000 mg/L Cd) contaminated by either Cu (50 mg/L), Fe (50 mg/L) or Cr (100 mg/L), the non-selective transport made the ED concentrate not reusable for electroplating baths [212].

However, metals from a mixture can be selectively separated in different channels by complexation–ED. A simulated Zn electroplating bath was prepared with 48.9 g/L Zn^{2+} and 1 g/L Fe^{3+}, and treated by testing different chelating agents [224]. These solutions were circulated through the diluate of a five-compartment ED stack with two concentrate channels (Figure 8a), obtaining the separation between Zn^{2+} and Fe complexes. The process was more efficient when heterogeneous IEMs and citric acid were used. A high retention of Fe (~92%) in the feed channel was caused by the generation of an electrically neutral citrate complex, while ~87% of Zn^{2+} was removed with $\eta \approx 85\%$. These results suggest that recovering concentrate solutions from contaminated baths can be feasible in conventional ED with one diluate and one concentrate. Further experiments confirmed these results [225], while finding problematic the selective separation for a Zn^{2+} solution contaminated by Cu^{2+}. This occurred due to a partial formation (~65%) of Cu–citric acid anions.

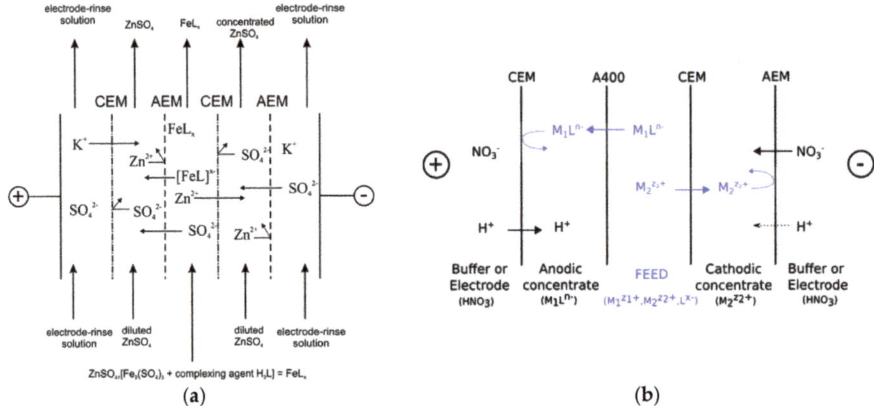

Figure 8. Schemes of complexation-enhanced ED: (a) Zn^{2+} recovery from a model Fe^{3+}-contaminated electroplating bath by Fe-citrate neutral complex retention; (b) selective separation of metal cations (Ag^+/Zn^{2+} or Cu^{2+}/Cd^{2+}). The feed solution contains initially the two metal ions $M_1^{z_1+}$ and $M_2^{z_2+}$, and the ligand L^{x-}, which forms complexes only with $M_1^{z_1+}$ (i.e., M_1L^{n-}) in situ. The A400 AEM allows for the transport of anion complexes of ~400D without fouling. Panel (a) is reproduced (adapted) with permission from [224], published by Elsevier, 2018. Panel (b) is reproduced with permission from [226], published by Elsevier, 2017.

By taking advantage of the actual formation of anion chelates, complexation–ED was carried out by a three-compartment configuration (Figure 8b) in order to separate metals from mixtures of Ag^+/Zn^{2+} or Cu^{2+}/Cd^{2+} [226]. The two ions of each mixture were transported to two different concentrate compartments. In fact, Zn or Cd formed anion complexes, instead Ag^+ or Cu^{2+} persisted as free ions. EDTA was found to be the best among various complexing agents, allowing for removal percentages higher than 99% from initial concentrations of 0.1–1 meq/L with E_{spec} between 0.28 and 0.55 kWh/m³, thus enhancing the process compared to previous results reported in [227]. BMED was suggested for separating the complexed cation and regenerating the ligand.

Ni and Co were separated by complexation with EDTA [178]. A Ni-EDTA negative complex was preferentially formed and retained, while Co^{2+} ions migrated through the CEM.

Waste mixtures with heavy metal ions can be purified by EDI processes. A five-compartment EDI device with electro-regenerated cation and anion IXRs in separated beds was tested with a real electroplating waste rinse with Ni^{2+}, Cu^{2+}, Zn^{2+}, Cd^{2+} and Cr^{3+} [46]. A similar unit was tested with a waste solution from a Zn electrolysis process containing Zn^{2+} and other metals [48].

Electrodialytic technologies have been studied for treating various other industrial waste effluents containing metal ions. H_2SO_4-$CuSO_4$ solutions with impurities (As(III), As(V) and Sb(III)) typical of Cu-electrorefining electrolytes were reclaimed via ED by separating and concentrating metal ions [228]. Similarly, a Cu–electrowinning model solution containing 50 g/L H_2SO_4 and 9 g/L Cu^{2+} (as $CuSO_4$

salt) with 0.5 g/L Fe^{2+} impurities was reclaimed by removing 96.6% of copper and 99.5% of iron with $E_{spec} \approx 1$ kWh/kg [229]. Cr(VI) was recovered as H_2CrO_4 by BMED from chromite ore processing residue (chromate production byproduct) [230]. The waste effluent contained a mixture of ~3728 mg/kg Cr(VI) and ~2650 mg/kg Cr(III), along with other metal ions (Fe, Al, As, etc.). The BM-AEM configuration was used, and MF membranes were placed in the wastewater compartment to protect the BM and the AEM from clogging. Recoveries of ~90% were obtained with η of 2.3% E_{spec} of 395 kWh/kg.

Selectrodialysis (SED) with MVC was used with synthetic solutions simulating an acidic metallurgical wastewater (pH = 2.3) with 47 mM $CuSO_4$, 146.8 mM $ZnSO_4$, and 31.6 mM Na_2HAsO_4 [231] (Figure 9). The process recovered 80% of Cu^{2+} and 87% of Zn^{2+} in a solution, and 95% of As(V) in another solution, at η of ~38% for Cu^{2+} and Zn^{2+} and E_{spec} of ~2.6 kWh/kg for their salts. The solution rich in Cu^{2+} and Zn^{2+} was pure by 99.8% (over 80% due to Zn^{2+}), while the product rich in As(V) contained a comparable concentration of Zn^{2+}. To solve this problem, the authors suggested recirculating the As(V) product to the feed.

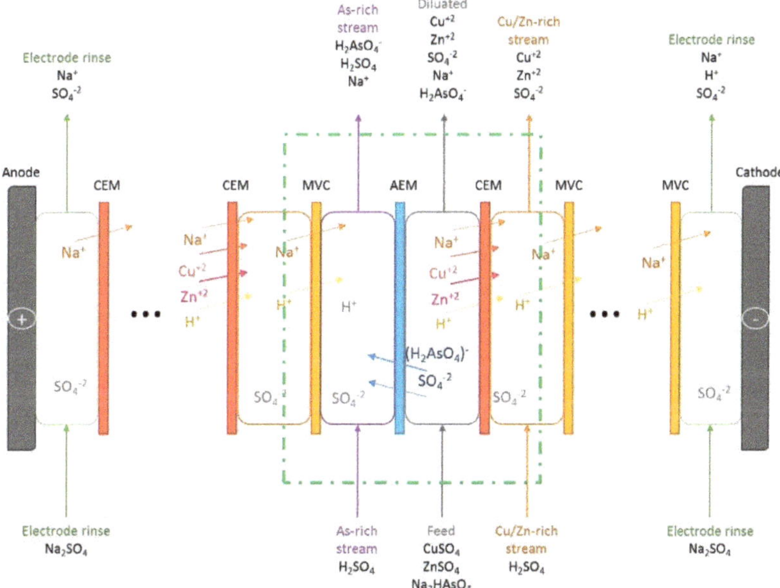

Figure 9. SED configuration for recovering Cu^{2+} and Zn^{2+} from acidic metallurgical wastewater containing As(VI). Reproduced (adapted) with permission from [231], published by Elsevier, 2018.

ED arrangements either with or without MVAs were proposed for reclaiming alkaline gold mine wastewater containing heavy metals (copper and zinc), sodium and cyanide [232].

EDI is suitable for purifying nuclear power plants' primary coolants, which contain low concentrations of Co^{2+} [233,234]. Among several arrangements tested with model solutions (e.g., 0.34 mM), a five-compartment EDI module was developed by using a layered bed within the diluate to prevent the precipitation of metal hydroxide, remove both anions and cations, and control the pH (Figure 10a) [233]. Starting from this EDI arrangement, a stack with four-compartment repetitive units was assembled (Figure 10b). Removals of over 99% at $\eta \approx 30\%$ and $E_{spec} = 14$ kWh/m^3 were obtained. Other experiments conducted with a four-compartment EDI device removed up to ~99.9% of Cs^+ from model waste solutions (e.g., 50 mg/L) [235]. Different radionuclides (Cs^+, Sr^{2+} and Co^{2+}) in traces were removed by 77.1–99.7% [236]. Th^{4+} was removed at rates of up to ~99% (from 30–90 mg/L) in experiments optimized by response surface methodology [237]. Overall, EDI processes are very promising for treating low radioactive effluents.

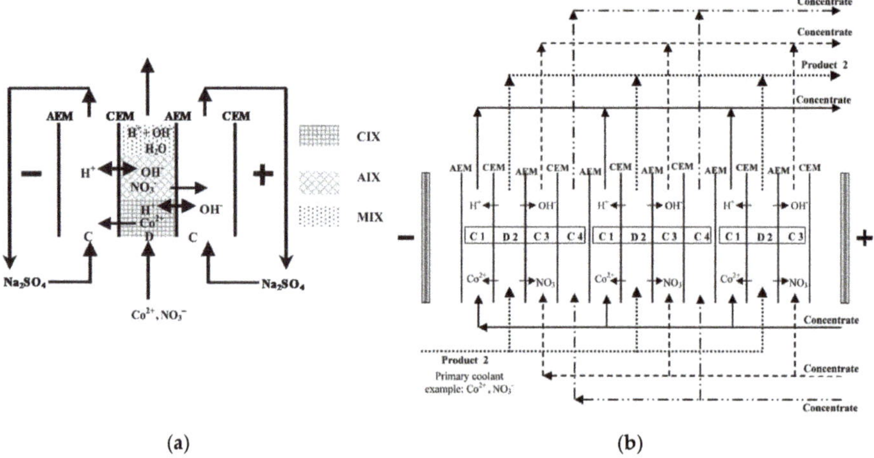

(a)   (b)

Figure 10. EDI configuration with diluate compartment filled by layered bed for the treatment of a Co^{2+} solution simulating the primary coolant from nuclear power plants: (**a**) sketch of the bed packing; (**b**) stack with three repetitive units with four channels each. Reproduced (adapted) with permission from [233], published by Elsevier, 2004.

4.2. Regeneration of Acid/Base, Salt Conversion

Many manufacturing processes produce large quantities of acidic/alkaline waste streams, and their neutralisation is commonly practiced for disposal. In other cases, spent alkaline/acidic solutions result in waste salt streams. In all cases, economic and environmental benefits can be drawn from reuse/recycling approaches. From this perspective, electrodialytic processes (mainly ED and BMED) can be used for treating different industrial effluents, such as acidic wastewaters with heavy metal ions from pickling ant other processes (Section 4.2.1), waste solutions without heavy metal ions (Section 4.2.2), spent alkaline solutions from flue gases chemical absorption (Section 4.2.3), and wastewaters with organic matter, including organic acids (Section 4.2.4).

4.2.1. Effluents with Heavy Metal Ions

Waste acidic effluents are produced form pickling and other processes of metal manufacturing and metallurgical industry. In particular, pickling is a surface treatment that removes impurities (oxides, rust, and others) before metal pieces go through painting, plating, etc. The main application is steel acid pickling. Pickle liquors contain sulphuric, hydrochloric, nitric or hydrofluoric acid, which react with oxides, thus dissolving metal ions. They are regarded as being spent once the acid concentration diminishes by 75–80%, and the metal concentration rises to 150–250 g/L [238]. Pickling processes produce large quantities of spent solutions. For instance, steelwork plants generate ~3 × 10^5 m^3/y waste pickle liquors in the only Europe.

The regeneration (recovery and purification) of pickling operations effluents can be accomplished by several methods, including ED for acid concentration and metal separation [238,239]. For example, 60–70% of H_2SO_4 was recovered from a pickling rinse water (9100 ppm) with a ten times increased acid/iron concentration ratio (from 7.4:1 to 74.6:1) [186]. Optimal operating conditions and selective membranes are crucial [240]. Proton leakage through AEMs, which limits the acid concentration [241], can be alleviated by purposely developed proton-blocking membranes [240,242–248]. On the other hand, the passage of metal ions across the membranes may impair the concentrate purity [249] and cause fouling [240]. However, using MVCs that retain multivalent metal cations allows for the acid recovery with high purity [250].

After neutralisation of the spent pickling solution and precipitation of metals, BMED can regenerate the acid stream in combination with an ED salt concentration step [251]. A BMED-ED integrated pilot has been used since 1987 [34] in an industrial treatment plant at the Washington Steel Corporation facilities (Pennsylvania) [37], where a pickling solution with mixed acids (8–15 wt% HNO_3 and 2–5 wt% HF) was used. The process scheme is depicted in Figure 11, and can be described as follows [37]. Metals in the spent liquor were removed by neutralisation/precipitation (KOH dosage) and filtration. The KF/KNO_3 solution obtained (1.1–1.5 M, with metal ions at concentration < 1 ppm) went through the salt compartment of BMED, which produced the base used for neutralisation, and the mixed acids (HF + HNO_3) recycled into the pickling bath. ED recovered water (for filter cake washing) and salt from the BMED diluate (0.3–0.5 M) stream. The base was diluted with a fraction of the BMED diluate. The BMED and ED stacks were assembled with 25 and 15 cell units, respectively, totalling 2.33 m^2 and 1.4 m^2 of membrane area. During a long-term run with 240 L/day waste acid, η was ~80% for acid and base, and remained quite stable over time, while E_{spec} was ~0.25 kWh/L acid product (180 L/day). 93% of F$^-$, 99% of NO$_3^-$, and 96% of K$^+$ were recovered. The economic analysis for a scaled-up system (6 × 10^6 L/y) found high investment costs, but with a 4-year payback period.

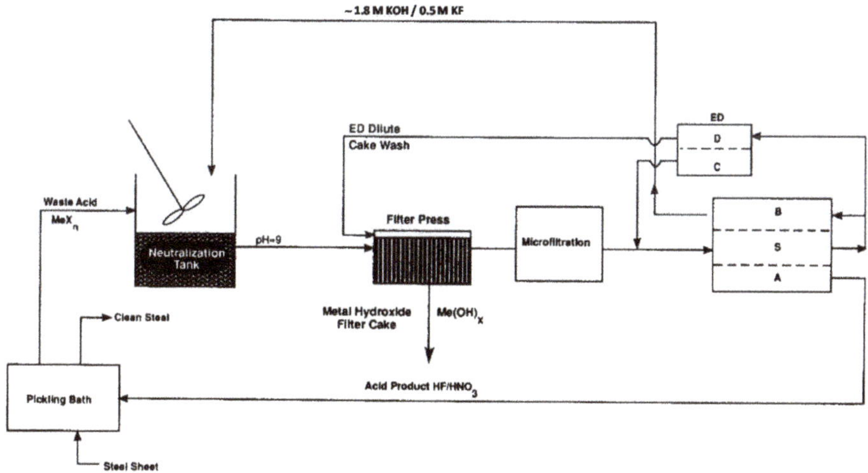

Figure 11. Flowsheet of the pickling liquor recovery by BMED-ED coupling. Reproduced (adapted) with permission from [38], published by Elsevier, 1991.

Several metallurgical processes produce spent acidic solutions with metal ions, and the reclamation via ED or BMED has been tested. ED concentration of a spent solution from a metallurgical industry, containing Ni^{2+}, Cu^{2+} and other ions, recovered more than 80% of H_2SO_4, showing the impact of the membranes (MVMs, proton-blocking AEMs) [252]. Similarly to the example previously reported for waste pickling effluents, BMED can produce acid and base from waste mixtures of heavy metals and salts, after pre-treatment removing metals. From a Ni washing wastewater, crystallisation (fluidized pellet reactor) removed up to 74% and 94.4% of Ni^{2+} and Ca^{2+}, respectively (with filtration), thus minimizing scaling in the BMED [253]. The feed salt solution of BMED contained ~45 g/L Na^+ and ~80 g/L SO_4^{2-} as main ions, with ~10 mg/L Ni^{2+}, ~16 mg/L Ca^{2+}, and other minor components. η was 69% and 80% for acid (H_2SO_4) and base (NaOH), respectively, while E_{spec} was 5.5 kWh/kg acid and 4.8 kWh/kg base. In a long-term test, acid and base at 1.76 N and 2.41 N, respectively, were obtained starting from 0.2 N, with small scaling.

An alternative way for reclaiming acidic streams from metal finishing is the conversion to organic acid via ion substitution ED [254]. The wastewater feed maintains metal ions retained by an MVC, and releases protons and anions to the organic salt stream and the inorganic salt stream, respectively (Figure 12). From a model waste acid (0.4 M HCl, 0.1 M $FeCl_2$) and a 0.3 M sodium acetate solution, acetic acid at high purity was produced (0.2 mM Fe^{2+}) with average η of 91%. A proton selective composite CEM was then developed, showing a possible process enhancement [255].

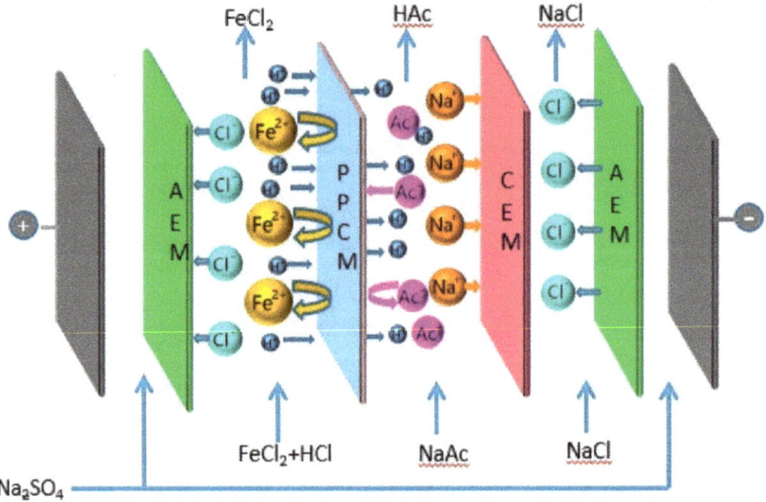

Figure 12. Schematics of ion substitution ED with three-compartment repeating units for the conversion of metal finishing waste acid into organic acid. Reproduced with permission from [255], published by Elsevier, 2019.

Other ED applications regard spent acids produced by Zn hydrometallurgy [256–258]. Again, in order to prevent acid and metal leakage, proton-blocking AEMs and MVCs are crucial elements for performing feasible recovery processes via ED [256]. MVCs were prepared by different methods [259,260], reaching $P^{H^+}_{Zn^{2+}}$ of 34.4 [260]. An interesting alternative is provided by replacing CEMs with NF membranes (Figure 13) [261]. Homemade NF membranes were used for acid recovery from a Zn^{2+}-containing synthetic solution (diluate feed with 0.5 mol/L H_2SO_4 and 0.23 mol/L $ZnSO_4$, concentrate with 0.05 mol/L H_2SO_4). The modified ED system exhibited better performance compared to the stack equipped with MVCs, increasing the permselectivity $P^{H^+}_{Zn^{2+}}$ from 15 to 354.

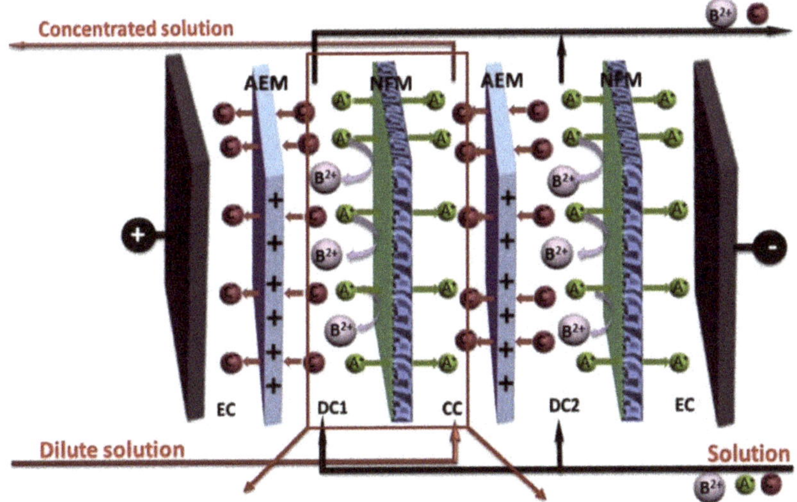

Figure 13. Modified ED equipped with NF membrane in place of CEM. Monovalent cations A^+ (e.g., H^+) can move across the NF membrane, while divalent cations B^{2+} (e.g., Zn^{2+}) are retained. Reproduced with permission from [261], published by Elsevier, 2016.

BMED of acidic raffinate from Cu ore hydrometallurgical processing was proposed [262]. From the raffinate (11,800 mg/L Fe, 336 mg/L Zn, 135 mg/L Cu, etc.) heavy metals were separated (from ~70% to ~99%) as precipitates in the base, and SO_4^{2-} (45.2 g/L) was transported to the acid (by ~86%), thus recovering H_2SO_4, albeit with some impurities. With E_{spec} below 0.1 kWh/L and high values of η, the treated raffinate reached a metal concentration below 100 mg/L, thus being reusable as leachate.

ED tests were performed to recover nitric acid from rinsing-wastewater from aluminium anodizing industry (acidity of 4085 mg/L $CaCO_3$) [263]. Despite some issues of Al precipitation and leakage, most of the waste acid was recovered with a conductivity removal of ~86–91% and E_{spec} values of ~0.11–0.3 kWh/mol acid. Acid recovery can be obtained also by combined membrane processes. HCl from acidic wastewater produced by aluminium foil industry was recovered by integrating diffusion dialysis and ED (model solution with 1.35 mol/L HCl and 0.15 mol/L $AlCl_3$) [264] or BMED (model solution with 4.7 mol/L HCl and 0.59 mol/L $AlCl_3$) [265], showing that cost-effective schemes can be devised. In the former combined process, up to ~75% of HCl was recovered with a metal leakage of ~12%. In the latter, after acid recovery by diffusion dialysis, the dialysate was fed to the base compartment of the BMED, thus allowing for the aluminium recovery.

Experiments showed the ED effectiveness for acidic wastewater with various metal ions (including Cu, Fe, Zn, Cd and As) from the chalcopyrite ($CuFeS_2$) mining industry [266]. An effluent from the SO_2 wet purification process was cleansed from metal ions by IX, then was fed into the ED diluate to concentrate H_2SO_4 in the concentrate (95–98% recovery from a feed concentration of ~17 g/L) and to provide reusable water.

Sulphuric acid and sodium hydroxide were produced by BMED from IX spent regenerant (containing ~0.75 M H_2SO_4, ~0.55 M Na_2SO_4, and metal ions) coming from hydroxy acids liberation from alkaline kraft black liquor [267]. Feeding acid and base compartments with initial concentrations of 0.1 M, the BMED provided a solution with 1 mol/L H_2SO_4 at 95% purity, and a solution with 0.79 mol/L

NaOH at 93% purity, thus presenting a promising perspective to reduce the chemicals consumption in the overall process.

A rare example of ED application for base recovery from an alkaline solution was reported in a study on a synthetic wastewater of 0.1 mol/L Na_2WO_4 and 1 mol/L NaOH [268]. Composite AEMs were prepared and tested, showing OH^- recovery ratios up to ~65%, E_{spec} of ~7 kWh/kg, and low tungstate leakages (5–14%).

4.2.2. Effluents without Heavy Metal Ions

BMED can recover acidic/alkaline solutions from several saline wastewaters produced in industrial processes. In rayon production plants, BMED can restore the acidity of spin baths by converting part of the Na_2SO_4 from spent baths into H_2SO_4, and produce NaOH reusable in cellulose dissolution [37,38] (Figure 14). The crystallisation of Na_2SO_4 produced Glauber salt. After purification and dissolution in water, the solution passed through the acid compartment of a two-cell BMED unit or through the salt compartment of a three-cell BMED unit (receiving a spent bath portion in the acid compartment). η values of 80–95% were reported. For a production of 10,000 Mt/y NaOH, a payback period of 2–5 years was estimated.

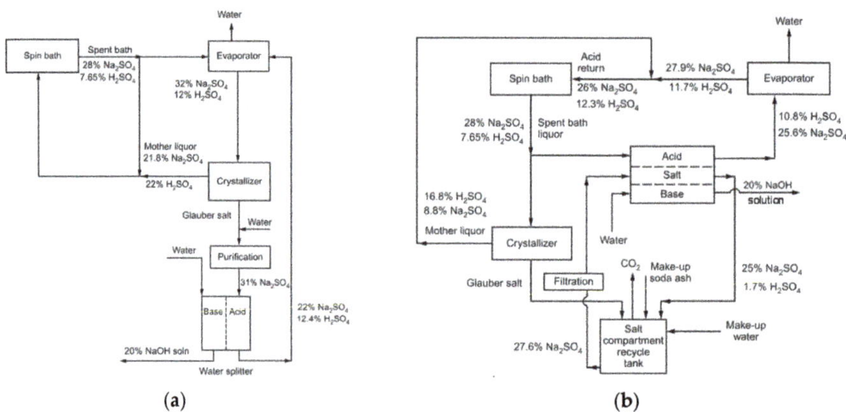

(a) **(b)**

Figure 14. Rayon process flowsheet with BMED: (**a**) Two-compartment configuration (BM-CEM); (**b**) Three-compartment configuration. Panel (**a**) is reproduced with permission from [37], published by Elsevier, 1988. Panel (**b**) is reproduced with permission from [38], published by Elsevier, 1991.

BMED was cost-effective for desalinating cooling tower blowdown and producing acid and base, which could be reused on-site, e.g., for IXRs regeneration [269]. From NaCl synthetic solutions (48–390 mM, with the lower part of the range being representative of cooling tower blowdown), 73–81% of salt was converted into acid/base, at η higher than 75% and E_{spec} of 0.02–0.09 kWh/mol. The maximum estimated cost (12.6 \$/kmol, under the assumption that the total cost is 1.7 times the energy cost) was less than the minimum cost of purchase (21 \$/kmol).

The technical feasibility of BMED was proven for recycling several other saline wastewaters. Tests on BMED stacks were performed with NH_4NO_3 nuclear fuel processing effluents [270,271] or $NaNO_3$ dying industry effluents [272] to produce HNO_3 and NaOH. Wastewaters from UF_6 production can be recycled as HF and KOH [251]. Phosphogypsum ($CaSO_4$) by-product from phosphoric acid production can be converted by NaOH into $Ca(OH)_2$ and Na_2SO_4, thus splitting the salt into base and sulphuric acid via

BMED [273]. Other applications were proposed to convert NH_4Cl into HCl and NH_3 [274,275], NH_4HCO_3 into NH_3 and CO_2 [276], Na_3PO_4 into H_3PO_4 and $NaOH$ [277], $NaBr$ into HBr and $NaOH$ [278,279], $Na_2SO_4/(NH_4)_2SO_4$ into H_2SO_4 and $NaOH/NH_3$ [280], $NaCl/KCl$ into HCl and $NaOH/KOH$ [281], boron into boric acid [282–284].

A SED-BMED coupled process was developed for waste salt mixtures ($NaCl$ and Na_2SO_4, originated from dye synthesis) conversion into $NaOH$ and separated acids (HCl and H_2SO_4) [285] (Figure 15). Different feed solutions were used, with sulphate concentration from 26 to 840 mM and chloride concentration from 63 to 497 mM. SED with MVAs fractionated the salts into two product streams at purity of ~90% for Cl^- and over 90% for SO_4^{2-}. From these solutions, the BMED processes yielded pure $NaOH$ and acid solutions rich in HCl or H_2SO_4 by 87% or 93%, respectively. E_{spec} was ~6.4 kWh/kg product on average for the SED with solutions at medium to high concentration, while it was ~5 kWh/kg $NaOH$ for the BMED.

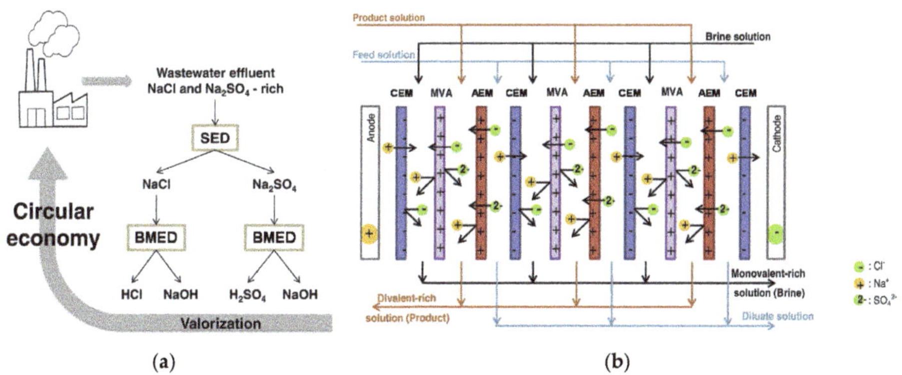

Figure 15. Valorisation of salt mixture wastewater by SED-BMED integration: (**a**) Scheme of the process; (**b**) Sketch of SED with MVAs. Reproduced (adapted) with permission from [285], published by Elsevier, 2016.

4.2.3. Spent Solutions from Chemical Absorption of Flue Gases

Chemical absorption through wet scrubbers is used in treatment lines for waste gases produced by combustion at power plants (e.g., coal-fired) and by other processes. BMED or ED can be adopted for regenerating spent alkaline or acidic solutions from flue gases chemical absorption, thus recycling the absorbent for the scrubbing tower. Two different processes were developed to recover spent alkaline absorbents for SO_2. One of them was based on the three-compartment BMED fed by the Na_2SO_4 solution from the stripper to produce $NaOH$, which is reused for absorption [8,286]. The other process was developed by the two-compartment BMED unit with BM and CEM [286,287], and took the name of Soxal^{TM} as an industrial process [38] (Figure 16). The spent solution is an $NaHSO_3/Na_2SO_4$ mixture that is converted into a regenerated stream of Na_2SO_3 (in the base channel), and into a stream with SO_2 (in the acid channel) that is then stripped. These regeneration processes led to significant economic advantages, exhibiting $\eta \approx 90\%$ and $E_{spec} \approx 1.3$ kWh/kg base [9].

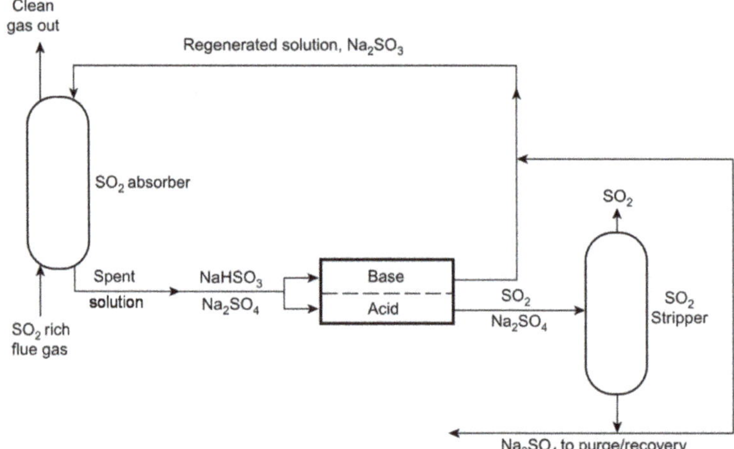

Figure 16. BMED Soxal^{TM} process for flue gas desulfurisation. Reproduced (adapted) with permission from [38], published by Elsevier, 1991.

The coupling of IX with BMED was tested to treat limestone-gypsum wet flue gas desulfurisation wastewater after chemical precipitation of Ca^{2+} and Mg^{2+} [288]. A synthetic wastewater (35–140 g/L NaCl and Na_2SO_4 mixed salts at mass ratio of 2:1, and 40–250 mg/L Ca^{2+} or Mg^{2+}) was softened by chelating IXRs to remove residual hardness, thus preventing scaling. The BMED obtained 99.3% pure acid and 99.0% pure base. The former will regenerate the saturated IXRs, and the latter will be dosed for the precipitation step.

BMED powered by solar organic Rankine cycle was tested with saline wastewater from flue gas desulfurisation in fluid catalytic cracking [289]. Experimentally assisted simulations were performed for a process in which the NaOH absorption spent solution contains HSO_3^- and SO_3^{2-} that are oxidized (in an aeration tank) to SO_4^{2-}. Thus, the Na_2SO_4 solution was treated by BMED. Simulation results showed that the salt solution was converted into H_2SO_4 (7.6 wt%) and NaOH (6.4 wt%), by reducing the salt content in the wastewater from 8.0 wt% to 0.37 wt% with $\eta \approx 52\%$ and $E_{spec} \approx 2.7$ kWh/kg salt.

A two-step ED treatment removed fluoride and chloride from ammonia-based flue gas desulfurisation slurry [290]. The slurry was pre-treated by MF and IX to remove fly ash and metals. A synthetic solution (10,000 mg/L F^-, 20,000 mg/L Cl^-, and 50% $(NH_4)_2SO_4$, by dissolving NH_4F, NH_4Cl, and $(NH_4)_2SO_4$) was also used for comparison purposes. After the first stage, Cl^- was almost completely transported to the concentrate (tap water) with $\eta \approx 54\%$ and $E_{spec} \approx 0.9$ kWh/kg, while a small amount of F^- was removed, mainly remaining with ammonium and sulphate in the slurry (diluate). A double second stage was then carried out to treat the two outlet solutions from the first stage. One was fed with the previous concentrate in the diluate channels to further separate Cl^-, another was fed with the previous diluate in the diluate channels to separate F^- and purify the slurry. The MVAs were crucial for SO_4^{2-} retention. A solution with Cl^- purity larger than 95% and a solution with F^- maximum purity of 51.4% were obtained.

ED can be cost-effective for regenerating spent alkanolamine effluents used for H_2S absorption, by removing inorganic and organic degradation by-products that form heat stable salts [291,292]. The estimated cost was 14.6 \$/ton with E_{spec} of 39.4 kWh/ton for a spent amine wastewater from the H_2S desulfurisation stripper of a thermoelectric factory (20.38 wt% N-methyldiethanolamine and 2.54 wt% salts, 36 L/day [291]). The selective removal of heat stable salts along with the minimisation of N-methyldiethanolamine loss was attained by developing ED or EDI stacks equipped with three-compartment repeating units [293]. This configuration comprised: CEM, concentrate, AEM, diluate (with or without anion-exchange resin), AEM, NaOH solution. Hydroxyl ions of the base compartment migrated to the diluate and reacted with binding amine, and thus neutral amine was regenerated. Moreover, the hydrolysed (cationic) amine was retained in the diluate. This solution was depleted in anions migrating to the concentrate. The spent solution coming from an H_2S desulfurisation stripper in an integrated gasification combined cycle power plant contained 21.06 wt% N-methyldiethanolamine and 5.19 wt% heat stable salts, at pH of 9.4 and conductivity of 12.32 mS/cm. Salts were removed by ~94%, 86% and 65% in the three-compartment EDI, three-compartment ED, and conventional ED, respectively, which exhibited losses of amine of ~3.8%, 5.6% and 21.1%, with E_{spec} of 71.7, 66.6, and 56.25 kWh/m^3 wastewater and estimated total cost of 0.88, 0.92, and 1.04 US\$/kg heat stable salt (treatment of 48 L/h). Moreover, AEM fouling was reduced in the EDI.

Similarly, ED can regenerate spent alkanolamine absorbents for CO_2. Pilot-scale studies were conducted with a spent solution at 30 wt% monoethanolamine, showing stable performances during long-term operations [294]. The effect of CO_2 loading (from 0 to 0.2 mole/mole amine) on heat stable salts (48 meq/L) removal from monoethanolamine-based solvent (30 wt%) was studied by two-stage ED [295] (Figure 17). An increase in recovery of salts was observed as the CO_2 concentration decreased, due to the smaller content of amine charged species and their lower competitive transport. An optimum CO_2 loading of 0.1 mole/mole amine and the associated E_{spec} of 25.9 MJ/kg solvent were estimated, considering that the change in CO_2 loading would require additional power for further solvent regeneration in the stripping column.

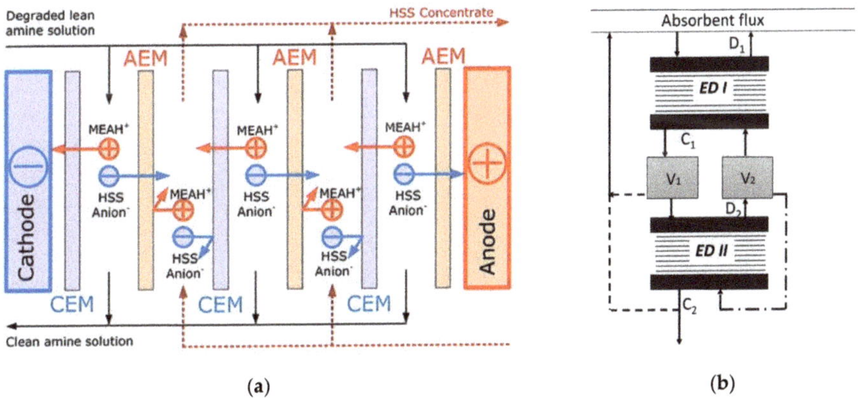

Figure 17. ED for heat stable salt removal from monoethanolamine solvent: (**a**) principle of the process; (**b**) flow scheme of the two-stage ED, where the concentrate stream after the first stage goes to the second stage to reduce monoethanolamine loss. Reproduced from [295].

A BMED stack with BM-CEM configuration recovered CO_2 and regenerated NaOH from model carbonate solutions [296], recording η values of 46–80% and E_{spec} values of 1–3 kWh/kg CO_2. The economic analysis highlighted the importance of the membrane cost for the process competitiveness. A coupled system was developed with the BM-AM two-cell BMED configuration and a hollow fibre membrane contactor aiming at regenerating spent absorbers (1 M monoethanolamine, piperazine or NaHCO$_3$), removing heat stable salts and separating CO_2 [297]. From BMED, the alkaline stream was recirculated into the flue gas absorber, while the acidic stream was recirculated into the membrane module for CO_2 separation. The developed mathematical model predicted $E_{spec} = 2$ MJ/kg CO_2, but the actual consumption in the experiments was 3–4 times higher due to the low η (~40%).

4.2.4. Effluents with Organic Matter

Salt or acidic wastewaters may have organic compounds. BMED or ED have been studied for acid/base recovery and organic matter separation, despite such solutions are complex and bring possible issues of organic fouling.

In a BMED stack used to regenerate NH_4^+ and H_2SO_4 from glutamate wastewater, CEM scaling at the base side was caused by Ca^{2+} and Mg^{2+}, along with minor fouling on the other surface [298]. However, acid–ultrasound cleaning restored the membrane properties. Glyphosate recovery and HCl/NaOH production were obtained by BMED of alkaline glyphosate (12.8 g/L, with ~175 g/L NaCl) neutralisation liquor from pesticide industry [299,300]. Glyphosate was recovered by 98.2%, and the NaOH solution (~1.45 M with ~96.5% purity) was produced with maximum η of 80.8% and minimum E_{spec} of 2.15 kWh/kg [300]. Prospecting the base reuse for CO_2 sequestration, the overall balance estimation was positive only if employing renewable energies.

A three-stage BMED process was developed to remove aniline (1000–3000 ppm) and salt (0.1 M NaCl) from a model wastewater and to simultaneously capture CO_2, through the reaction of the amine group with carbon dioxide that results in positively charged amine [301]. Aniline was completely transported to the base compartment (at η up to 80% and E_{spec} of ~3 kWh/kg), and the desalination exceeded 94% (at $\eta = 90\%$ and $E_{spec} \approx 1$ kWh/kg). The possible use of conventional ED for aniline-H_2SO_4 wastewater was suggested by performing an analysis of mass transfer and current–voltage characteristics [302].

Measures for counteracting the adverse effects of leakage currents (parasitic currents flowing through the manifolds) were suggested in a BMED process fed by high ammonium chloride organic wastewater (2.46 M NH_4^+, 1.95 M Cl^-, 6 g/L carbocystein) [303]. Experiments and simulations showed that, to enhance the process efficiency and reduce overheating phenomena, a proper stack design should be devised, increasing the relative resistance of the parasitic pathways. In particular, using low-resistance membranes, thin spacers, sufficiently long slots in the spacer gasket, and implementing a two-stage scheme can be fruitful to this aim.

BMED with BM-CEM configuration reclaimed waste solutions with polymeric bonding agents and sorbed heavy metals from, e.g., polymer-enhanced UF [304]. Bonding agents were regenerated (84–95%) in the acid compartment by protonation, while heavy metals (Cu^{2+}, Ni^{2+}, Co^{2+} and Pb^{2+}) were transported to the base compartment and separated by formation of hydroxides.

Spent caustic reclamation for NaOH regeneration via BMED was demonstrated by testing a BM-CEM unit [305]. From a spent caustic with 0.44 M NaOH, 0.29 M Na_2CO_3, and 0.048 M Na_2SO_4, optimal conditions yielded a 0.11 M base at η approaching 100%, E_{spec} of ~8 kWh/kg NaOH, and an estimated cost of 0.97 US\$/kg (process capacity of 18.9 kg/year), without observing effects due to oil.

ED concentration was reported for HCl from waste effluents originating from hydrolysis of palm oil by-products [306], and for NaOH from cellulose mercerisation wastewater [307].

ED technologies have been widely studied for production of organic acids, with development for some industrial applications [19,308]. New opportunities are derived from the recovery of organic acids from wastewaters via ED [309–315], or BMED [316–318] or both [319,320], including systems combined with biotechnologies.

Naphthenic acids were recovered by BMED from sodium naphtenate solutions [321]. Naphthenic acids are valuable chemical raw materials, which negatively affect the quality of petroleum distillates. They are removed by alkaline extraction, thus generating a solution with salts of naphthenic acids. The BMED with

BM-CEM-CEM three-chamber unit cell formed insoluble naphthenic acids (salt compartment fed with 18 wt% sodium naphtenate, 24 wt% NaOH, 1.5 wt% oil), which are separated in a sodium naphthenate reservoir filled with Raschig rings (Figure 18). The same process was conducted by introducing a cation-exchange resin (EDI) and sodium sulphate in the salt chamber, obtaining (i) a reduction in the electrical resistance, (ii) gains in limiting current density, η (up to ~80%) and E_{spec} (0.38 kWh/L), and (iii) a total conversion.

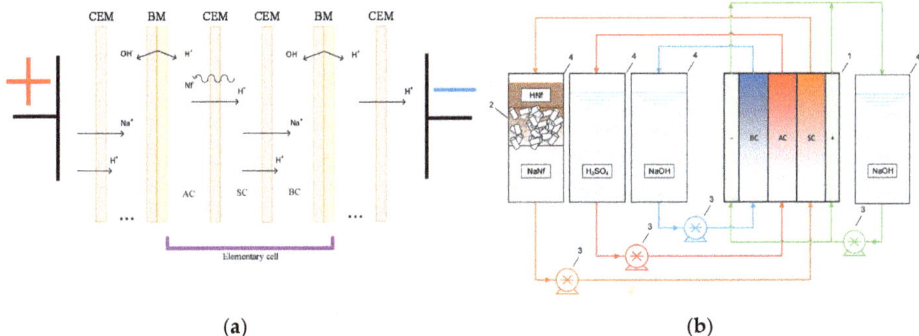

(a)   (b)

Figure 18. BMED recovering naphthenic acids from sodium naphtenates: (**a**) BM-CEM-CEM configuration; (**b**) flow scheme with electrodialyzer (1), Raschig rings (2), pumps (3) and tanks (4). Reproduced (adapted) with permission from [321], published by Elsevier, 2019.

4.3. Desalination

Several industrial processes originate salty wastewater, which needs to be desalinated before its reuse or discharge. To this end, the use of ED has been studied for the following main types of wastewater from industrial activities: produced water from oil and gas extraction (Section 4.3.1), wastewater from refineries and petrochemical industries (Section 4.3.2), drainage wastewaters from coal mining (Section 4.3.3), and wastewater from power plants (Section 4.3.4).

4.3.1. Oil and Gas Extraction

Operations of oil and gas extraction produce large volumes of effluents. The amount of water produced worldwide as a by-product of oil and gas production is ~250 × 10⁶ barrels/day, i.e., approximatively triple the produced oil [322]. During some extractions of oil and gas, water can be brought to the surface, as present in (or nearby) the hydrocarbons reservoir, or because intentionally injected with additives to enable the withdrawal. Water is pumped during extraction of unconventional gases, such as coal seam gas and shale gas. The former (known also as coal bed methane), which is adsorbed to the coal surface, is released up to ~1 km underground by effect of a depressurisation applied by wells pumping water from the seams. The latter is extracted from greater depths via hydraulic fracturing, i.e., rocks fracturing by injecting a high-pressure liquid (water with chemicals) from vertical or horizontal wells, thus producing flow-back water. The produced water is also generated during oil recovery enhanced by the polymer flooding technique, which consists of injecting a high-viscosity aqueous solution with soluble polymers that improves the oil sweep efficiency by a less mobile phase.

The produced water composition is strongly site-specific. Total Dissolved Solids (TDS) may amount from some to ~300,000 mg/L [322], being 200–40,000 mg/L in coal seam gas produced water [323] and more in shale gas produced water [324,325]. Polymer flooding produced water falls in the lower part of this range.

Additionally, produced waters contain oil, organic compounds, suspended solids, heavy metals and natural radioactive materials. Therefore, treatments (e.g., biological and physico-chemical processes) are needed before reuse/discharge. Membrane processes [322–327], including ED for desalination, can be adopted.

Two ED applications are documented in [328]. After de-oiling and removal of dissolved organics (floatation and fluidized bed reactors), the produced water from a conventional well was desalinated, attaining a TDS removal of ~89% from 9100 ppm. Coalbed methane produced water (TDS up to 27,000 ppm) was recovered by 80–90% by mobile ED units, and was reused for fracturing.

ED experiments with simulated produced waters (TDS from ~4400 mg/L to ~97,600 mg/L through the diluate, 25 g/L NaCl through the concentrate) were carried out in order to evaluate the attainment of standards for different reuses (targets of 500–5000 mg/L TDS) [329]. At low feed concentrations, regardless of the composition, it was possible to attain the concentration targets with E_{spec} of ~1.2 kWh/m^3, while at high concentrations it was not feasible due to exaggerated E_{spec} values (more than 20 times higher) or even to unattainable targets. Several lab-scale tests exhibited promising results, even for hypersaline solutions [330–332]. Cost analyses (based on experiments with NaCl solutions) and models showed that ED is a cost-effective method for brackish water [333], but, if optimized, it can be competitive also for high salinity feeds [331]. However, the system behaviour should be characterized by tests with real feeds, where other ions are present.

With simulated produced waters from shale gas fracking (3% or 6% NaCl with 1000 or 4000 mg/L Ca^{2+}), scaling at the cathode chamber was mitigated by an MVC end-membrane (Ca^{2+} flux decreased by 47–73%), thus obtaining a current density increase of ~40% [334]. Both simulated and real produced waters from shales were then used [335]. After pre-treatment (NaOH dosage, settlement and MF) and, in some cases, dilution, the field samples were partially desalinated by ED (feeds with 25,000–44,600 mg/L TDS, removals up to ~60%) showing similar performance compared to simulated effluents. Operations with a periodic pulse polarity-reversal enhanced the ion migration by temporarily disrupting the stagnant layer of Ca^{2+}, Mg^{2+} and Ba^{2+} retained by the MVC. Nevertheless, the occurrence of some precipitation of $Fe(OH)_3$ on any IEM suggested to boost the pre-treatment.

For reusing polymer flooding produced water, TDS must be reduced at 500–1000 ppm because higher concentrations could lessen the viscosity of the solution. ED desalination can be applied (Figure 19), and has been demonstrated by pilot-/large-scale plants [336,337]. Serious fouling issues occur, and its mechanisms were investigated along with fouled membranes characterisation [338–341], proposing chemical cleaning strategies [337,342]. Synthetic solutions at 5000 mg/L (brackish water) or 32,000 mg/L (seawater) TDS and with 1.0 g/L partially hydrolysed polyacrylamide (HPAM) were reusable after ED desalination (concentrate feed with 5 g/L NaCl) with small replenishment of polymer (~25% was withhold in the stack) [343]. η was 85–99% and E_{spec} was 0.5–6 kWh/m^3. Moreover, a preferential removal of divalent ions was feasible, especially at low current densities [344]. Separating multivalent ions is desirable to allow polymer-flooding produced water to be reused, since they have the most significant effects in reducing the solution viscosity (calcium and magnesium) and could lead to scaling and reservoir souring. Tests with different solution compositions showed that fouling issues were associated mostly to HPAM adsorption on AEM and formation of a gel layer favoured by divalent cations [345]. However, the gel layer was significantly removed by application of current reversal and use of foulant-free solution. The minimisation of the gel layer formation was then obtained by applying pulsed electric fields [346]. Oily compounds (synthetic solution with 53.3 mM NaCl plus HCO_3^-, SO_4^{2-}, Ca^{2+} and Mg^{2+}, 250 mg/L HPAM and 2 mg/L crude oil) increased slightly membrane fouling, but made the HPAM gel layer less stable. The best condition (1 s/1 s of pulse/pause) led to a reduction of ~35% in E_{spec} (~0.6 kWh/m^3).

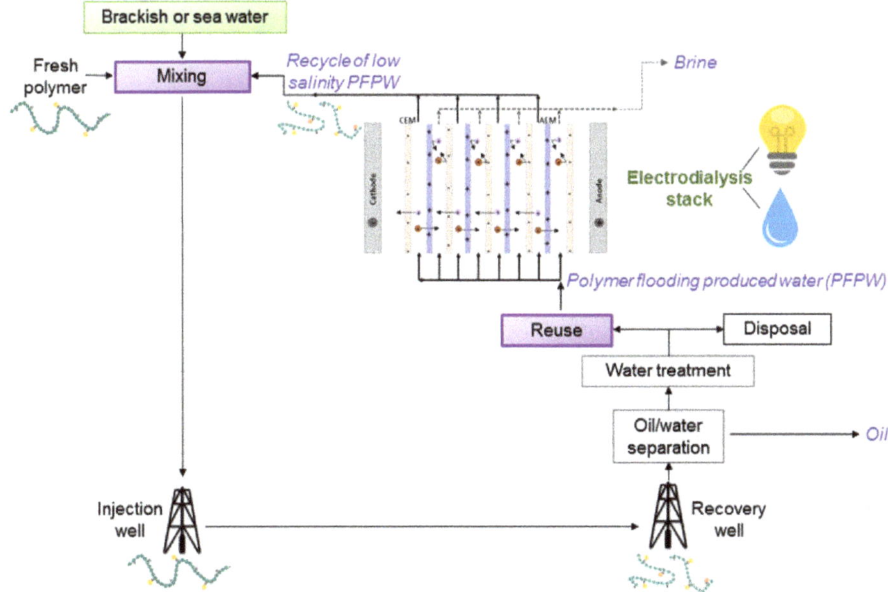

Figure 19. Scheme of ED desalination of polymer flooding produced water. Reproduced from [343], published by Elsevier, 2018.

One study proposed reverse electrodeionisation (REDI) for simultaneous energy recovery and water reuse by controlled mixing between treated fracturing produced water and fresh water (needed for replenishment) [347]. The REDI diluate contained IXR-wafers to lessen the electrical resistance. The highest P_d was 0.9 W/m^2 along with a $P_{d,net}$ of 0.79 W/m^2 by using produced water at 162 mS/cm (~130 g/L, after NF) coupled with fresh water at 1 g/L, without observing fouling. Assuming the use of 5×10^6 gallons and 60% water recovery from drilling, rough economic calculations show an average increase in revenue over 300,000 $/(year·well). However, the profit will depend strongly on the process (mainly capital) cost.

4.3.2. Refineries and Petrochemical Industries

Petroleum refineries use water for many processes [348,349]. However, cooling processes are responsible for ~90% of overall water consumption [349]. Refinery wastewaters, including cooling tower blowdown, are sent to treatment plants [350], and ED can be adopted for desalination.

A pilot EDR was installed (late 1980s) at STANIC Industria Petrolifera (Livorno, Italy) [351]. The effluent from the biological treatment (~1500 ppm TDS) was desalinated by ~90%, thus providing cooling tower makeup and boiler feeding, along with a concentrate suitable for discharge, and kicking off the construction of a 180 m^3/h full-scale plant.

A pilot EDR was used as pre-desalination step followed by RO for a tertiary effluent (~1150 mg/L TDS) from petrochemical industry [352]. Two ED modules (75 cell pairs per each, with a total area of 28.8 m^2) were used either in series or in parallel mode. The EDR step removed up to ~90% of TDS, while the EDR-RO hybrid system achieved overall removals above 90% for several physico-chemical parameters, with 41% water recovery (75% in EDR, and 50% in RO). Water recovery can be boosted by hybrid schemes in which ED treats the RO brine retentate (see Section 6.1.3).

4.3.3. Coal Mines

Coal mining drainage waters are brines with high salt content that must be reduced to allow water reuse. An industrial EDR application at Tutuka power station (South Africa) was upgraded to a more than doubled capacity (13,200 m^3/day) to desalinate mine water as well (2500 mg/L TDS) [353] (see Section 4.3.4).

EDR tests with coal mine effluent at low concentration (~2100 mg/L TDS) were performed at single pass with a diluate velocity ten times the concentrate velocity [354]. Despite the high super-saturation level of calcium sulphate and calcium carbonate in the concentrate, crystallisation was avoided by the insufficient residence time, thus preventing scaling. Water recovery of ~90% and salts removal of ~70% were obtained. ED brine valorisation by salt production via two-stage ED (prior to evaporation-crystallisation) was studied with a coal mine solution with 32.8 g/L Cl$^-$, finding E_{spec} values in the order of 10 kWh/m^3, and producing a sufficiently pure concentrate [355]. Then, a combined NF-ED-RO system was proposed [356], supported by additional experiments [357]. An alternative way of coal mine brine valorisation could be represented by energy recovery through RED [358]. Artificial solutions simulating coal mine brine (111 g/L NaCl) and fresh water (0.56 g/L NaCl) produced a $P_{d,max}$ of 0.87 W/m^2 and a corresponding $P_{d,max,net}$ of 0.71 W/m^2. An investment cost of 3 \$/kWh was estimated by assuming a peak power of ~1 W/m^2 (low-resistance membranes), showing that the economic feasibility is strongly dependent on the membrane cost.

Sulfide minerals (pyrite, FeS$_2$) can be oxidized when in contact with water and oxygen, thus resulting in acid mine drainage that contains sulphate, iron and other (heavy) metals. Samples collected from different locations in a carboniferous, with pH < 3 in most cases and different compositions (conductivity from 1155 to 15,300 μS/cm, SO$_4^{2-}$ from ~500 to ~8000 mg/L, various cations) were desalinated by ED (after settling and MF) achieving removals of 97–99%, thus recovering the diluate [359]. Iron precipitation was observed on CEMs, thus long-term operations could require pre-treatment to prevent scaling. A reduction in membrane resistance was found at higher current densities with solutions of Fe$_2$(SO$_4$)$_3$, attributing this behaviour to the FeSO$_4^+$ dissociation into more mobile Fe^{3+} and SO$_4^{2-}$ ions at the boundary layer [360]. Moreover, a significantly different selectivity was observed between homogeneous and heterogeneous CEMs immersed in mixtures with Na$_2$SO$_4$.

4.3.4. Power Plants

Power plants' cooling tower blowdown can be desalinated by ED. The industrial EDR with 7-year operation at Tutuka power station cited in Section 4.3.3 was accomplished in ZLD approach [353]. The plant, upgraded (13,200 m^3/day) also to treat mine water, received a feed with 2500 mg/L TDS with ~50% CaSO$_4$ saturation. After pre-treatment with HCl dosage for scaling inhibition, chlorine dosage against organics, and coagulation–filtration for suspended solids removal, the EDR plant recovered water by 75% at η of 86%, with attractive costs and long membrane life.

Cooling tower blowdown (conductivity from 2.3 to 3.5 mS/cm, flow rate of 2.3 m^3/h) was treated by including EDR desalination in the pilot facility (lamella separator, UF, MF, EDR) in Terneuzen, The Netherlands [361]. The ED stack comprised four hydraulic stages and two electrical stages. Normalized parameters were introduced for pressure drop, IEM resistance and η to control, monitor and optimize the process. In a 2-month operation, η was stable.

4.4. Treatment of Other Wastewaters

This section reports studies on ED methods (for separation, desalination, concentration, regeneration or energy recovery) applied to other industrial wastewaters that have not been presented above. Tables 2 and 3 regard effluents with and without organic matter, respectively.

Table 2. Other studies on industrial wastewaters without organic matter (adapted from [23]).

Wastewater	Treatment Process	Main Remarks	Ref.
NaCl + Na$_2$SO$_4$ solution, 0.01 M each	ED with MVAs	Layer-by-layer composite AEM, $P^{Cl^-}_{SO_4^{2-}} = 11.5$	[312]
NaCl + Na$_2$SO$_4$ solution, 0.05 M each	ED with MVAs	MVAs with hydrophobic alkyl side chain, max $P^{Cl^-}_{SO_4^{2-}} = 13.1$, long-term stability	[362]
NaCl + MgCl$_2$ or LiCl + MgCl$_2$ solutions, 0.1 M each	ED with MVCs	Zwitterion structure MVCs, $P^{Na^+}_{Mg^{2+}} = 58.4$ and $P^{Li^+}_{Mg^{2+}} = 6.5$	[363]
NaCl+MgCl$_2$ solution, 0.1 M each	ED with MVCs	Zwitterion structure MVCs with hydrophobic alkyl side chain, max $P^{Na^+}_{Mg^{2+}} = 25.3$	[364]
Model solutions with two salts with the same counter-ion among NaCl, Na$_2$SO$_4$, MgCl$_2$, MgSO$_4$ and NaNO$_3$, 0.01 M each	ED or ED with MVMs	Max separation efficiency ~68% for cations by MVMs (comparable to NF), but lower for anions	[365]
NaCl + Na$_2$SO$_4$ solution, 8 mM each	SED	SO$_4^{2-}$ purity > 85%, $\eta \approx 50\%$	[39]
MgCI$_2$ + Na$_2$SO$_4$ solutions, 0.3–0.5 M each	EDM	$\eta > 100\%$, $E_{spec} \approx 0.9–1.6$ kWh/kg, MgSO$_4$ purity ~98%	[40]
Na$_2$SO$_4$ solutions, 0.01 M/0.3 M	RED-alkaline polymer electrolyte water electrolysis	$V_{OC} \approx 12$ V (200 cell pairs), $P_{d,max} = 0.04–0.11$ W/m^2 by changing solutions velocity and temperature, H$_2$ production 50 cm^3/(h·cm^2)	[366]
Catalyst plant model wastewater $\left(Na^+,\ Cl^-,\ Mg^{2+},\ Ca^{2+},\ SO_4^{2-}\right)$, 25.8 g/L TDS	Two-stage ED with MVCs	On-line membrane modification, $P^{Ca^{2+}}_{Na^+}$ reduced from 0.36 to 0.11, $P^{Mg^{2+}}_{Na^+}$ from 0.81 to 0.12, $\eta = 75–92\%$, 1 g/L diluate, stable long-run, but larger water transport, membrane resistance, and E_{spec} (up to ~35% more)	[367]
Photovoltaic industry simulated wastewater, 120–180 mg/L NaF and/or 750–2000 mg/L NaNO$_3$	ED	With single salt, max removal efficiency ~60% and 75% for F$^-$ and NO$_3^-$ in 6 min, under optimal conditions $E_{spec} = 0.25–0.36$ kWh/m^3; with mixture, ion competition affected only F$^-$ removal	[368]
F$^-$ solutions: single salt at 25–200 mg/L, binary and ternary mixtures with 100 mg/L F$^-$ + Cl$^-$ and/or SO$_4^{2-}$ at same equivalent concentration	ED	High removal efficiencies, $E_{spec} = 0.02–0.49$ kWh/m^3, Cl$^-$ affected F$^-$ separation, SO$_4^{2-}$ did not	[369]
Synthetic secondary effluent of graphite industry, 10–30 mg/L NaF, 6 g/L NaCl	ED	Response surface methodology, F$^-$ removal 99.69% with $E_{spec} = 0.76$ kWh/m^3 under optimal conditions	[370]
B artificial wastewater, 25–100 mg/L; binary or ternary mixtures with 100 mg/L B + Cl$^-$ and/or SO$_4^{2-}$ at same or doubled equivalent concentration	ED	Max removal of B ~80%, enhanced at high pH (10.5) due to a predominance of B(OH)$_4^-$, hindered by Cl$^-$ and not by SO$_4^{2-}$, $E_{spec} = 0.02–1.24$ kWh/m^3	[371]
Acidic model solution from B-selective sorbents regeneration, 0.2 M HCl or 0.1 M H$_2$SO$_4$ + 1.0 or 5.2 g/L H$_3$BO$_3$	Two-stage ED with pH increase	Regenerating acid (HCl or H$_2$SO$_4$) recovered in the concentrate ~90%, ~93% of H$_3$BO$_3$ (non-ionic) retained within the diluate and concentrated in the 2nd stage after alkalinisation, reusable solutions	[372]
B-containing industrial landfill leachate, 62.8–76.5 mg/L B (+ SO$_4^{2-}$, Cl$^-$, Ca^{2+} and Mg^{2+})	Two-stage ED with pH increase	Desalination in the 1st stage 80%, B(OH)$_4^-$ removal in the 2nd stage 97% under alkaline conditions, max $\eta = 25–28\%$, estimated cost 1.27 \$/m^3	[373,374]
Model nuclear power plant effluent, 60–400 mg/L H$_3$BO$_3$	Three-compartment EDI	Max removal ~45%, optimal pH = 10	[375]
NH$_4$NO$_3$ model wastewater from fertilizer production, 0.012 M	ED	Thin heterogeneous IEMs vs. commercial ones: higher limiting current density due to larger back-diffusion and electroconvection; lower alkalisation due to lower water dissociation	[376]
Synthetic solutions with single acid or salt: H$_2$SO$_4$, HNO$_3$, NH$_4$NO$_3$, NaCl, LiCl, Na$_2$SO$_4$, 0.06–0.3 M	ED or BMED-ED	ED concentrator without flow through concentrate chambers, acid concentration 1.16 M at $\eta = 89\%$ for BMED and 26% for ED, $E_{spec} = 0.83$ kWh/mol SO$_4^{2-}$	[377]
Alkaline liquid from bauxite solid residue (Bayer process) washing (2.4 g/L Al^{3+}, +K$^+$, Na$^+$, F$^-$, SO$_4^{2-}$ ...)	ED with aeration	NaOH recovery, NaAl(OH)$_4$ separation, TDS and OH$^-$ removal 61.3% and 76.6%, $\eta = 60\%$, $E_{spec} = 11.15$ kWh/kg	[378]
Synthetic or real wastewater from mineral carbonation for CO$_2$ sequestration, 0.05–1.0 M (NH$_4$)$_2$SO$_4$, 0.05–0.54 M (NH$_4$)HSO$_4$, (+MgSO$_4$, NH$_3$, Fe(II), Fe(III) ...)	BMED	Different setups for regenerating rock-derived solutions after leaching or after carbonation, $E_{spec} = 1.7–350$ MJ/kg NH$_4^+$	[379]
Model solution from Li-ion waste batteries, Li$^+$ and Co^{2+} 0.02 M each	BMED with complexation	Co-EDTA chelated anions and Li$^+$ separated in the acid and base compartments, respectively, removals 99%, but Co absorption in AEM; metal recovery enhanced in semi-batch operation for the feed	[380]

Table 3. Other studies on industrial wastewaters with organic matter (adapted from [23]).

Wastewater	Treatment Process/Exp. Device	Main Remarks	Ref.
Solutions with octanoic acid or anionic surfactants; alkaline bleach plant filtrate from sulphate pulp mill, 1370 mg/L COD	IEM resistance measurement cell	Slight inorganic fouling on CEM by bleach plant filtrate, significant organic fouling on AEM by all solutes	[29]
Solutions with carboxylic acids (propanoic, octanoic and decanoic acid); alkaline bleach plant filtrate from sulphate pulp mill, 1850 mg/L COD	IEM resistance measurement cell, ED	No CEM fouling, AEM fouling due to organic anions, especially compounds with longer chain, and at higher currents	[381]
Solutions with 16 charged or neutral trace organic contaminants, 0.1 mg/L with 100 g/L NaCl	ED	Adsorption governed by electrostatic interactions, transport mostly diffusion driven, migration of charged components only at very low NaCl concentration	[382]
Solutions with NaCl, Na_2SO_4 or $MgCl_2$ and acetic acid, phenol or glucose, 0.8 eq/L salts and 0.1 M organics	ED	Phenomenological model: convection-diffusion of neutral organics affected by steric effects and ion hydration	[383]
Solutions with NaCl, Na_2SO_4 or $MgCl_2$ and acetic acid, phenol, glucose or acetate	ED	Phenomenological model: transport of several organics larger with SO_4^{2-} than with Cl^-, opposite trend for phenol	[384]
Wastewater from bisphenol A diphenyl phosphate production, 4.5–4.8% total salt (NaCl and sodium phenolate), pH = 13.2–13.5, diluted with pure water	RED-ED	Ultrapure water fed into the RED diluate, V_{OC} up to 1.65 V (10 cell pairs) and $P_{d,max,net}$ up to 1.12 W/m^2 in RED at dilution ratio 1.0:0.5, E_{spec} lower than that of standalone ED (17.65 vs. 25.32 kWh/m^3) with 27.4% pre-desalination in RED	[385]
NaCl-glycerol solution, 1.11–1.67 M NaCl and 0.06–0.6 M glycerol	ED	7 membrane pairs tested, phenomenological model: $C_3H_8O_3$ electro-osmotic co-transport 38–64%, osmotic co-transport 16–41%, diffusion 9–28%, low glycerol/NaCl flux at low glycerol/NaCl and NaCl concentrations	[386]
Simulated dairy wastewater, 10 mM citrate, 1 mM lactate, 30 mM NaCl …	ED	Guanidinium groups in AEM as functional moiety binding oxyanions, enhanced transport of phosphate and citrate	[387]
Diluted effluent from sodium dithionate processing, 35 g/L HCOONa, 30 g/L $Na_2S_2O_3$ …	ED with MVAs	Recovery of HCOONa 69%, with 87% purity, η = 70%, E_{spec} = 96 kWh/m^3	[388]
Steel manufacturing wastewater (Cl^-, SO_4^{2-}, Na^+, Mg^{2+}, Ca^{2+}), 2.8–4.0 mS/cm, 36–72 mg/L COD	Sand filtration-EDR	Water recovery 75%, desalination 92%, concentrate COD below discharge limit, E_{spec} = 0.85 kWh/m^3, operation cost 0.146 \$/m^3	[389]
Secondary effluent from spinning processes, chemical industries, and metal processors (Cl^-, SO_4^{2-}, Mg^{2+}, Ca^{2+}, NO_3^-, PO_4^{3-} …), 7.3 mS/cm, 41.5 g/L COD	Sand filtration-EDR	Lower techno-economic efficiency compared to fiber filtration-UF-RO	[390]
ZnO washing wastewater (Na^+, K^+, Cl^-, Ca^{2+} and SO_4^{2-}), ~0.35 M, 1.2 mM TOC	ED with MVMs	Overall $\eta \approx$ 80%, divalent ions retained, thus scaling prevented, stable long-term performance of pilot plant with removal target of 50% (before evaporation)	[281]
Kraft pulp mill dissolved electrostatic precipitator dust (Cl^-, CO_3^{2-}, SO_4^{2-}, Na^+, K^+), 137 g/L TDS (0.1 wt% TOC in the dust)	ED with MVMs	Selective removal of Cl^- at η = 60–78% and $E_{spec} \approx$ 1 kWh/kg, organics in the dust recycled with the sulphate-rich diluate, no fouling, accumulated dust simply flushed, successful long-term operation, operation saving of 800 \$/1000 ton Kraft pulp	[391]
Paper mill effluent, 6046 mg/L TDS, 390 mg/L COD	MF-ED	Max TDS removal ~90%, water recovery 80%, $E_{spec} \approx$ 0.5 kWh/m^3, concentrate usable as biomass	[392]
Primary textile effluent, 2,980 mg/L TDS, 220 mg/L COD	UF-ED	Desalination ~96%, E_{spec} = 0.9 kWh/m^3, reusable water	[393]
Model textile effluent with 1 g/L reactive blue 194 and 40 g/L Na_2SO_4	Tight UF-based diafiltration-BMED	Pre-concentration at a factor of 8 and diafiltration with 8 diavolumes, UF permeate with low dye content (2.7 mg/L) and 21.06 g/L Na_2SO_4, 99.5% dye recovery, ~99% salt conversion into 99% pure 0.29 M acid and 0.4 M base without fouling, E_{spec} = 4.2 kWh/kg	[394]
Model textile effluent with 0.25 g/L Remazol Brilliant Blue R and 50 g/L Na_2SO_4	BMED	Effect of zeta potential of dye molecule on fouling, fouling controlled by the identification of a "critical salt concentration" below which desalination cannot proceed due to fouling, η =39%, desalination 74%, 72% of Na^+ and 66.9% of SO_4^{2-} converted into base and acid, respectively	[395]
Tannery unhairing effluent, pH = 12, 576 mg/L S^{2-}, 23,289 mg/L COD, 436 mg/L Ca^{2+}, 429.6 mg/L Cl^-	ED with protective UF membrane on AEM	Anti-fouling solution against proteins and peptides, desalination 56%, 90% of organics retained within the diluate, thus water recycling	[396]

Table 3. *Cont.*

Wastewater	Treatment Process/Exp. Device	Main Remarks	Ref.
Almond processing treated wastewater (electrocoagulation and electrooxidation), 7.2 mS/cm, 296 mg/L TOC	ED	Concentration factor of 10 in the concentrate, diluate target 0.5 mS/cm, TOC removal ~70%, water recovery 94%, no fouling, scale-up at pre-industrial scale, E_{spec} = 1.1–2.9 kWh/m^3	[397]
Waste brine from olive pickling process, 103.3 mS/cm, pH = 3.5, 8033 mg/L dissolved organic carbon; coupled with storm water, 3.6 mS/cm	RED	V_{OC} = 1.37 V (~70% of the ideal one, 10 cell pairs), $P_{d,max}$ = 0.59 W/m^2 enhanced (with respect to NaCl solutions at the same conductivities, 0.44 W/m^2) by pH gradient and organic acids (lower resistance)	[398]
Lysine fermentation effluent, 152 mS/cm, 17,800 g/L NH$_4^+$, 71,000 mg/L SO$_4^{2-}$, 102,300 ppm TOC	MF-ED	Separation of 73.1% NH$_4^+$ and 83.5% SO$_4^{2-}$, E_{spec} = 106 kWh/m^3, pulsed electric field effective against fouling, demineralized waste usable as animal feed, concentrate as fertilizer	[399]
Bio-refinery effluents: molasses effluents, lignocellulosic stream, sugar cane juice, 3.2–72.4 mS/cm, 38–380 g/L COD	ED	Salt removal 96% and 63% from lignocellulosic and molasses effluents, lower from sugar cane juice, low COD loss (< 6.3%), η = 69–104%, E_{spec} = 0.44–1.59 kWh/kg salt	[400]
Bio-refinery effluents: synthetic salt mixtures with sorbitol, molasses effluent	ED	Simplified process model, predictions in good agreement with experimental results	[401]
Vinasse from a distillery producing ethanol from sugar can juice, 30,500 mg/L COD, 11.5 mS/cm	UF-ED with MVMs	K$^+$ recovery 72%, E_{spec} = 9 kWh/m^3, η = 54%, concentrated stream for fertigation, diluate stream for fertigation or biogas production (anaerobic digestion)	[402]
Solutions of 3-chloro-1,2-propanediol (α-monochlorohydrin or 3-MCH) (model effluent from biodiesel production or other sources), 10 or 30% wt + 0.1 M KCl	BMED	Recovery of glycidol by dehydrohalogenation caused by OH$^-$ in the base compartment, selectivity 96%, η = 64%, glycidol distilled with 75.6% yield	[403]
Model antibiotic effluent with 0.95 g/L penicillin, 1 g/L SO$_4^{2-}$ and 1 g/L bovine serum albumin	ED with UF membrane, 3-comp. (AEM-UF-CEM)	Penicillin recovery ~20%, removal of SO$_4^{2-}$ from feed and antibiotic product 90%, no fouling, E_{spec} = 0.058–0.082 kWh/g, estimated profit 6850 \$/ton produced penicillin (8 L/day wastewater)	[404]
Effluent from anaerobic digester–decanter, 13,800 mg/L COD, 1700 mg/kg total N, 1800 mg/kg Cl$^-$, 2,900 mg/kg Na$^+$ ...	ED	Separation 70–96% for monovalent ions, < 50% for divalent ions, E_{spec} = 6–11 kWh/m^3 for water recovery 50–95%	[405]
Simulated supernatant of excess sludge mixed with influent from anaerobic-aerobic biological treatment, 100 mg/L P *	ED or ED-BMED	ED : PO$_4^{3-}$ recovery 95.8%, ED-BMED: 0.075 M H$_3$PO$_4$ recovered, $\eta \approx$ 70–80%, E_{spec} = 5.3–29.3 kWh/kg	[406]
Effluent of upflow anaerobic sludge blanket reactor from potato processing, 2.5 mM phosphate $\left(+\text{K}^+, \text{NH}_4^+, \text{Cl}^- ...\right)_*$	SED-struvite precipitator	6.8 mM phosphate in SED product (from 0.8 mM initial product, struvite effluent), average overall $\eta \approx$ 70%, desalination 95%, phosphate recovery 93%, E_{spec} = 16.7 kWh/kg phosphate	[407]

* These effluents did not contain organic matter; however, as they come from biological treatment, they could have, in general, residual organics.

5. Municipal Wastewater and Other Effluents

Desalination via ED can make treated municipal wastewater reusable, as shown by several field plant applications (Section 5.1). As an alternative, it could be used as a low-salinity solution coupled with seawater for recovering salinity gradient energy (Section 5.2). Additionally, ED methods are under study to recover nutrients (as well as water) and, in some cases, volatile fatty acids (VFAs) from treated wastewater and related/similar effluents (Section 5.3). Another ED application studied is the regeneration of liquid desiccant solutions for air conditioning (Section 5.4).

In addition, ED methods have been proposed for desalinating drainage wastewaters for agricultural reuse [408], and the experimental screening/optimisation has been studied [409].

5.1. Desalination of Municipal WWTP Effluents

Secondary or even tertiary effluents from wastewater treatment plants (WWTPs) are normally not reusable in irrigation, aquifer recharge, or industrial processes. When the salinity of treated effluents is relatively high, it can be suitably reduced by ED. MF is often used before ED to remove suspended solids and microorganisms. Pre-treatments, EDR operation and cleaning procedures against fouling can maintain or restore, at least partially, IEMs properties.

A rapid sand filtration-activated carbon-EDR plant of ~1100 m^3/day capacity supplied treated wastewater to Moody Gardens plants and fishponds (Galveston, Texas) [410]. Other EDR plants (Middle East) were cited in the same paper. In Gran Canaria, a pilot MF-EDR plant was tested for irrigation water reuse [411]. The EDR achieved reductions of conductivity and TDS of 89% and 74%, respectively, from a feed with 2.7 mS/cm average conductivity and 1565 mg/L average TDS. Among the other physico-chemical parameters, the following removals were obtained: 79% ammonia, 88% nitrate, 59% phosphate, 83% BOD_5, 40% COD, 50% faecal coliforms, from feed concentrations of 24 mg/L, 50 mg/L, 56 mg/L, 18 mg/L, 65 mg/L, 4 colonies/100 mL, respectively. The chance of slightly reducing capital costs compared to RO (by ~6%) was shown.

In another plant with UF and RO for irrigation reuse (Las Palmas, Gran Canaria), the pilot EDR produced 100–140 m^3/day of desalted water (<500 mg/L) with 82–90% recovery [412]. Metal membrane MF and ED provided a stable effluent quality over a 6-month testing, reducing by more than 90% most physico-chemical parameters, including nutrients [413]. A pilot plant with 144 m^3/day capacity consisted of 500 μm pre-filtration, coagulation-disinfection ($Fe_2(SO_4)_3$ and NaClO), 15 μm multimedia filtration and EDR [414]. EDR performed a desalination of ~70% (1104 mg/L TDS in the EDR feed), thus providing an effluent for horticultural reuse with TDS below the 375 mg/L target identified by guidelines. E_{spec} was 1 kWh/m^3 (60% of which was in EDR) and 82% of wastewater was recovered with an estimated operating cost of 18 \$cents/m^3. A further benefit from ED is the hypochlorite production in the anolyte, to be used then for disinfection [415]. Pilot MF-RO and MF-EDR plants were compared for a tertiary effluent containing endocrine disrupting chemicals, and pharmaceuticals and personal care products, showing that only RO was capable of removing them, as expected [416]. The same conclusion was drawn from another study, which exhibited moderate removals of some compounds (48–58%) and low removals for other compounds [417].

ED desalination was tested for 930 h during one year with a macrophytes pilot system effluent [418]. The conductivity reduced from ~0.67 mS/cm to 0.2 mS/cm, obtaining an effluent appropriate for reuse (e.g., in cooling towers) with only slight fouling. Further tests were conducted with the secondary effluent from a small WWTP for a university campus sewage [419]. The conductivity of ~1 mS/cm reduced to 0.05–0.1 mS/cm, with removals above 80% for cations and 70% for anions, and with E_{spec} values of 0.104 kWh/m^3 increased by fouling to 0.119 kWh/m^3. Other physico-chemical features (colour, turbidity, COD, BOD, etc.) were mildly cut down, thus providing an effluent suitable for fish farming after simple pH correction, but requiring some further treatment to reduce BOD and turbidity in case of urban and agricultural reuse.

Different schemes coupled ED and forward osmosis (FO). For example, FO extracted water from a secondary effluent (0.05 M salt concentration) to provide it to a draw NaCl solution then sent to the ED stage [420] (Figure 20). Ions, organic and inorganic substances were rejected in the FO retentate, while the draw stream enriched in water (diluted from 0.5 M to 0.2 M) by osmosis went to the ED. This yielded high-quality water (0.81–0.88 mS/cm) and the draw solution for the FO. ED driven by photovoltaic energy exhibited E_{spec} of 4.98–5.57 kWh/m^3 and η of 78.2–100%, while the estimated cost for a system producing 130 L/day potable water was ~3–5 €/m^3.

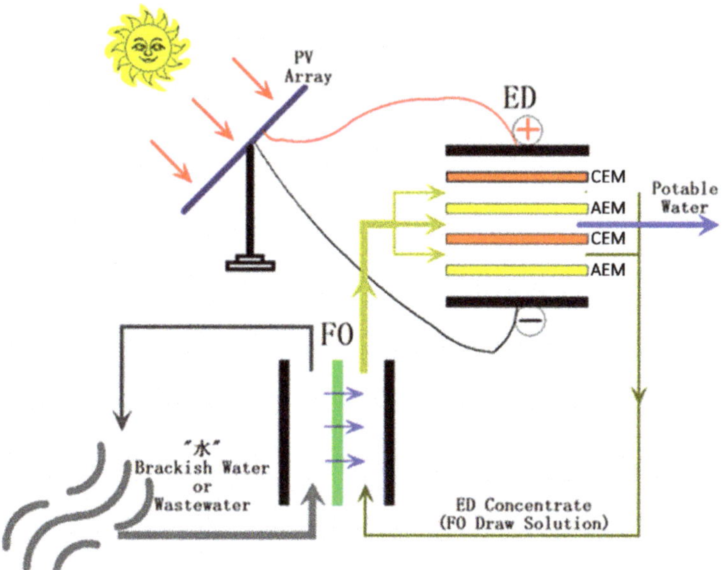

Figure 20. Scheme of FO–ED integrated process for wastewater (or brackish water) reclamation. Reproduced (adapted) with permission from [420], published by American Chemical Society, 2013.

An osmotic membrane bioreactor–ED system treated a synthetic primary effluent (300 mg/L COD, 0.51 g/L salts, 1.1 mS/cm) [421]. FO and biological degradation by activated sludge occur in the bioreactor, which thus suffers from salt accumulation caused by osmosis to the draw side and contrary solute flux. This would result in increased costs for draw replenishment, as well as in discharge issues and microbial growth inhibition. In this study, ED desalted the treated wastewater, by achieving a salinity build-up mitigation (conductivity maintained at 8 mS/cm), which allowed for (i) an increase by 6 times of the biological treatment duration (24 days), and (ii) the waste salt recovery, thus providing the concentrate draw solution. η was 41.6–76.2% and E_{spec} 1.88–4.01 kWh/m^3. In another hybrid process, ED mitigated the salinity build-up in the FO (submerged module) feed (secondary wastewater with 29.3 mg/L COD, ~0.5 mS/cm) by using a fertilizer draw solution with 0.5–2 M $(NH_4)_2HPO_4$ [422]. A diluted fertilizer was recovered by FO from wastewater, while 96.6% of the fertilizer lost by reverse flux (63 mg/L ammonium and 83 mg/L phosphate) to the feed was recovered through ED, which also returned the desalinated feed to the FO module. E_{spec} of the system was 0.72–1.49 kWh/m^3.

5.2. Energy Recovery

WWTP effluents can be used as diluate coupled with salty waters as concentrate in RED stacks recovering energy. The most abundant high-salinity solution is represented by seawater, thereby implying possible applications in coastal areas.

RED experiments with artificial NaCl solutions (diluate 0.002–0.08 M, concentrate 0.6 M) showed that the optimal diluate concentration maximizing the power output ($P_{d,max}$ of 0.39 W/m^2) was in the range 0.01–0.02 M [423], which often corresponds to the concentration range of WWTP effluents from biological treatment. By increasing the temperature from 25 °C to 60 °C, $P_{d,max}$ increased by 60%, suggesting that co-locating RED with a thermal power plant where the solutions are pre-heated (e.g., cooling tower seawater) can boost the energy recovery. Moreover, the seawater–WWTP effluent RED process may be used as a pre-desalination step before RO, by providing great potential of reducing the energy consumption of seawater desalination plants [424]. The RED unit may also be operated under "assisted" conditions, in which the applied electrical current overcomes the short-circuit current, thus requiring a lower membrane area [425]. Simulation results showed that hybrid RED-RO systems either with assisted or conventional RED yielded cost savings compared to standalone SWRO [426]. However, a critical issue affecting these RED-RO systems is the potential contamination of the seawater by organic micropollutants that may be adsorbed and transported from the impaired water [427]. In contrast, this problem is less important in RO-RED schemes (Section 6.3), where the RED process is fed by the RO reject brine and the WWTP effluent.

Several studies tested RED with real WWTP effluents and seawater. After filtration by a 10 μm filter, a treated wastewater at conductivity of 0.44 mS/cm (~0.002 M) coupled with seawater (46.2 mS/cm) produced a $P_{d,max}$ of 0.15 W/m^2 [398]. The low concentration of the diluate led to a high electromotive force (V_{OC} of 1.66 V with 10 cell pairs, 70% permselectivity), but limited the power density due to the high electrical resistance. The presence of natural organic matter (NOM) in the WWTP effluent (16.3 mg/L dissolved organic carbon) increased the resistance, causing a reduction in $P_{d,max}$ of ~17% with respect to a model solution lacking NOM.

Higher values of $P_{d,max}$, i.e., up to 0.38 W/m^2, were obtained by testing a pilot plant [428]. The WWTP effluent was treated by dual-media filtration, bag filter (50 μm) and cartridge filters (5 μm), while the seawater underwent only the 5 μm filtration. The RED feed solutions had conductivity of 1.3–5.7 mS/cm and 52.9–53.8 mS/cm, respectively. The transport of inorganic solutes and NOM (4.3 mg/L dissolved organic carbon in the treated wastewater) was investigated. Over 12 days, the power density was on average ~20% lower than the highest one. It was observed a slight increase of the IEMs resistance, which can be attributed to fouling and effects of divalent ions. However, the reduction of P_d was mainly caused by precipitates clogging the cathode chamber, where the wastewater was used as electrolyte (high pH due to the hydrogen evolution reaction). Organics at low molecular weight were transported towards the seawater compartments. Therefore, attention should be paid to this aspect in case RED is followed by RO. Pressure drops increased continuously over 12 days up to almost 3 and 1.5 times in the wastewater and seawater compartment, respectively, due to spacer-filled channels clogging at the inlet regions. This may affect significantly the net power (not calculated). Therefore, suitable pre-treatments and cleaning procedures should be adopted.

The same $P_{d,max}$ (~0.38 W/m^2) was recorded by another pilot RED fed with treated water (anaerobic-oxix activated sludge process) at conductivity of 1.0–2.5 mS/cm and seawater at 50 mS/cm [429]. Actually, the seawater solution was obtained by mixing a desalination brine with the treated water. Suspended particles were removed from both streams by cartridge filter and fibre filter (10 μm pore size). V_{OC} was 28.6 V (200 cell pairs), 20% lower than the theoretical one due to effects of divalent ions and reduction of driving force along the channels. A 5 wt% Na_2SO_4 electrolyte was fed to the electrode compartments, obtaining 0.9 L/h H_2 with ~100% efficiency by cathode reduction. Without pre-filtration, the performance was stable over 300 h, but a subsequent reduction of power density was observed (up to 80% after 800 h). Cleaning without chemicals removed clogging from the disassembled stack and restored its performance. However, cleaning in place methods should be developed with short interruptions of the RED process.

Filtration pre-treatments of a domestic WWTP effluent (1.13 mS/cm) were compared, i.e., 100 μm filtration, rapid sand filtration or river bank filtration [430]. Seawater (48.35–58.38 mS/cm) was pre-treated with sand filtration, bead filtration and UV. During 40-day RED testing, the pressure drop increased by only 0.09–0.18 bar from the initial value of 0.03–0.04 bar when using 100 μm filtration and rapid sand filtration, respectively. With an almost stable $P_{d,max}$ of ~0.25 W/m^2, $P_{d,max,net}$ was 0.23 and 0.22 W/m^2, respectively. Instead, the RED operation with river bank filtration or without pre-treatment exhibited high pressure drops (~0.6 bar on average) and low $P_{d,max,net}$ (~0.06 W/m^2), despite alkaline and acidic cleanings. Inlet and outlet regions of spacer-filled channels were critical points of biofilm development even at the seawater side (biological growth during storage).

The pre-treatment with polyaluminium chloride coagulant (and 0.45 μm filtration) was tested for reclaimed water (~0.5 mS/cm) [431]. Filtered seawater (~48.5 mS/cm) was the RED concentrate. Polyaluminium chloride residue affected the RED performance by increasing the CEM resistance and, thus, by reducing the power density. However, the optimized dosage removed up to 50% of the organic matter (from ~6.5 ppm TOC) and resulted in a $P_{d,max}$ of ~0.42 W/m^2, increased by 20% with respect to that obtained with filtration only. Multivalent ions and NOM, instead, reduced $P_{d,max}$ by ~20% with respect to that produced by model solutions. The long-term operation should be tested.

A secondary effluent was treated by coagulation-flocculation, decantation and 10 μm filtration [432]. The treated stream (1.8 mS/cm, ~0.008 M NaCl) was used with 1 μm-filtered and UV-disinfected seawater (54.7 mS/cm, ~0.5 M NaCl) for RED energy recovery, showing a stable performance over 480 h. The treatment before RED and the slight increase of salinity after RED provided a water quality acceptable for reuse. V_{OC} was 3.45 V (20 cell pairs) and $P_{d,max}$ was 1.43 W/m^2 without observed fouling. This is the highest value of P_d recorded so far with this kind of streams. However, no data on pressure drop nor on $P_{d,net}$ were provided.

RED was performed with both compartments fed by treated wastewater [433]. Similarly to the previous cases, the low-salinity solution was a reclaimed urban WWTP effluent (from membrane bio-reactor pilot plant) at conductivity of 0.6–1.8 mS/cm (corresponding to ~0.004–0.016 M NaCl). Instead of seawater, the concentrate stream was a fish canning factory wastewater, treated by a pilot aerobic granular sludge sequence batch air-lift reactor, 5 μm MF, and acidification at 4 < pH < 5, which had a conductivity of 47.0 or 87.5 mS/cm. During long-term experiments of 29 days, fouling and increase of pressure drops (clogging) were observed. Backwashing with short pulses (1–2 s) at high fluid velocity (10 cm/s) was performed. An alkaline solution through the compartment of the fish wastewater was more effective than other solutions, while distilled water through the compartment of the reclaimed WWTP effluent was sufficient. Periodic ED pulses reduced the absorption of foulants, maintaining almost constant the stack resistance. However, fouling issues did not vanish, especially those involving AEMs. Under the best conditions tested, V_{OC} was almost constant around 1.6 V (10 cell pairs), while $P_{d,max}$ declined from ~0.9 to ~0.6 W/m^2. $P_{d,max,net}$ corrected by excluding the effect of the blank resistance declined from ~1.3 to 0.1 W/m^2. Therefore, measures against fouling and clogging should be further studied. Some detrimental effects of divalent ions were observed.

The integration between membrane distillation (MD) and RED was proposed to recover water and energy from urine in off-grid applications [434]. From real urine feed (12.65 mS/cm, 207 mg/L $NH_4^+ - N$, 6.33 g/L COD), the MD produced a retentate with doubled conductivity (24.1 mS/cm) and a permeate at 0.21 mS/cm. These streams were then used as feeds for an RED unit for partial remixing with energy recovery. $P_{d,max}$ (~0.2 W/m^2) was comparable to that produced by NaCl solutions (0.32 W/m^2). By increasing the temperature from 22 to 50 °C, $P_{d,max}$ could be increased by 70%, as shown by experiments with synthetic solutions, thus prospecting the use of waste heat. In RED tests with recirculation, ~47% of the Gibbs free energy was extracted. In optimized systems, the energy efficiency could be enhanced even compatibly with a good quality of the final diluate.

5.3. Recovery of Nutrients and VFAs

ED methods can recover nutrients from wastewater, thus lowering the ecological impact of discharge (eutrophication) and producing fertilizers. In some cases, VFAs are other valuable components that can be recovered along with nutrients. Studies have been performed on treated municipal wastewater

(Section 5.3.1), excess sludge sidestreams (Section 5.3.2), separately collected human urine (Section 5.3.3), and waste effluents from animal farming (Section 5.3.4).

5.3.1. Municipal WWTP Effluents

Municipal WWTP effluents can provide nutrients, i.e., ammonia, phosphate, nitrate, and potassium, thus producing fertilizers. To this aim, ED systems can concentrate P-based nutrients before precipitation/crystallisation of struvite, i.e., $(NH_4)MgPO_4 \cdot 6(H_2O)$, or calcium phosphates [435].

Several studies have tested SED units with MVAs to fractionate and concentrate phosphate (Figure 21). Experiments with synthetic wastewater (3–7 mM KH_2PO_4 and 13–17 mM NaCl) achieved a phosphate removal of 62.3% with a product concentration of 16 mM at 44% purity, and a Cl^- removal of 87% [436]. Values of η for $H_2PO_4^-$ through the AEM and for Cl^- through the MVA were 26.6% and 63%, respectively. These outcomes were obtained at pH = 12 in the product, as multivalent phosphates (HPO_4^{2-} and PO_4^{3-}) predominate under alkaline conditions. Crystallisation with $CaCl_2$ in a pellet reactor produced hydroxyapatite and brushite with 82.7% efficiency, thus demonstrating the feasibility of the integrated process.

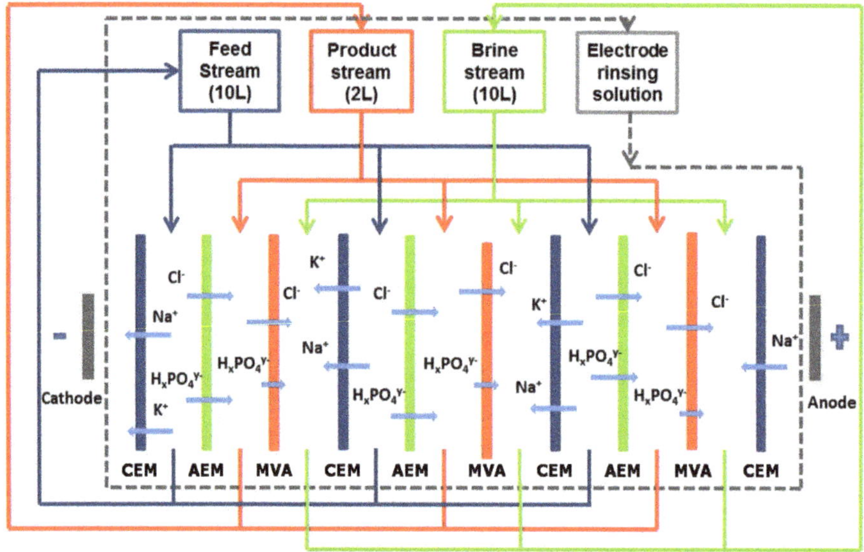

Figure 21. SED stack equipped with MVAs to fractionate and concentrate phosphate. Reproduced (adapted) with permission from [437], published by Elsevier, 2015.

With a more realistic synthetic municipal effluent containing an ion mixture (1 mM KH_2PO_4, and 2 mM NO_3^-, HCO_3^-, SO_4^{2-}, Ca^{2+} and Mg^{2+}), although the process required a longer operation, 41.9% of phosphate was removed and concentrated by 161% in the product, and acid cleaning removed scaling [437]. Other experiments with a synthetic secondary effluent (NaCl, $NaNO_3$ and Na_2HPO_4 with 355 mg/L Cl^-, 30 mg/L N, 10 mg/L P) exhibited recovery efficiencies of 56.97–64.28% for N in the brine and 67.42–73.67% for P in the product, with overall η of 56.7–61% and E_{spec} of 1.63–2.92 kWh/m³ [438]. The pH in the tank with the product was adjusted at 10.2–10.5. However, at high values of applied voltage, lower concentration rates of P suggested water dissociation as a likely cause of pH reduction.

Even conventional ED is able to recuperate nutrients [439]. To accomplish a selective separation, a two-stage underlimiting/overlimiting ED was proposed [440]. A model macrophyte-treated wastewater (0.022 g/L $Na_2HPO_4 \cdot 7H_2O$, 0.011 g/L $NaH_2PO_4 \cdot H_2O$, 0.481 g/L Na_2SO_4) was circulated through both compartments. In the first stage, ions were concentrated by cycles where the diluate was changed once reached 50% desalination (49.8% Na^+, 46.7% $H_xPO_4^{3-x}$, 42.6% SO_4^{2-}) to avoid pH reduction that would occur at higher desalination percentages. $H_xPO_4^{3-x}$ reached 0.118 g/L (concentration factor of ~10), which is satisfactory for an efficient precipitation/crystallisation. To segregate phosphate, a concentrated solution was treated with a second stage at overlimiting currents promoting water dissociation and phosphate protonation-deprotonation. Na^+ and SO_4^{2-} were removed by 97.7% and 94.2%, respectively, while phosphate transfer was significantly hampered, by retaining it by 81.3% in the diluate. Studies for membrane characterisation and transport mechanisms elucidation can help to enhance such ED applications [441,442].

A pilot ED was assembled with Mg anode to provide Mg^{2+} to the concentrate and precipitate struvite [443]. Synthetic wastewater with 34.6 mg/L $NH_4^+ - N$, 10 mg/L $PO_4^{3-} - P$ and 300 mg/L NaCl was used as initial solution for concentrate, anode and diluate. The concentrate chambers were connected with the anode, so that the product solution exiting the concentrate flowed through the anode and vice versa. At the optimal pH of 8.8 and with multiple cycles in the diluate, 65% of phosphate was removed as struvite, the diluate had on average less than 4 mg/L $PO_4^{3-} - P$ (below 0.5 mg/L at the end of several cycles), and the concentrate had 30 mg/L $PO_4^{3-} - P$. The cost of the Mg anode was 31.27 \$/kg P.

5.3.2. Excess Sludge Sidestreams

ED techniques were proposed also for recovery of fertilizers from excess sludge sidestreams (supernatant, centrate, filtrate) of municipal WWTPs. An economic analysis based on an ED simulator estimated a total cost of 0.392 \$/kg N (29.5 m^3/day capacity), 65% of which due to operation cost (E_{spec} of 2.36 kWh/kg), showing the convenience of ED compared to other conventional or novel processes [444]. As Table 3 reports, either ED or ED-BMED recovered phosphate or phosphoric acid from a synthetic solution modelling the supernatant of excess sludge mixed with the influent [406], and these treatment processes may be intended also for municipal effluents. In other experiments, an integrated system with ED, struvite precipitation and ammonia stripping was developed by testing synthetic sludge anaerobic digestion sidestreams (dewatering by, e.g., centrifuge or belt filter press) [445]. The feed contained 200 mg/L P and 600 mg/L N. After concentration via ED, nutrients were recovered by the struvite reactor, while the ammonia excess was recovered via gas stripping at 40 °C. Overall removal percentages were ~86% for P and ~92% for N.

Lab-scale [446] and pilot-scale [447] ED experiments were conducted with a real anaerobic digester supernatant (centrifuge centrate) to concentrate NH_4^+ and K^+. A crystallisation/precipitation pre-treatment was performed to recover struvite and prevent scaling (Figure 22). The ED feed (232 mg/L K^+, 1003 mg/L Na^+, 768 mg/L Cl^-, 835 mg/L $NH_4^+ - N$, etc., 351 mg/L COD) was pumped with single pass through the diluate (23% reduction of conductivity) and with recirculation through the concentrate, reaching concentration factors of ~8 for $NH_4^+ - N$ and K^+ with average overall η of 76% and E_{spec} of 4.9 kWh/kg $NH_4^+ - N$ [447]. The treatment was competitive and provided a product usable as fertilizer. However, improvements were needed, especially in terms of increase of product recovery (affected by water flux and ions back diffusion) and elimination of unwanted ions like Cl^-. The transport of pharmaceuticals (10 or 100 µg/L) was then studied [448]. Nutrients were concentrated by a maximum factor of ~5, while less than 8% of pharmaceuticals were transported to the concentrate product. However, the lower concentrations usually present in real effluents do not hinder the product use as fertilizer.

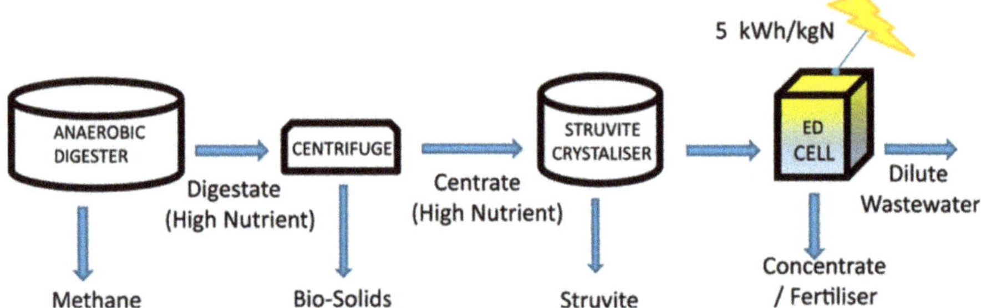

Figure 22. Conceptual scheme of the recovery of materials from anaerobic digestion of WWTP excess sludge, including ED concentration of nutrients from centrate after struvite precipitation for fertilizer production. Reproduced (adapted) with permission from [447], published by Elsevier, 2018.

A solution prepared with 6.6 g/L NH_4HCO_3 (1.5 g/L NH_4^+), simulating sludge reject water, was concentrated by ED with dynamic current [449]. The current density was controlled during the batch process by applying values equal to a fraction of the instantaneous limiting current density (related to the diluate electrical conductivity). This reduced the operational run time by 75% compared to the operation with fixed current, thanks to reduced effects of osmosis and back-diffusion. A more efficient separation was thus obtained: the concentration factor increased from 4.5 to 6.7, while E_{spec} remained unchanged at 5.4 MJ/kg N removing 90% of NH_4^+ at η = 83–96%.

Dissolved ammonia (1.7 or 4.0 g/L) in anaerobic digestion centrate solutions was converted by acid stripping through a liquid–liquid hollow fibre membrane contactor into ammonium salts (NH_4NO_3 or $NH_4H_2PO_4$ at 5.1 wt% or 10.1 wt% in nitrogen), which were then concentrated by ED [450]. Under optimal conditions (10.1 wt%), the ED produced a liquid fertilizer at 15.6 wt% NH_4NO_3-N, with E_{spec} of 0.21 kWh/kg and η of ~93%.

A treatment integrating BMED, struvite precipitation, and multi-stage membrane capacitive deionisation (MCDI) was demonstrated for recovering phosphorus and ammonium from a synthetic supernatant (12.5 mM NH_4^+ and 2.5 mM PO_4^{3-}) [451]. BMED was used only to alkalify the wastewater, thus facilitating struvite precipitation, while MCDI was employed to separate excess ammonia in a small volume. E_{spec} of the BMED was ~1 kWh/m^3. The integrated process removed ~100% of phosphorous and ~77% of ammonia, recovering ~81% of high-quality effluent and ~19% of concentrated stream meeting reuse standards.

VFAs are other products extractable from excess sludge. To recover them, as well as nutrients, thermally hydrolysed waste activated sludge was fermented anaerobically, and after screening and MF, the permeate was concentrated by ED [452]. The 6.15 g/L total VFAs present in the MF permeate were concentrated by ED to 19.82 g/L, corresponding to 92% of transferred mass, while 0.92 g/L NH_4^+ and 0.16 g/L PO_4^{3-} were concentrated to 3.02 and 0.45 g/L, respectively. After that, struvite precipitation was performed to remove the excess ammonium and phosphate, and fermentation was conducted to produce polyhydroxyalkanoates. ED and precipitation significantly enhanced their accumulation in the fermentation broth (from 24 mg/L to 165 mg/L), thus offering a cost-effective valorisation process.

ED techniques can also recover lactic acid from the organic fraction of municipal solid waste hydrolysate [453].

5.3.3. Human Urine

Human urine comprises up to 91–96% water, with urea ($CO(NH_2)_2$) being the major solute fraction (50% of TOC) and with other organic and inorganic components, including P and K [454]. Anthropogenic urine represents a small fraction of domestic wastewater. Nevertheless, nutrients are present in it, along with micropollutants (endocrine disrupting compounds and pharmaceuticals). Therefore, suitable treatment processes have been proposed for separately collected urine to separate and concentrate nutrients (fertilizers) from micropollutants, including ED [455]. IEMs enable ion passage, while retaining neutral organics, proteins and microorganisms. Therefore, ED produces a nutrient-enriched concentrate stream (starting from water with or without salt), and a waste urine stream (diluate).

After MF, in experiments with lab-scale [456] and pilot-scale [457] ED units, maximum concentration factors of nutrients were 3.3 and 4.1, respectively, with urine desalination of 85–99%. The urine contained 4.85 g/L $NH_4^+ - N$, 0.23 g/L $PO_4^{3-} - P$, 1.96 g/L Na^+, 3.83 g/L Cl^-, 1.72 g/L K^+, 0.67 g/L SO_4^{2-}, and 4.36 g/L COD in the former study, while it contained 2.9 g/L $NH_4^+ - N$, 0.18 g/L $PO_4^{3-} - P$, 1.6 g/L Na^+, 3.0 g/L Cl^-, 1.4 g/L K^+, 0.7 g/L SO_4^{2-}, and 3.6 g/L COD in the latter study. The compartments of concentrate were filled at the start of the experiment; then, the concentrate flow was generated only by water transport. η values up to 50% were observed for ions, suggesting that part of the current was transferred by charged fractions of COD [456]. During a 90-day operation, spiked micropollutants (mixture of pharmaceuticals and hormones) were adsorbed and, after saturation, partially permeated the membranes, reaching the concentrate [456]. However, natural levels of micropollutants (e.g., ~100 µg/L ibuprofen) in the feed would allow much longer operations (e.g., 400 days) without permeation [457]. Membranes cleaned after 195 operating days exhibited a desalination rate enhancement of 35%. Ozonation removed completely micropollutants. Fertilisation by the concentrate showed good performances [457]. However, the ED method would be economically feasible only for large-scale plants, while other processes are affordable for developing countries [458].

A system with precipitation, anaerobic nitrification and ED concentration was developed and operated for ~7 months [459] (Figure 23). The first two treatment processes minimized scaling and biofouling in the ED step. The influent was with 20% or 40% urine in water (151 mg/L $NH_4^+ - N$, 0.4 mg/L $NO_3^- - N$, 42 mg/L $PO_4^{3-} - P$, 449 mg/L Na^+, 753 g/L Cl^-, 401 mg/L K^+, 31 mg/L Ca^{2+}, 11 mg/L Mg^{2+}, 14 mg/L SO_4^{2-} in the 20% solution). After precipitation by NaOH dosage for Ca^{2+} and Mg^{2+} removal, the biological treatment (moving bed biofilm) oxidised organics and stabilised N via ammonification–nitrification, which converts urea (volatile and thermally unstable) into nitrate. The ED transferred 70%/80% of ions in 15%/20% of the starting volume by treating an influent with 20%/40% of urine, respectively (concentration factors of 3–5), with E_{spec} of 4.3 kWh/m³. The P-rich solids from precipitation and the ED concentrate produced fertilizers; the ED diluate could allow water recovery.

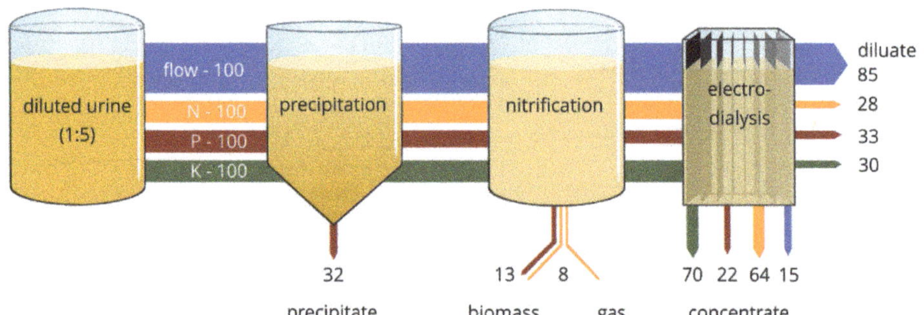

Figure 23. Flowsheet with mass balances of the pilot plant for urine-water solution treatment. Reproduced with permission from [459], published by Elsevier, 2018.

5.3.4. Animal Farming

The augmentation of intensive livestock production caused by massive meat consumption implies several environmental impacts, including those associated with inappropriate disposal of raw or digested animal manure (e.g., eutrophication). Animal manure is a slurry which contains faeces, straw, urine, and water, but sometimes the solid fraction is separately collected [460]. The nutrient content (P, K, N) could allow fertilisation by animal manure. Additionally, biogas can be produced via anaerobic digestion [461]. Unfortunately, the direct agricultural use is limited by transportation costs and foul odours release. However, separation and concentration of nutrients from raw or digestate manure can solve these problems, and ED techniques (after solid–liquid separation) have been studied to this aim, especially for effluents from pig farms, which account for ~36% of the meat market [462].

After vacuum filtration, ammonium contained in liquid swine manures (3.29 or 5.14 g/L NH_3-N, 2.09 or 2.52 g/L K^+, 0.2 or 0.24 g/L P, 20.66 or 40.32 g/L soluble COD, VFAs, solids, etc.) was concentrated into a 1 g/L KCl feed concentrate, by testing different IEMs [463]. Total ammonia reached ~14.4 g/L (of which free ammonia, NH_3, represented ~6%) with $\eta \approx 75\%$. Most of the remaining current was used for K^+ transport (3.8 concentration factor), while a small quantity of phosphorus was transferred (0.45 concentration factor). Volatilisation of free ammonia caused a 17% loss of total ammonia, whose concentration was also limited by water transport. Coupling RO with ED did not bring improvements. Fouling phenomena impaired the IEMs properties resulting in a decline of ED performance [464]. An alkaline–acidic cleaning was fully successful for the CEMs, but restored 80% of conductivity of AEMs, which were likely affected by permanent organic fouling (dark coloration). After cleaning, current density and desalination rate were 95% and 91% of the values measured with new IEMs. To limit losses of ammonia by volatilisation, its transfer to an acid solution by air stripping of the concentrate was tested [465]. Total ammonia nitrogen was concentrated by ~7 times (from a feed at 3.2 g/L) with $\eta = 64$–71%. However, the acidic trap recuperated only 14.5% of NH_3 present in the concentrate. Since the residual total ammonia in the swine manure was 1.2 g/L, the NH_3 recovery could be enhanced at higher pH values.

Nutrients were recovered by EDR treatment of pig manure digestate [466]. The raw manure was completely digested. Then, pre-treatments prior to ED included: acidification for P extraction from solids, 0.4 mm sieving, flocculation, centrifugation. The generated effluent (2637 mg/L $NH_4^+ - N$, 492 mg/L $PO_4^{3-} - P$, 3259 mg/L K^+, 8894 mg/L Cl^-, 691 mg/L Na^+, 1021 mg/L Ca^{2+}, 555 mg/L Mg^{2+}, 2.8 g/L soluble COD, VFAs, solids, etc.) was the diluate feed. All ammonium and most of phosphate (84%) were removed, obtaining a concentrate product at 4.2 g/L $NH_4^+ - N$ and 0.7 g/L $PO_4^{3-} - N$, with overall η of 46.8–61.3% and

E_{spec} of 0.13 kWh/L. At every polarity reversal (15 min), acidic cleaning was performed. Reversible fouling was removed, but irreversible effects of organic fouling were observed as well. However, they became stable, likely because the foulants transport inside the membrane was hindered by superficial fouling, thus suggesting the feasibility of long-term operations. The antibiotics fate in the EDR process was studied with spiked (from 50 µg/L to 5 mg/L) pre-treated pig manure, by investigating sorption and migration mechanisms and fouling formation [467]. The main conclusion was that antibiotics can be transported to the concentrate, thus presenting the possible need for further treatment.

The ED stack with Mg anode for struvite precipitation discussed at the end of Section 5.3.1 was also tested with a solution that was 10 times more concentrated (100 mg/L $PO_4^{3-} - P$) simulating swine wastewater digestate [443]. The final diluate had on average ~20 mg/L $PO_4^{3-} - P$, the concentrate product had ~100 mg/L $PO_4^{3-} - P$.

The simultaneous fractionation of cations and anions into several streams by SED offers an interesting approach to recover nutrients from digested swine manure [468]. The tested SED stack (Figure 24a) was built with repeating units of four membranes, i.e., AEM, CEM, MVC and MVA, and four channels, i.e., feed, cationic product (Mg^{2+} and Ca^{2+}), brine (K^+ and NH_4^+) and anionic product (PO_4^{3-} and SO_4^{2-}). The initial feed was prepared with NaH_2PO_4 (40 mg/L P), NH_4Cl (500 mg/L N) Na_2SO_4 (100 mg/L SO_4), KCl (400 mg/L K), $MgCl_2$ (60 mg/L Mg), $CaCl_2$ (100 mg/L Ca) and NaCl (3.192 mM). The other initial solutions were with 0.1 M NaCl. The feed conductivity was practically reduced to zero, obtaining fractionations from ~33% (K^+) to ~90% (PO_4^{3-}), η from ~2% (SO_4^{2-}) to ~30% for NH_4^+, and E_{spec} from ~1 kWh/kg NH_4Cl to ~0 kWh/kg NaH_2PO_4. The two divalent ion products were mixed, obtaining phosphate precipitation by NaOH dosage.

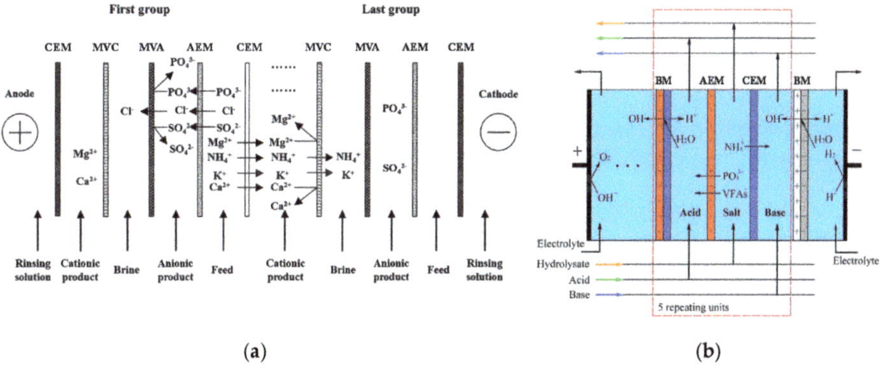

Figure 24. Schematics of ED methods for animal farming effluents: (**a**) four-compartment SED of swine manure digestate; (**b**) BMED of pig manure hydrolysate. Panel (**a**) is reproduced (adapted) with permission from [468] published by Elsevier, 2019. Panel (**b**) is reproduced (adapted) with permission from [469], published by Elsevier, 2018.

BMED recovered nutrients and VFAs from pig manure hydrolysate (after acidification and solid–liquid separation, the effluent contained ~4.0 g/L $NH_4^+ - N$, ~1.8 g/L $PO_4^{3-} - P$, 9.54 g/L VFAs, 52.68 g/L COD, residual solids, etc.) [469]. The salt tank was filled with the effluent, while the acid and base tanks with deionized water (Figure 24b). Preliminary experiments with model solutions exhibited low values of recovery efficiency, η and purity, due to ion diffusion. The migration of Cl^- and SO_4^{2-} (as well as NH_4^+) was faster than other ions and constant until reaching low concentrations in the feed compartment. Instead, PO_4^{3-} and acetate remained in the feed compartment until that time, and then started to migrate. Therefore, a two-stage process enhanced significantly the performance. Once reached the inflection point of voltage (galvanostatic mode) corresponding to the start of PO_4^{3-} and VFAs transport, the produced acid and base were substituted with demi water. Two acid and two base products with higher purity were thus generated: acid I contained mainly strong acidic ions (~30 g/L Cl^-), acid II had high concentrations of acetate and PO_4^{3-} (~15 and 5 g/L, respectively, with recovery efficiency of 87% and 77%), base I and II contained mainly NH_4^+ (~10.5 and 3.2 g/L, respectively, with recovery efficiency of 60%) with Cl^- impurity. Air stripping of base I recovered NH_3, while acid I acidified the pig manure. P and VFAs may be extracted from acid II by further processes.

5.4. Regeneration of Liquid Desiccant Solutions for Air Conditioning

Liquid desiccant air conditioning has emerged as an efficient and energy-saving alternative to vapor compression air conditioners in buildings. In particular, liquid desiccant can dehumidify air by absorption. Membrane processes can replace conventional thermally driven evaporation methods in the regeneration (concentration) of hypersaline liquid desiccant solutions, by avoiding droplet carry-over and by lowering energy consumption [470]. Several studies have addressed the ED regeneration of LiCl or LiBr liquid desiccant solutions.

The application potential and the benefits over thermal regeneration were first shown by theoretical studies, where ED was powered by photovoltaic energy [471]. A two-stage ED was more advantageous, by saving more than 50% of energy compared to a single stage under optimized conditions [472,473]. An experimental setup was then developed, obtaining a maximum difference in LiCl mass concentration in the regenerated desiccant solution between start and end of ED of 0.03 wt% (experiments conducted with initial regenerate concentration of ~21–23 wt%) [474]. Low η values were observed (<55%), which is not surprising, given the hard conditions where the IEMs have to work, i.e., high concentrations and high concentration gradient. η values spanned over a large range (21–65%) in other experiments, showing the important role played by the concentrate (regenerate) –diluate concentration difference [475]. The same occurred towards E_{spec} and the overall coefficient of performance of the liquid desiccant air-conditioning system with ED regeneration, COP, defined as the refrigerating capacity divided by the power consumption ($E_{spec} \approx 1.4–16$ kWh/m^3, COP $\approx 0.3–4$). However, at high current densities and concentration differences, η and COP were governed by liquid desiccant concentration and flow rate, membrane properties and stack design. Experiments and simulations showed that higher values of concentration difference, initial regenerate concentration and applied current, reduced η (20–70%) [476]. However, the concentration difference had lower effects on the COP and energy efficiency. In contrast, the increase of the initial concentration from 27 wt% to 35 wt% reduced η, but enhanced the COP from 4 to 6.2 ($E_{spec} \approx 8.3–8.7$ kWh/m^3).

In other experiments, η values between 55.17% and 73.54% were found [477]. With initial concentrations of 23.96 wt% and 28.77 wt% for spent and regenerate solution, respectively, when a concentration difference of 5.86 wt% was reached, a concentration decrease in the regenerate was even observed. Negative effects derived from osmosis and electro-osmosis from diluate to concentrate, and salt back diffusion. Further tests and simulations showed that water transport was more significant than salt transport, and increased as the applied current increased and the initial solution concentration decreased [478]. It was concluded that minimizing the concentration difference between the two solutions can improve the ED regeneration. Its performance was then investigated and optimized by the Taguchi method, and the percentage contribution of each factor was evaluated by analysis of variance [479]. The optimal initial concentration was 27.5 wt% (without concentration difference, as expected). The applied current was the main parameter affecting the energy consumption, but had mild effects on the concentration increase. Instead, it was mostly affected by the concentration difference between the two solutions (accounting for ~78% of the effects) and, at a lesser extent, by the initial regenerate concentration. Compared to average values, optimal conditions led to an energy saving of 31.7% ($E_{spec} \approx 25$ kWh/m^3) and to a concentrative effect increased by 9.4% (regenerate at 36.22% wt).

A hybrid method combining continuous ED and thermal regeneration of the spent solution by a low-temperature heat source was developed [480] (Figure 25). The technical feasibility of the system was assessed by modelling validated by ED experiments. Simulations of one week's summertime weather in Darwin, Australia, showed that the two outlet concentrations from the ED regenerator were maintained at 29.93–30.17 wt% and 26.70–26.85 wt%, respectively (from inlet concentrations of 29.80–30.05 wt% and 26.80–26.95 wt%). The computed water removal in the low-temperature thermal regenerator and water absorption in the dehumidifier amounted at 128.6 kg and 126.6 kg, respectively. The largest part of energy for desiccant regeneration was consumed by ED (85%, corresponding to E_{spec} of 22 kWh/m^3), while the COP was on average equal to 0.5, thus suggesting that high-performance IEMs are crucial for the system efficiency. Experiments were designed by Taguchi's method, and results were analysed by analysis of variance, showing that increasing inlet temperatures from 30 to 45 °C and from 20 to 30 °C for the spent and regenerate solutions, respectively, resulted in an enhancement of regenerate concentration increase of 19% and to a similar energy saving [481]. These tests confirmed that a higher initial concentration of the regenerate causes higher E_{spec} and a lower concentration increase, which is considerably affected also by the initial concentration difference.

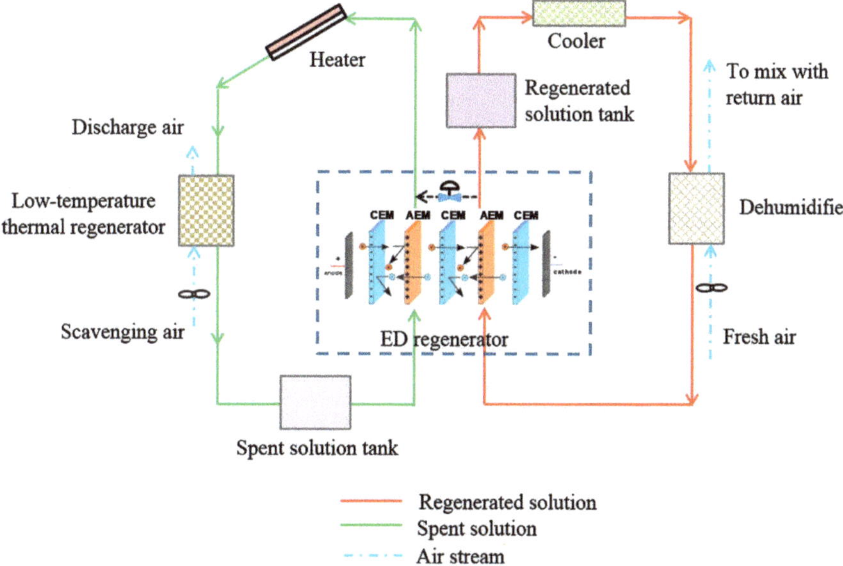

Figure 25. Scheme continuous ED regeneration integrated into a liquid desiccant dehumidification system. Reproduced (adapted) with permission from [480], published by Elsevier, 2019.

The ED regeneration of LiBr solutions showed lower values of η (9–31%) compared to LiCl solutions, because of the higher concentration (~45 wt%) [482]. As the operating conditions were let to vary, η and the COP changed significantly. However, a maximum COP of 4.26 was reached. A mathematical model simulated LiCl, LiBr and CaCl$_2$ solutions regeneration, confirming that initial concentration and applied current are important factors in all cases [483]. Concentrations of 15, 25 and 15 wt% were suggested for LiCl, LiBr and $CaCl_2$ solutions, respectively ($\eta \approx 70\%$ and $E_{spec} \approx 0.1$ kWh/mol).

6. Waste Brine from Desalination or Ion Exchange

Desalination technologies provide an important contribution in response to water scarcity. The global desalination capacity reached $\sim100 \times 10^6$ m^3/day [484,485], by producing $\sim142 \times 10^6$ m^3/day waste brine [485]. Therefore, brine management is a critical issue facing the desalination industry. RO holds $\sim$60–70% of the desalination market share [485,486] and has several applications. About 50% of RO plants produces potable water from brackish or sea water, $\sim$40% provides ultrapure water to industries, and some installations reclaim polluted or waste effluents, or process food [487]. Brackish water RO (BWRO) has water recoveries of 85–90%, while seawater RO (SWRO) of 35–50%, limited by high osmotic pressure [487]. RO retentate is usually discharged or evaporated. Similarly, environmental impacts are related to the disposal of IXRs regeneration spent brines and ED desalination brines. However, novel brine management methods at low environmental impact are oriented by (near) ZLD strategies towards waste disposal minimisation and resources (water and others) recovery [488–490]. Membrane processes are under study for this aim [1,488–490]. In particular, ED techniques may recover water and salt (Section 6.1), acid and base (Section 6.2), or energy (Section 6.3).

6.1. Water and Salt Recovery

ED has been studied to recover water and salt (diluate and concentrate, respectively) from BWRO brine (Section 6.1.1), SWRO brine (Section 6.1.2) or wastewater RO (WWRO) brine (Section 6.1.3) in ZLD approaches. Moreover, ED has been proposed for the regeneration of spent NaCl brines from IX (Section 6.1.4).

6.1.1. BWRO Brine

ED of BWRO brine can boost water recovery (diluate product) and facilitate the salts separation (concentrate product) in ZLD desalination.

A pilot EDR treated BWRO brine, and a gypsum precipitator for the ED concentrate protected the stack, since $CaSO_4$ was near to saturation [491]. The diluate salt concentration decreased from $\sim$340 mN to $\sim$20 mN, allowing for a total water recovery of 97–98%. Other performances were: concentrate concentration increase to 10% (from 1.5%), $\eta = 60$–75%, $E_{spec} = 7$–8 kWh/m^3. Despite the slow transport, the attainment of silica saturation level limited the brine concentration. Concentration and water recovery were similar for the BWRO-ED system depicted in Figure 26 (feed concentration $\sim$0.3%, retentate $\sim$1%) [492]. Scaling was minimized by acidification, EDR operation, and alleviation of sparingly soluble salts super-saturation via crystallisation, settling and MF or UF. Wind-aided intensified evaporation increased finally the TDS concentration over 30%. η was of 81%, E_{spec} was of 5–6 kWh/m^3, and the estimated cost was 0.408 €/m^3 for treating 100 m^3/h feed with 98% water recovery, thus showing the process competitiveness.

Pilot tests were conducted with a commercial-size ED unit (two electrical stages and four hydraulic stages) fed by BWRO brine ($\sim$10 g/L TDS, water recovery of 82.5%) [493]. ED recovered 55% of its feed, raising the overall recovery to 92.1%. MVMs obtained by modification of commercial IEMs changed the product composition. Compared to original IEMs, the MVMs ED achieved the same conductivity reduction (up to 60% of 19.5 mS/cm), while requiring higher E_{spec} (up to $\sim$70% more, $\sim$4 kWh/m^3). Efficient concentrative ED operations are designed by multi-stage configurations within more complex schemes. For example, two- or three-stage batch ED with concentrate split concentrated a 3.5 wt% NaCl solution to 17.9 or 20.6 wt%, with average $\eta = 82.9\%$ or 84.8%, and $E_{spec} = 0.31$ or 0.45 kWh/kg salt (18.83 and 27.06 kWh/m^3), respectively [494].

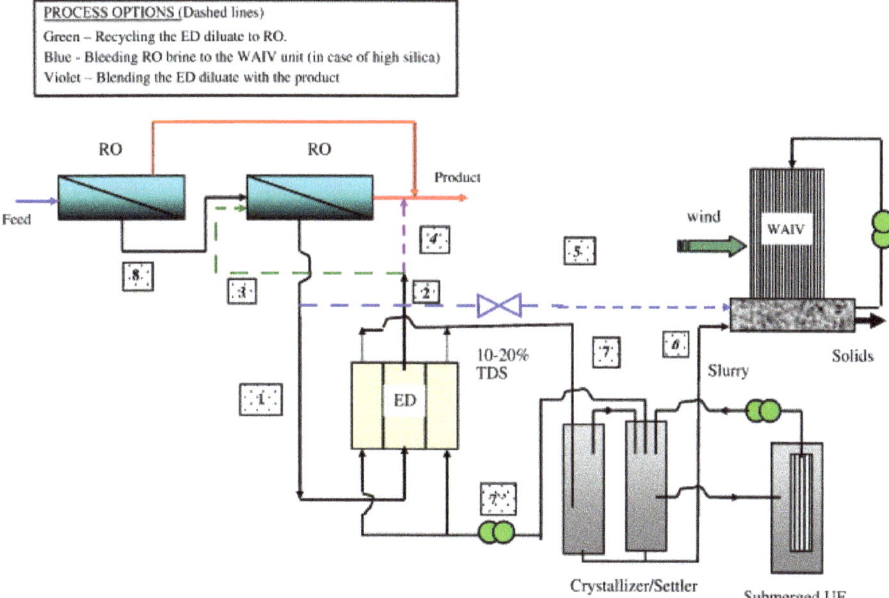

Figure 26. BWRO-ED desalination process. EDR treats the RO retentate (1). The EDR diluate product (2) can go through the second RO step (3) or, if it is sufficiently desalted, is conveyed with the permeate (4). The EDR concentrate (7) is sent to the seeded crystallizer/settler, and to the submerged UF module, thus the crystal-free concentrated stream (7′) recirculates to the EDR. The bleed stream (6) is evaporated by the WAIV unit. Reproduced with permission from [492], published by Elsevier, 2010.

6.1.2. SWRO Brine

SWRO brine treatment via ED recovers water and/or a high-concentration NaCl stream usable to produce coarse salt (via evaporation-crystallisation) or for other purposes, e.g., as raw material for the chlor-alkali industry.

ED stacks assembled with different IEMs were used with a simulated SWRO brine (10.5% TDS) [495]. A multi-stage operation (Figure 27a) produced two solutions with concentrations of up to 27.13% and 470 mg/L. The concentrated brine was suitable for producing edible salt, the diluate (water recovery of ~68%) was suitable more for industrial use than for drinking.

A real SWRO brine (70 g/L TDS) was concentrated through a pilot ED unit (single-pass diluate) equipped with MVMs, tested for 24 months to produce a feed for the chlor-alkali industry [496]. η ranged from 80% to 92% for Cl$^-$. The concentrate stream was even depleted in most multivalent ions because of water transport. Instead, the Cu and Ni concentration increased due to the transported monovalent Cl-complexes. Better performances were observed at higher temperatures (27 °C), which led the salt concentration to a maximum of 245 g/L. E_{spec} values comparable with the target of ED plants producing edible salt from seawater (0.12 kWh/kg salt, > 200 g/L) were obtained at lower concentrations (185 g/L). Instead, a process enhancement would be needed to meet the requirements of electrolysis for chlorine production (300 g/L, high purity) along with competitive costs. Experimental data from the same pilot plant validated a modelling tool [497]. Model predictions highlighted again that the RO-ED system was competitive in the edible salt market, but did not reach the targets for industrial use. Among different synthetized MVCs, the best permselectivities measured with a model SWRO brine were $P^{Mg^{2+}}_{Na^+} = 0.09$ and $P^{Ca^{2+}}_{Na^+} = 0.8$, which can provide concentrates at purity sufficient for the chlor-alkali process [498].

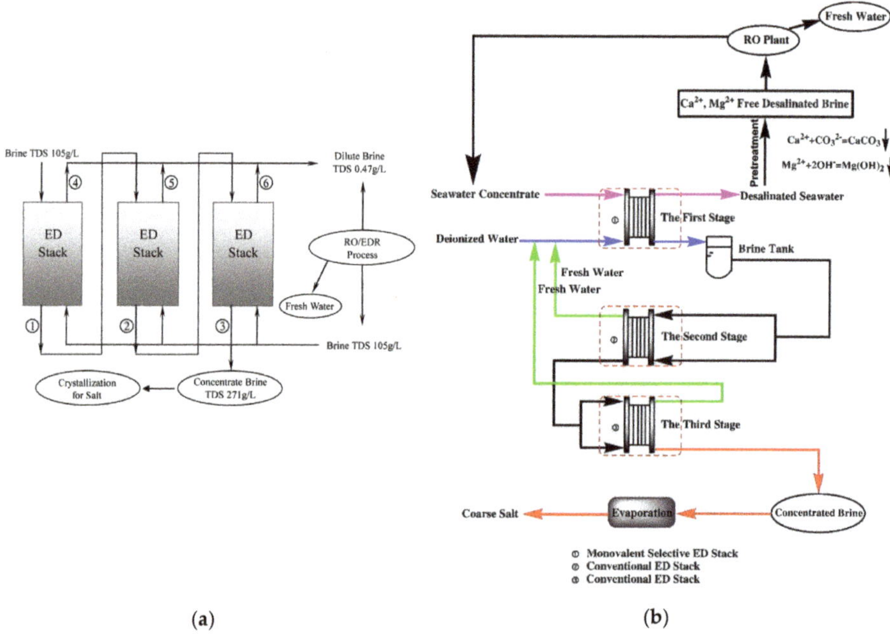

Figure 27. Flowcharts of three-stage ED systems for water reclamation and salt production from SWRO brine: (**a**) conventional ED; (**b**) ED with MVMs in the first stage, followed by conventional ED in the second and third stage. Panel (**a**) is reproduced (adapted) with permission from [495], published by Elsevier, 2014. Panel (**b**) is reproduced (adapted) with permission from [499], published by Elsevier, 2017.

A three-stage ED was developed for recovering water and salt from SWRO brine (~45 g/L TDS, 60 mS/cm) [499] (Figure 27b). The first ED was performed with MVMs to retain divalent ions in the diluate feed and transfer NaCl to the concentrate (deionized water). The produced diluate (70% desalination) can return to RO after removal of divalent ions, while the concentrate went through two concentrative steps with conventional IEMs. The conductivities of the concentrate products were 42.4, 73.2 and 105 mS/cm, and coarse salt 85% pure in NaCl was produced by final brine evaporation. The water recoveries relative to the initial diluate of each stage were 90%, 86% and 82%, obtaining fresh water from the last two stages. The total E_{spec} was of 2.3–2.4 kWh/kg NaCl.

Another strategy for enhancing the salt purity consists of using an NF pre-treatment, reducing the divalent ions concentration [500]. From an artificial SWRO brine (66.8 g/L TDS), the NF rejected Ca^{2+}, Mg^{2+} and SO_4^{2-} by 40, 87 and 100%, respectively, recovering water by 54.3% as permeate (58.7 g/L TDS). The ED tests achieved a maximum concentration of 160 g/L NaCl with some impurities (5 g/L) at E_{spec} of 1.4 kWh/kg NaCl. The final ED diluate (~95% of the NF permeate) had a minimum concentration of 25.25 g/L TDS, hence it could go to the RO. Similarly, the NF retentate could go to further recovery processes.

Theoretical approaches, including process simulators, techno-economic estimations and thermodynamic evaluations, can drive the design of optimized systems. Several hybrid schemes and multi-stage operations are viable, prospecting promising developments [501–504]. A techno-economic analysis found that the salt production costs of an SWRO-ED-crystallizer plant (RO feed at 50 m^3/h) may be competitive (61–111 \$/ton salt), i.e., lower by 19–55% than those of conventional standalone ED-crystallizer systems [505]. However, the process main aim (brine minimisation or salt production) and the site-specific features (water, salt and electricity prices) may require different strategies. For example, in another analysis, feeding the ED diluate with SWRO brine instead of seawater reduced the water production costs by 87% (i.e., from 27 to 3.5 \$/m^3), but increased the salt production costs by 26% (i.e., from 135 to 170 \$/ton salt), by considering an SWRO plant capacity of 150,000 m^3/day [506]. Optimal currents further reduced the water costs (3.0 \$/m^3), but increased E_{spec} by 26% to 12.7 kWh/m^3. The complete salt production scheme (SWRO-ED-crystallizer) was competitive with the SWRO desalination only in some Middle-East countries, where the salt price is higher than 104.5 \$/ton.

EDM of BWRO [507] or SWRO [41] brine was proposed. The complete conceptual scheme of SWRO-EDM is depicted in Figure 28. EDM separates ions from SWRO brine (diluate 1) into two high-solubility salt solutions products (the initial feed is SWRO permeate), one with Na$^+$ salts (concentrate 1) and another one with Cl$^-$ salts (concentrate 2), following the metathesis MX + NaCl → NaX + MCl, where NaCl is supplied by the substitution solution (diluate 2). The diluates recirculation to RO increases water recovery, crystallizer I recovers Na$_2$SO$_4$ and NaCl from concentrate 1, crystallizer II recovers NaCl from concentrate 2 after Ca^{2+} and Mg^{2+} precipitation. In the EDM experiments, initial solutions were an SWRO retentate at 50 g/L TDS, an NaCl substitution solution with the same conductivity, and deionized water (concentrates). To find optimal operating conditions, preliminary experiments with partial concentration were performed, exhibiting E_{spec} values lower than 1 kWh/kg salt. Then, a maximum concentrates concentration of ~200 g/L TDS was attained (while desalinating the diluate products at 5 mS/cm) with negligible co-ion leakages and scaling. However, 170 g/L was the concentration recommended to preserve the process efficiency. The feasibility of the overall process scheme should be assessed by further studies.

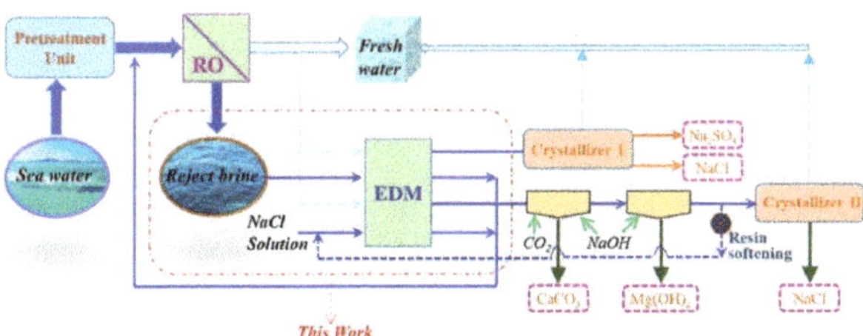

Figure 28. Conceptual scheme of SWRO-EDM for water and salts recovery. Reproduced with permission from [41], published by Elsevier, 2019.

6.1.3. WWRO Brine

ED was studied to recover further water from RO brines produced in industrial/municipal WWTPs (WWRO brine).

At an industrial WWTP of ~1900 L/min (heavy metal clarification, settling UF, RO), a six-stack EDR plant treated ~265 L/min WWRO retentate with TDS at 3000–5000 ppm [508]. The EDR was designed for E_{spec} of 4 kWh/m^3 and a cost of 0.33 \$/m^3. It reclaimed 85% of the WWRO reject brine as diluate (550 ppm), which recirculated to the treatment line, while the remaining 15% waste concentrate (45,000 ppm) went to evaporation ponds.

A WWTP receiving mainly a domestic effluent was endowed with UF and RO tertiary treatment to reuse water in groundwater recharge [509]. A pilot ED stack was tested to desalinate the WWRO retentate and recirculate it into the biological step (after ozonation preventing organics accumulation), while discharging the ED concentrate (Figure 29). The WWRO retentate had ~4.8 mS/cm conductivity, with high scaling potential (Ca^{2+}, Mg^{2+} and carbonates), which was lowered by acidification-decarbonation with HCl. Batch or feed and bleed experiments provided a diluate with 75% desalination (average overall $\eta \approx$ 70–85%), while a long-run test (42 h) attained a 69% desalination, thus the effluent could be recirculated, though it needed a TOC reduction. The ED addition enhanced water recovery from 75% (standalone RO) to 95%. An ED operational cost of 0.19 €/m^3 was estimated, 20% of which went from E_{spec} of 0.9 kWh/m^3 [510]. The capital cost was actually prohibitive (~15 €/m^3) for a plant capacity of 300 L/h, but the significant abatement with a full-sized plant could make the system feasible. The membrane processes produced significant CO_2 emissions, but if they were driven by renewable energy, the total emissions could be lower than those from conventional methods. The precipitation/crystallisation in a pellet reactor with fluidized bed showed that 80% of Ca^{2+} removal made the ED operation stable ($\eta \approx$ 70%) without any scaling [511]. The integration of a phytoremediation pre-treatment (willow field for nutrients and organics removal) with ED recycled effectively the treated WWRO concentrate in the WWTP [512]. However, again, an oxidation step after ED was needed.

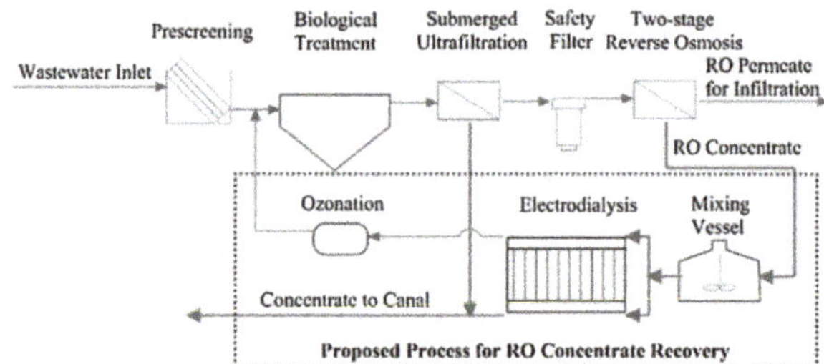

Figure 29. Scheme of the WWTP with RO-ED water recovery. Reproduced with permission from [509], published by Elsevier, 2011.

A pilot plant integrating a membrane bioreactor with RO and EDR was developed for treating municipal landfill leachate (15 mS/cm, 2250 mg/L COD) [513]. The RO had a water recovery of 84% and a rejection > 95% for most components. The EDR produced a diluate product at 30–65 mS/cm and recovered 67% of water, leading to an overall water recovery over 93%.

Recovery of nutrients and/or organics via ED from RO concentrate of food industry wastewater was assessed [514]. Experiments with model solutions (salts mixture, ~20 mM as total concentration) without or with organic compounds (120 g/L TOC) showed that monovalent ions and multivalent ions can be separated from each other under suitable operating conditions (low currents, for example), especially by using MVAs. However, separating nutrients (NO_3^- and $H_xPO_4^{y-}$) from other ions (Cl^- and SO_4^{2-}) was not feasible. The organics fate was strongly affected by molecular weight and charge, as the transport was slower for larger molecules and even slower for zwitterions and uncharged compounds. With a real WWRO retentate (90 mM Cl^-, 4.5 mM SO_4^{2-}, 70 ppm TOC), ions were almost completely removed, while more than 85% of the organics was retained within the diluate.

When RO treats wastewater, organic micropollutants (e.g., pesticides) may interact with the IEMs functional groups and may by adsorbed by the IEMs, with important implications such as fouling/poisoning and release during cleaning [515].

Some ED applications were studied for desalinating RO brine from petrochemical industry wastewater treatment, showing interesting results despite the scaling and fouling potential. A pilot EDR exhibited removal efficiencies above 90% for Cl^- and alkalinity, and of 76% for TDS, from a WWRO brine with 1104 mg/L TDS, and this RO-EDR process enhanced significantly the water recovery (87.3%) [516] compared to the EDR-RO scheme [352] (41%, see Section 4.3.2), thus offering the chance of reuse (cooling towers). Another EDR study achieved TDS removal of 50% (from 8663 mg/L TDS in the feed) with 85% of water recovery that reduced the brine volume by ~6.5 times [517]. No organic fouling was observed; however, the stack resistance increased during operation due to scaling at the concentrate side. A comparison among ED, NF and IX assessed the separation of NaCl and natural organic matter (NOM) [518]. The WWRO retentate had 6.9 g/L TDS and 35 mg/L NOM. ED removed ~90% of NaCl and recovered ~97% of water, retaining more NOM (e.g., TOC by 65%) when using MVMs. All the tested technologies exhibited results opening new opportunities to recover solid NaCl from brines in ZLD perspective.

The response surface methodology modelled and optimized the ED for RO concentrate (1950 mS/cm) reclamation in coal-fired power plants, finding a reduction in conductivity of 75.3% with E_{spec} of 0.11 kWh/m^3 and ~50% of water recovery with optimized conditions [519].

An IX-RO-ED treatment process for industrial Li-containing wastewater was developed [520] (Figure 30). The effluent (1268.9 mg/L Li$^+$, 17.87 mS/cm) from a Li-ion batteries production plant was softened to prevent scaling. Then, it was concentrated by RO and a two-stage ED, thus obtaining fresh water (RO permeate) with increased recovery (ED diluate recycle) and a concentrate solution suitable for Li$_2$CO$_3$ precipitation by Na$_2$CO$_3$ addition. Under optimal conditions, the RO retentate (60 mS/cm) was split into diluate/concentrate ED feed solutions with 3:1 volume ratio, obtaining water recoveries of 67.51%, 78.73%, and 69.44% in the RO step, the first and the second ED, respectively (E_{spec} in ED of ~30 and 50 kWh/m^3). A final LiCl concentration of ~87 g/L was reached with average η of 67.52%, total E_{spec} of 0.772 kWh/kg and a total cost of 0.47 $/kg (process capacity of 282 kg/year).

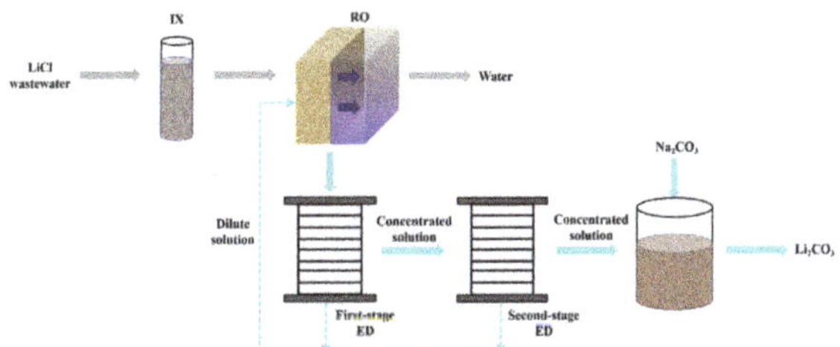

Figure 30. Scheme of the IX-RO-ED system for concentrating LiCl from industrial lithium-containing wastewater. Reproduced with permission from [520], published by American Chemical Society, 2019.

6.1.4. IX Spent Brine

Studies on ED regeneration of spent brines from IX treatment of surface water have exhibited promising results. A two-stage pilot ED treated an NOM-containing spent brine from anion exchange resin regeneration [521] (Figure 31). Electrostatic interactions and formation of metal-organic complexes caused the NOM removal by IX and the transfer of these compounds to the regeneration brine. The spent NaCl brine (8530 mg/L Na$^+$, 9050 mg/L Cl$^-$) contained NOM at 700 mg/L as dissolved organic carbon, but also NO$_3^-$ (113 mg/L), SO$_4^{2-}$ (3350 mg/L) and HCO$_3^-$ (2660 mg/L). The first ED step was conducted with MVAs, while the second one was with standard membranes. Both the ED steps used fresh water (RO permeate) in the concentrate, while the spent brine flowed through the diluate of both stages. As a result, with a removal of 85% for Cl$^-$ and 65% for Na$^+$, the ED stage 1 produced a monovalent salt solution (concentrate) with sufficient quality to be reused for IX regeneration, while the ED stage 2 produced a multivalent salts solution (concentrate, with predominance of Na$_2$SO$_4$) and a 470 mg/L NOM solution (diluate). The potential reuse of the NOM solution should be further studied, since its quality was impaired by heavy metals.

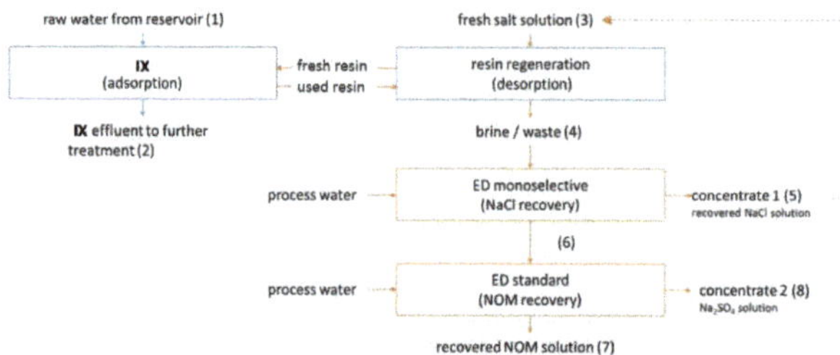

Figure 31. Process scheme of IX water production and two-stage ED treatment of brine. (1) Surface water from IJssel lake (the Netherlands), (2) ceramic MF and advanced oxidation, (3) NaCl regenerating solution, (4) spent regeneration brine, (5) concentrate from monovalent selective ED, (6) diluate from monovalent selective ED, (7) NOM-containing diluate from conventional ED, (8) multivalent salts-containing concentrate from conventional ED. Reproduced (adapted) from [521], published by Elsevier, 2019.

A simpler process was adopted for an IX spent brine with NaCl, sulphate and NOM concentrations of 90, 1, and 1 g/L (as dissolved organic carbon), respectively [522]. A single ED step (spent brine as diluate, 2 g/L NaCl solution as concentrate) with MVMs was effective, yielding a pure NaCl solution with 88.8% removal, negligible membrane fouling, and E_{spec} of 2 kWh/kg salt.

6.2. Salt Conversion into Acid and Base

BMED can valorise brines by salt transformation into acid and base. Moreover, the desalted feed may improve water recovery (ZLD approach). Studies on this application are presented for BWRO brine and SWRO brine in Section 6.2.1, and for WWRO brine and IX spent brine in Section 6.2.2.

6.2.1. BWRO Brine and SWRO Brine

The cost-effectiveness of acid/base production was demonstrated with artificial solutions (up to 390 mM NaCl) representing also BWRO concentrates [269] (see Section 4.2.2).

Several studies have been devoted to concentrated solutions from seawater desalination. A model SWRO brine was prepared with a salts mixture (~61 g/L) without Ca^{2+} and Mg^{2+}, and BMED produced mixed acids and bases up to 1 M [523]. The feed was desalinated up to 80% (from 70 to 12 mS/cm), but its pH decreased to ~2 due to proton leakage. η values between 50% and 80% indicated co-ion leakages as well, and impurities (e.g., SO$_4^{2-}$ in the acid) made the products unsuitable to reach commercial chemicals quality standards. However, they could be intended for uses not requiring high purity. The employment of nanocomposite MVAs synthetized by commercial AEMs coating, led only to a 10% reduction of SO$_4^{2-}$ in the acid compared to the original AEMs (~6 mM) [524]. However, the MVA was stable over more than 90 h operation ($P_{Cl^-}^{SO_4^{2-}} \approx 0.8 =$ constant, against ~1.08 for the AEM). By using a 1 M NaCl feed and a photovoltaic solar array simulator, a BMED process powered by photovoltaic energy was characterized [525].

In overflow configuration, the manipulation of flow rate for pH control resulted in a drop in E_{spec} from 7.3 kWh/kg acid (reference case at constant current) to 4.4 kWh/kg at variable current. An acid stream at a constant concentration of ~1 M HCl was produced for 30 h. Other BMED experiments produced up to 3.31 M HCl and 3.65 NaOH (from 1 M NaCl feed) [526]. E_{spec} was in the range 21.8–41.0 kWh/kg HCl at constant currents, and 26.7–43.5 kWh/kg HCl at variable currents. Azeotropic distillation was simulated as post-concentration step, producing 11.4 M HCl (35 wt%, commercial level) with overall E_{spec} of ~40–60 kWh/kg HCl. A life cycle assessment included environmental burdens associated with brine disposal and carbon footprint [527]. It was shown that renewable energies can be crucial for an overall process sustainability. However, though photovoltaic energy reduced the carbon footprint of the BMED-distillation process, it was still higher than that of industrial production from H_2-Cl_2 reaction, excluding the contribution of the transportation [526].

A real brine from seawater desalination (~42 g/L) pre-treated for Ca^{2+} and Mg^{2+} removal (precipitation by NaOH and CO_2) was used [528]. Preliminary tests were conducted with model solutions for optimisation purposes, finding η = 50–74% and $E_{spec} \approx 7.5$ kWh/kg. With the real effluent, the BMED produced continuously ~1 M acid and base without visible fouling. These products were usable at the desalination plant. For other industrial uses necessitating high quality standards, they would require purification. The low-salinity diluate product enhanced water recovery (direct reuse or recirculation to RO).

An SWRO brine (~60 g/L) was cleaned from Ca^{2+} and Mg^{2+} by NF and chemical precipitation, then the effluent (52 g/L) went to BMED [529]. The feed was almost fully desalted, and the NaCl conversion into HCl and NaOH was over 70%, obtaining ~1 M products at η of ~77% and E_{spec} of ~2.6 kWh/kg NaOH. Improvements were achieved with a combined process where an SWRO brine (~70 g/L) was purified and concentrated (~100 or ~200 g/L) by ED with MVMs, and then was converted (1.6 or 2 M acid, 1.2 or 2 M base) by BMED [530]. The diluate produced by BMED was at ~20 mg/L, the conversion was between 46% and 84%. The ED E_{spec} was 0.055–0.217 kWh/kg NaCl, the BMED one was 1.82–3.62 kWh/kg NaOH (η = 55–88%). The associated operating cost exceeded the market prices of the products. Therefore, in situ uses and/or stringent brine management regulations could justify the process.

BMED equipped with MVMs (bipolar membrane selectrodialysis, BMSED) was used with SWRO model brines at 70 or 105 g/L [531] (Figure 32). The process, which used commercial membranes, was highly selective ($P^{Na^+}_{Ca^{2+}}$ = 3.6–10.6, $P^{Cl^-}_{SO_4^{2-}}$ = 31.0–67.5) and thus produced high purity (approaching 99.99%) HCl and NaOH solutions at concentration up to 1.9 and 2.2 M, respectively. The final salt water was still at high concentration (~50 g/L) and could return to RO.

6.2.2. WWRO Brine and IX Spent Brine

BMED for RO concentrates was actually first proposed for a WWTP effluent (2590 mg/L TDS, 9 mS/cm) softened by IX [532]. Mixed acids and mixed bases at ~0.2 N concentration were produced, along with a diluate at conductivity below 2 mS/cm. For an RO capacity of 37,850 m^3/day, the estimated process cost of ~0.7 \$/m^3 was lower than those of conventional disposal (e.g., evaporation pond) or thermal ZLD processes.

BMED was proposed for waste neutralisation brine (20.4 mS/cm) from acid and base effluents regenerating IXRs used for surface water desalination [533]. The process concept included an ED step with MVCs followed by IX to concentrate and soften the brine and increase water recovery (ED diluate recycle) before BMED. Acid and base products from BMED were reusable for IXRs regeneration, and the produced diluate could recirculate to the ED concentrate. The saline water conductivity increased to 40 mS/cm by ED. BMED desalinated this solution up to ~5–10 mS/cm, and produced 0.9 M acid and base at η of 47% and E_{spec} of 6.25 kWh/kg HCl. Higher concentrations were reached with worse performance, due to co-ion leakage and current leakage (shunt currents).

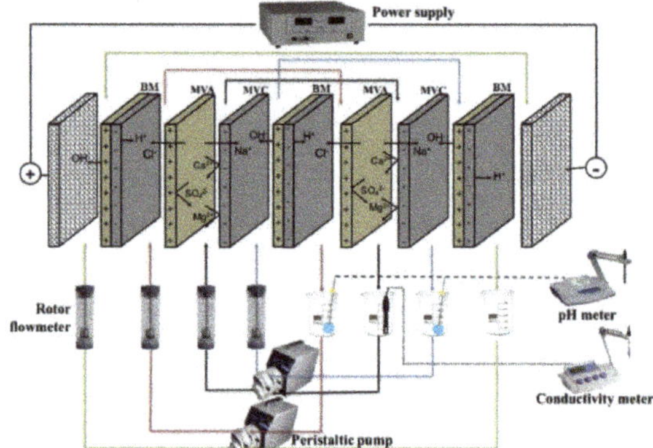

Figure 32. Scheme and experimental set-up of BMSED for acid/base recovery from SWRO brine. Reproduced with permission from [531], published by Elsevier, 2018.

6.3. Energy Recovery

The valorisation of desalination waste brines can be done in the form of energy recovery through salinity gradient power technologies [534,535]. Brines from RO desalination (of seawater, typically) can be employed as high-salinity feed (concentrate) for RED to produce electrical energy, reducing the overall energy consumption of desalination. Moreover, the RO brine is partially desalted prior to discharge. RED units require also a low-salinity feed; thus, using reclaimed wastewater as RED diluate can enhance the energy recovery compared to that achievable when using the desalination plant influent, e.g., seawater. Figure 33a shows the RO-RED scheme in which seawater is desalinated by RO, and RED receives the SWRO brine and a secondary effluent [424]. This scheme avoids issues of seawater contamination by organic micropollutants that, instead, may occur in the opposite scheme, i.e., RED-RO (Section 5.2). However, the overall energy balance may be less favourable in the RO-RED configuration, despite the higher power produced by RED due to the higher-concentration concentrate. For example, model predictions of the energy consumption are ~1 kWh/m^3 for RO-RED and ~0.5 kWh/m^3 for RED-RO [424]. More complex schemes were assessed, i.e., with RED pre- and post-treatment or with brine recirculation, but similar energy performances were predicted. In all cases, the costs should be evaluated.

Using desalination brines as concentrate, energy recovery by RED has been evaluated with different feed solutions and within different desalination schemes. Figure 33 depicts some of them.

Some integrated schemes involve a third process, e.g., membrane distillation (MD, Figure 33b) or membrane capacitive deionisation (MCDI, Figure 33c), between RO and RED, enhancing water recovery and/or energy saving. In the ED-RED coupling shown in Figure 33d, RED uses the ED brine (seawater desalination) and treated wastewater. ED and RED offer the possibility of internal integration in one stack with four-compartment repeating units (Figure 33e) and could by coupled with other desalination processes (e.g., RO) by using their brines as high-concentration streams boosting the energy recovery.

Table 4 reports several studies on energy recovery via RED using desalination brines, including those illustrated in Figure 33. As shown in Table 4, $P_{d,max}$ spans in a wide range in the order of ~1 W/m^2. It was shown that the addition of RED can reduce the energy consumption of desalination. Promising results were exhibited by ED-RED couplings, thus posing the bases for the development of self-sufficient or low-energy consuming systems. In all cases, however, a critical issue is represented by the capital costs.

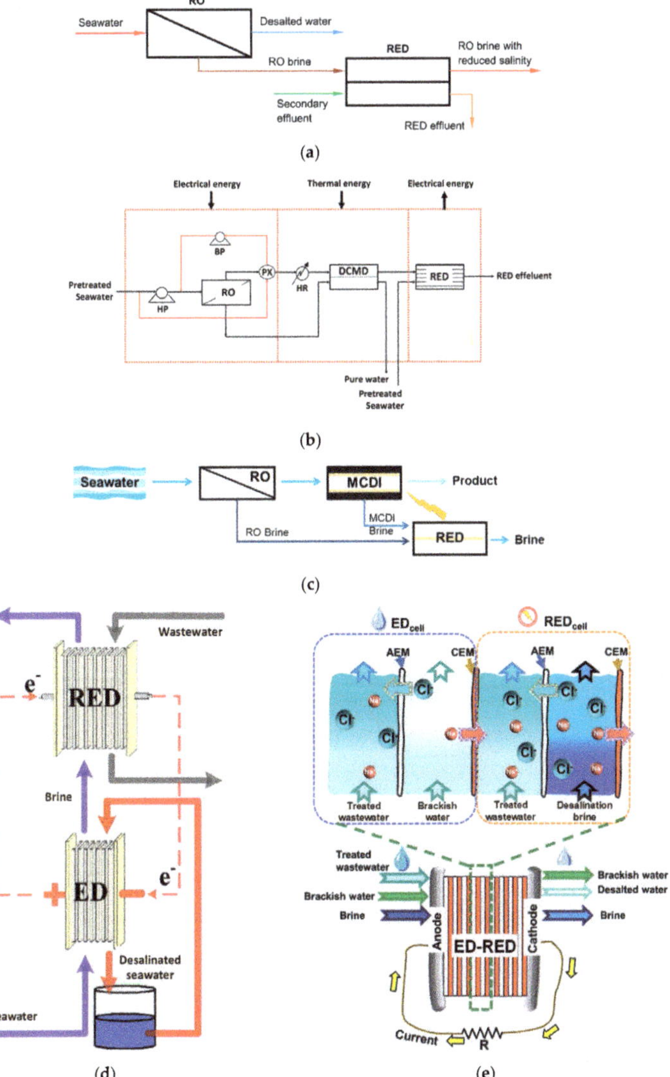

Figure 33. Desalination systems with RED using desalination brine for recovering energy: (**a**) RO-RED scheme in which RED receives the SWRO brine and a secondary effluent; (**b**) RO-MD-RED scheme; (**c**) RO-MCDI-RED scheme; (**d**) ED-RED scheme; (**e**) ED-RED process integrated in single stack (four-channel repetitive unit, coupling an ED cell with an RED cell). Panels (**a**–**e**) are reproduced with permission from [424,536–539] (adapted), respectively, all published by Elsevier, 2013, 2019, 2019, 2017 and 2020, respectively.

Table 4. Recent experimental studies on energy recovery via RED from desalination brines.

High-Salinity Sol.	Low-Salinity Sol.	Performance	Ref.
SWRO brine 1 or 2 M NaCl	Secondary effluent 0.02 M NaCl	V_{OC} (5 cell pairs), $P_{d,max}$: 1 M–0.02 M → 0.90 V, ~0.48 W/m^2 2 M–0.02 M → 1.02 V, ~0.57 W/m^2	[423] Figure 33a
• SWRO brine 1.2 M NaCl • FO brine 2.4 M NaCl	• River water 0.01 M NaCl • Seawater 0.6 M NaCl	$P_{d,max}$, estimated maximum reduction of E_{spec}: 1.2 M–0.01 M → 1.48 W/m^2, 7.8%; 2.4 M–0.01 M → 1.86 W/m^2, 13.5%; 1.2 M–0.6 M → 0.09 W/m^2, 0.5%; 2.4 M–0.6 M → 0.37 W/m^2, 2.2%	[540]
BWRO brine 31.3 mS/cm (~0.4 M), 19.5 mg/L dissolved organic carbon	Brackish groundwater 8.3 mS/cm (~0.095 M), 4.1 mg/L dissolved organic carbon	V_{OC} = 0.53 V (10 cell pairs, 78% permselectivity), $P_{d,max}$ = 0.07 W/m^2; with NaCl solutions at the same conductivity, $P_{d,max}$ = 0.09 W/m^2 indicated effects of NOM and divalent ions	[398]
MD brine 4, 5 or 5.4 M NaCl (from 1 M feed, i.e., SWRO retentate)	Seawater 0.5 M NaCl	At 20 °C and 0.7 cm/s, V_{OC} = 1.23–2.1 V (25 cell pairs), $P_{d,max}$ = 0.45–1.1 W/m^2, water recovery 92%; at 10–50 °C, 5 M and 0.7 cm/s, V_{OC} ≈ 1.7 and $P_{d,max}$ ≈ 0.5–1.05 W/m^2; at 20 °C, 5 M and 1.1 cm/s, $P_{d,max}$ ≈ 1.1 W/m^2 and $P_{d,max,net}$ ≈ 0.67 W/m^2	[541] Figure 33b
• SWRO brine 1 M NaCl • MD brine 5 M NaCl	• Brackish water 0.1 M NaCl • Seawater 0.5 M NaCl	V_{OC} (25 cell pairs), $P_{d,max}$: 1 M–0.1 M → 2.1 V, 0.39 W/m^2; 5 M–0.1 M → 3.4 V, 1.5 W/m^2 ($P_{d,max,net}$ ≈ 1.2 W/m^2, H$_2$ production by alkaline polymer electrolyte water electrolysis cell 44 cm^3/(h·cm^2)); 1 M–0.5 M → 0.71 V, 0.05 W/m^2; 5 M–0.5 M → 1.9 V, 0.55 W/m^2	[542]
MD brine 2–5 M NaCl (from 1 M feed, i.e., SWRO retentate)	Seawater 0.5 M NaCl	At 20–60 °C, water recovery 75–95%, V_{OC} = 1.26–1.95 V (25 cell pairs), $P_{d,max}$ ≈ 0.22–1.1 W/m^2, exergetic efficiency 49% under best conditions, electrical energy consumption (1.3 kWh/m^3) reduced by 23% and E_{spec} (4.4 kWh/m^3) reduced by 16.6% through RED inclusion	[536] Figure 33b
SWRO brine 1.1–1.5 M NaCl (RO with 30–50% water recovery from 43 g/L feed, i.e., high salinity seawater)	MCDI brine ~0.023 M NaCl (MCDI with 50–80% water recovery from ~0.85 g/L feed, i.e., SWRO permeate)	$P_{d,max}$ ≈ 2.45–2.83 W/m^2, E_{spec} = 2.0 kWh/m^3 reduced by ~39% compared to RO-RO and by ~17% compared to RO-RO-RED	[537] Figure 33c
• ED diluate 38.1 mS/cm (from ED of real seawater in salt production plant) • Seawater 48.7 mS/cm	Distilled water 0.2 mS/cm (from evaporation of ED brine in salt production plant)	V_{OC} (10 cell pairs), $P_{d,max}$: 38.1–0.2 mS/cm → 2.02 V, ~0.23 W/m^2; 48.7–0.2 mS/cm → 2.1 V, ~0.26 W/m^2	[543]
ED brine (from ED of simulated seawater at 30 g/L sea crystal)	Simulated wastewater 0.8 g/L NaCl, 0.1 g/L KH$_2$PO$_4$, 0.1 g/L NH$_4$Cl, 0.5 g/L glucose	Partial desalination of seawater (~60%) by consuming only the electrical energy produced by RED	[538] Figure 33d
Desalination brine 1.2 M NaCl	• Brackish water 0.02 M NaCl • Seawater 0.6 M NaCl	Partial desalination of brackish water (~75%) or seawater (~50%) by consuming only the self-produced electrical energy, the developed model predicted enhanced desalination by changing the operating conditions	[539] Figure 33e

7. Discussion, Conclusions and Outlook

ED and unconventional configurations of ED, i.e., BMED, SED, EDM and EDI, have great potential in desalination and valorisation strategies of wastewater for a broad range of applications to recover water and/or other valuable components. The main ones are metals, salts, acids and bases, nutrients, and organics. Energy recovery via RED is another possibility.

ED methods can be applied for effluents originating from various industrial processes (Section 4). In the separation of heavy metal ions (Section 4.1), such as Ni, Cu, Zn, Cr, Cd, and Pb, ED can provide solutions suitable for reuse, e.g., plating baths and rinse waters, including solutions with complexing agents (cyanide or organic acids), and tanning solutions. Two-stage operations (either both or one with MVMs) can improve purity, while EDI can reduce the energy consumption when treating diluted solutions (including low radioactive effluents). ED techniques (including complexation-enhanced ED) can also produce reusable solutions from heavy metal ions mixtures. BMED and SED have been poorly studied so far, but have exhibited promising results. Future research should be focused on experiments with real effluents, by assessing long-term operations, and should aim at scaling up the systems. Cost analyses are needed to assess the techno-economic feasibility.

For the regeneration of acid/base and salt conversion (Section 4.2), ED methods have been studied with a variety of industrial wastewaters. In the presence of heavy metal ions (waste solutions from pickling and other metallurgical processes), the use of proton-blocking AEMs and proton-selective CEMs is crucial for recovering acids by ED concentration. Interestingly, NF membranes instead of CEMs can increase significantly the proton/metal permselectivity. Further studies on the development and testing of special membranes would be beneficial. BMED is a valid alternative and can also convert salts into acids and bases, either with or without previous precipitation of heavy metals, if present. The regeneration of spent solutions from chemical absorption of flue gases (SO_2, H_2S, CO_2) is an interesting application of ED techniques, which may offer economic advantages. Various effluents containing organic matter were treated as well, finding only minor fouling issues in many cases. There are some commercialized systems and pilot plants. The scaling up should be extended to a wider range of applications.

Desalination (Section 4.3) via ED enables water reuse by treating salty wastewater from different industrial sources. Produced water of oil and gas extraction poses challenges related to high energy consumptions in the case of high-salinity feeds. Preliminary studies show that optimized systems can be competitive even for these solutions, but significant efforts are still needed in this direction. Dealing with fouling, cleaning procedures and EDR operation may have a partial effectiveness, as also shown by pilot plants. However, pulsed electric fields can minimize fouling phenomena. Therefore, this technique deserves further studies. Pilot installations have reclaimed wastewater of refineries and petrochemical industries, drainage water of coal mines, and wastewater of power plants. The treatment costs were shown to be attractive in some cases, but further economic analyses are needed. Energy recovery via RED is an interesting option, thus it should be explored more deeply in the future, by paying attention to the investment costs.

ED methods may be effective for various other industrial wastewaters (Section 4.4). Single salts or mixtures, waste effluents from pulp and paper manufacturing, textile processing, and bio-refining are just some examples. However, only a small number of studies have been conducted so far for each application. Therefore, further research is required to improve the performance (e.g., in terms of selectivity and energy consumption/recovery) and to develop techno-economically competitive systems for the various types of industrial wastewater.

There are several possible applications for municipal wastewater and other effluents (Section 5). In the desalination of municipal WWTP effluents (Section 5.1), ED can be a cost-effective treatment enabling water reuse (e.g., irrigation), as shown by several field plant applications. Interesting results were shown by some studies on ED coupled with FO for recovering high-quality water or for controlling salinity build-up, thus this coupling is deserving of more attention in the future.

Treated wastewater may be used as a low-salinity stream coupled with seawater or other waste effluents as high-salinity stream in order to recover energy via RED (Section 5.2). The presence of organics requires the development of cost-effective pre-treatments and cleaning methods against fouling and clogging, enabling long-term operations with stable net power. Some pilot installations have been tested, but techno-economic analyses are needed. The integration of membrane distillation with RED has been proposed to recover water and energy from urine, but further studies are needed to assess the process feasibility.

The recovery of nutrients (Section 5.3) via SED or ED is another option, but energy consumptions and costs still need to be evaluated. Fertilizers can be produced also by ED treatment of excess sludge sidestreams, human urine and animal farming effluents. Pilot installations were operated, and the processes may be competitive. Pre-treatments (in the case of human urine) or cleaning and polarity reversal (in the case of digested swine manure) minimized fouling phenomena, and thus long-term operations may be feasible. In a couple of studies, VFAs were recovered along with nutrients by ED (from excess sludge) or BMED (from pig manure). Promising results were obtained in a study on SED for the simultaneous fractionation of cations and anions into several streams from digested swine manure. These emerging applications deserve further studies.

A possible application of ED in the buildings sector is the regeneration of liquid desiccant solutions for air conditioning (Section 5.4). The high-concentration solutions (for example ~30 wt% LiCl) used for this process require the development of high-performance IEMs and a delicate process optimisation in order to limit detrimental effects due to undesired transport phenomena.

Resources can also be recuperated by treating waste brine from desalination or ion exchange (Section 6). Water and salt recovery (Section 6.1) has been demonstrated at pilot scale in various zero brine discharge systems. Scaling and/or fouling were controlled in all cases, without particular problems. ED exhibited competitive costs in recovering water from BWRO brine. The same applies for ED with MVMs recovering concentrated brines for edible salt production (evaporation–crystallisation) and diluted solutions (fresh or brackish water) from SWRO retentate, as shown also by long-term testing. However, the competitiveness with respect to a standalone SWRO depends on the local water, salt and electricity prices. Instead, NaCl recovery for industrial use (e.g., in the chlor-alkali industry) is not yet feasible. EDM has been proposed to separate salts, and it was found to be efficient. However, the overall process should be assessed. Some studies showed that water recovery by ED may be feasible also from WWRO brine (either municipal effluents or industrial effluents, e.g., from petrochemical sites). In the presence of organics (e.g., food industry wastewater), attention should be paid to their molecular weight and charge, which affect their transport. Water and salt (LiCl concentrated solution) were recovered from the effluent of a Li-ion batteries production plant, thus deserving further research to promote the industrial application. A couple of studies have investigated the treatment of surface water IX spent regeneration brines containing NOM. NaCl solutions that were reusable for IX regeneration were obtained via ED with MVMs, thus minimizing waste brine disposal. Further studies should be focused on the potential reuse of the NOM solution, the process sustainability and the long-term operation.

Salt conversion into acid and base from waste brines (Section 6.2) is a recovery method that can be performed by BMED. Moreover, the desalted solution may improve water recovery. In most cases, the treatment was tested with SWRO brine. Acid and base recovered have not reached the quality standards of commercial chemicals. This, in conjunction with the high energy consumptions, does not yet allow market entry. However, in situ use at the desalination plant is possible. The use of MVMs significantly enhanced the purity of the products. Future research should be intensified in this direction, as well as in the development of highly selective membranes, in the process optimisation, in the evaluation of post-concentration systems and in the scaling up.

Waste brines from desalination plants are suitable as concentrated solutions for energy recovery via RED (Section 6.3). Low-energy consuming systems were demonstrated in a few recent studies. However, capital costs may increase significantly, and thus economic analyses should reveal the actual feasibility. Other critical points for further research are the development of scaled-up systems, the testing with real solutions, and the evaluation of the net power density. Addressing all these aspects is necessary in order to attempt the implementation of integrated methodologies in real systems.

The application of ED techniques in wastewater treatment offers new opportunities for environmental protection and recovery of resources. Techno-economic challenges are still present, but great efforts have been made, mainly in the last 20 years, opening promising perspectives within efficient ZLD systems.

Some commercial applications and several pilot installations are accompanied by hundreds of studies with laboratory tests.

Some process limitations can be alleviated or even remediated. EDR operation, pulsed electric fields, pre-treatment and cleaning procedures against fouling can maintain or restore, at least partially, IEM properties. However, permanent fouling and poisoning of membranes may occur. Special membranes, e.g., proton-blocking AEMs, proton-selective CEMs, monovalent selective membranes, and even UF or NF membranes, can improve process selectivity and products purity. Nevertheless, energy consumptions may be high. Therefore, in addition to developing high-performing membranes (low resistance, high selectivity, low osmotic transport), optimizing system design and operation is essential to implement competitive processes. In this regard, novel concepts based on multi-stage ED configurations or integrated (electro-)membrane processes provide interesting technological solutions. Performance with real effluents, scaling up, long-term operation, overall sustainability, and techno-economic analysis have still to be assessed for several applications. An abatement in the membrane cost will be important to improving the process economics.

Please note that studies on similar or hybrid processes have been growing, i.e., EDI with configurations deviating from conventional ED stacks (important role of electrode chambers) [42,45–47,183], RED and fuel cell (Fenton)-RED with wastewater treatment at the electrode compartments [53,57,544–547], concentration gradient or pH gradient flow batteries [548–550], membrane electrolysis and electro-electrodialysis [551–560], hybrid liquid membrane-ED [561–563], decoupled ED [564], shock ED [565], (membrane) capacitive deionisation [566–574], membrane electrode redox transistor ED [575], bio-electrochemical systems [576–579] including microbial desalination cell [580–583], microbial desalination and chemical-production cell [584], and (Fenton) microbial RED [585–587]. Hence, knowledge acquired on common aspects will likely promote substantial advances in ED.

In light of all of the above considerations, a realistic scenario where ED techniques will conquer a wider market share for real applications can be prospected for a not far future.

Author Contributions: All the authors contributed to the writing and editing of this work. All authors have read and agreed to the published version of the manuscript.

Funding: This work was performed within the framework of the REvivED project (Low energy solutions for drinking water production by a REvival of ElectroDialysis systems) and the BAoBaB project (Blue Acid/Base Battery: Storage and recovery of renewable electrical energy by reversible salt water dissociation). These projects have received funding from the European Union's Horizon 2020 Research and Innovation program under Grant Agreements no. 685579 (www.revivedwater.eu) and 731187 (www.baobabproject.eu).

Conflicts of Interest: The authors declare no conflict of interest. The funders had no role in the design of the study; in the collection, analyses, or interpretation of data; in the writing of the manuscript, or in the decision to publish the results.

Abbreviations

AEL	Anion exchange layer
AEM	Anion exchange membrane
BM	Bipolar membrane
BMED	Bipolar membrane electrodialysis
BMSED	Bipolar membrane selectrodialysis
BWRO	Brackish water reverse osmosis
CEDI	Continuous electrodeionisation
CEL	Cation exchange layer
CEM	Cation exchange membrane
COP	Coefficient of performance
ED	Electrodialysis

EDI	Electrodeionisation
EDL	Electrical double layer
EDM	Electrodialysis metathesis
EDR	Electrodialysis reversal
EDTA	Ethylenediaminetetraacetic acid
FO	Forward osmosis
HPAM	Partially hydrolysed polyacrylamide
IEM	Ion exchange membrane
IX	Ion-exchange
IXR	Ion-exchange resin
MCDI	Membrane capacitive deionisation
MD	Membrane distillation
MF	Microfiltration
MVA	Monovalent selective anion exchange membrane
MVC	Monovalent selective cation exchange membrane
MVM	Monovalent selective ion exchange membrane
NF	Nanofiltration
NOM	Natural organic matter
RED	Reverse electrodialysis
REDI	Reverse electrodeionisation
RO	Reverse osmosis
SED	Selectrodialysis
SWRO	Seawater reverse osmosis
TDS	Total Dissolved Solids
UF	Ultrafiltration
VFA	Volatile fatty acid
WWRO	Wastewater reverse osmosis
WWTP	Wastewater treatment plant
ZLD	Zero liquid discharge

References

1. Tong, T.; Elimelech, M. The Global Rise of Zero Liquid Discharge for Wastewater Management: Drivers, Technologies, and Future Directions. *Environ. Sci. Technol.* **2016**, *50*, 6846–6855. [CrossRef]
2. Ahirrao, S. Zero Liquid Discharge Solutions. In *Industrial Wastewater Treatment, Recycling and Reuse*; Butterworth-Heinemann: Oxford, UK, 2014; pp. 489–520. ISBN 9780444634030.
3. Yaqub, M.; Lee, W. Zero-liquid discharge (ZLD) technology for resource recovery from wastewater: A review. *Sci. Total Environ.* **2019**, *681*, 551–563. [CrossRef] [PubMed]
4. Voulvoulis, N. Water reuse from a circular economy perspective and potential risks from an unregulated approach. *Curr. Opin. Environ. Sci. Health* **2018**, *2*, 32–45. [CrossRef]
5. European Comission. *Report from the Commission to the European Parliament, the Council, the European Economic and Social Committee and the Committee of the Regions on the Implementation of the Circular Economy Action Plan*; European Comission: Brusselles, Belgium, 2019.
6. Kirchherr, J.; Reike, D.; Hekkert, M. Conceptualizing the circular economy: An analysis of 114 definitions. *Resour. Conserv. Recycl.* **2017**, *127*, 221–232. [CrossRef]
7. Obotey Ezugbe, E.; Rathilal, S. Membrane Technologies in Wastewater Treatment: A Review. *Membranes* **2020**, *10*, 89. [CrossRef] [PubMed]
8. Strathmann, H. *Ion.-Exchange Membrane Separation Processes*, 1st ed.; Elsevier: Amsterdam, The Netherlands, 2004; ISBN 044450236X.

9. Tanaka, Y. *Ion. Exchange Membranes: Fundamentals and Applications*; Elsevier: Amsterdam, The Netherlands, 2007; Volume 12, ISBN 0927-5193.

10. Xu, T. Ion exchange membranes: State of their development and perspective. *J. Membr. Sci.* **2005**, *263*, 1–29. [CrossRef]

11. Nagarale, R.K.; Gohil, G.S.; Shahi, V.K. Recent developments on ion-exchange membranes and electro-membrane processes. *Adv. Colloid Interface Sci.* **2006**, *119*, 97–130. [CrossRef]

12. Ran, J.; Wu, L.; He, Y.; Yang, Z.; Wang, Y.; Jiang, C.; Ge, L.; Bakangura, E.; Xu, T. Ion exchange membranes: New developments and applications. *J. Membr. Sci.* **2017**, *522*, 267–291. [CrossRef]

13. Xu, T.; Huang, C. Electrodialysis-Based separation technologies: A critical review. *AiCHE J.* **2008**, *54*, 3147–3159. [CrossRef]

14. Zhao, W.Y.; Zhou, M.; Yan, B.; Sun, X.; Liu, Y.; Wang, Y.; Xu, T.; Zhang, Y. Waste conversion and resource recovery from wastewater by ion exchange membranes: State-of-the-art and perspective. *Ind. Eng. Chem. Res.* **2018**, *57*, 6025–6039. [CrossRef]

15. Strathmann, H. Electrodialysis, a mature technology with a multitude of new applications. *Desalination* **2010**, *264*, 268–288. [CrossRef]

16. Campione, A.; Gurreri, L.; Ciofalo, M.; Micale, G.; Tamburini, A.; Cipollina, A. Electrodialysis for water desalination: A critical assessment of recent developments on process fundamentals, models and applications. *Desalination* **2018**, *434*, 121–160. [CrossRef]

17. Sajjad, A.-A.; Yunus, M.Y.B.M.; Azoddein, A.A.M.; Hassell, D.G.; Dakhil, I.H.; Hasan, H.A. Electrodialysis Desalination for Water and Wastewater: A Review. *Chem. Eng. J.* **2020**, *380*, 122231.

18. Fidaleo, M.; Moresi, M. Electrodialysis Applications in The Food Industry. *Adv. Food Nutr. Res.* **2006**, *51*, 265–360.

19. Huang, C.; Xu, T.; Zhang, Y.; Xue, Y.; Chen, G. Application of electrodialysis to the production of organic acids: State-of-the-art and recent developments. *J. Membr. Sci.* **2007**, *288*, 1–12. [CrossRef]

20. *Electrodialysis and Water Reuse: Novel Approaches*; Moura Bernardes, A.; Zoppas Ferreira, J.; Siqueira Rodrigues, M.A. (Eds.) Springer: Berlin/Heidelberg, Germany, 2014; ISBN 9783642402494.

21. López-Garzón, C.S.; Straathof, A.J.J. Recovery of carboxylic acids produced by fermentation. *Biotechnol. Adv.* **2014**, *32*, 873–904. [CrossRef] [PubMed]

22. Gurreri, L.; Cipollina, A.; Tamburini, A.; Micale, G. Electrodialysis for wastewater treatment—Part I: Fundamentals and municipal effluents. In *Current Trends and Future Developments on (Bio-) Membranes-Membrane Technology for Water and Wastewater Treatment-Advances and Emerging Processes*; Basile, A., Comite, A., Eds.; Elsevier: Amsterdam, The Netherlands, 2020; pp. 141–192. ISBN 9780128168233.

23. Gurreri, L.; Cipollina, A.; Tamburini, A.; Micale, G. Electrodialysis for wastewater treatment—Part II: Industrial effluents. In *Current Trends and Future Developments on (Bio-) Membranes-Membrane Technology for Water and Wastewater Treatment-Advances and Emerging Processes*; Basile, A., Comite, A., Eds.; Elsevier: Amsterdam, The Netherlands, 2020; pp. 195–241. ISBN 9780128168233.

24. Gurreri, L.; Battaglia, G.; Tamburini, A.; Cipollina, A.; Micale, G.; Ciofalo, M. Multi-physical modelling of reverse electrodialysis. *Desalination* **2017**, *423*, 52–64. [CrossRef]

25. Battaglia, G.; Gurreri, L.; Airò Farulla, G.; Cipollina, A.; Pirrotta, A.; Micale, G.; Ciofalo, M. Membrane Deformation and Its Effects on Flow and Mass Transfer in the Electromembrane Processes. *Int. J. Mol. Sci.* **2019**, *20*, 1840. [CrossRef]

26. Battaglia, G.; Gurreri, L.; Airò Farulla, G.; Cipollina, A.; Pirrotta, A.; Micale, G.; Ciofalo, M. Pressure-Induced Deformation of Pillar-Type Profiled Membranes and Its Effects on Flow and Mass Transfer. *Computation* **2019**, *7*, 32. [CrossRef]

27. Battaglia, G.; Gurreri, L.; Cipollina, A.; Pirrotta, A.; Velizarov, S.; Ciofalo, M.; Micale, G. Fluid–Structure Interaction and Flow Redistribution in Membrane-Bounded Channels. *Energies* **2019**, *12*, 4259. [CrossRef]

28. Pawlowski, S.; Crespo, J.; Velizarov, S. Profiled Ion Exchange Membranes: A Comprehensible Review. *Int. J. Mol. Sci.* **2019**, *20*, 165. [CrossRef]

29. Lindstrand, V.; Sundström, G.; Jönsson, A.S. Fouling of electrodialysis membranes by organic substances. *Desalination* **2000**, *128*, 91–102. [CrossRef]

30. Strathmann, H.; Krol, J.J.; Rapp, H.J.; Eigenberger, G. Limiting current density and water dissociation in bipolar membranes. *J. Membr. Sci.* **1997**, *125*, 123–142. [CrossRef]

31. Mareev, S.A.; Evdochenko, E.; Wessling, M.; Kozaderova, O.A.; Niftaliev, S.I.; Pismenskaya, N.D.; Nikonenko, V.V. A comprehensive mathematical model of water splitting in bipolar membranes: Impact of the spatial distribution of fixed charges and catalyst at bipolar junction. *J. Membr. Sci.* **2020**, *603*, 118010. [CrossRef]

32. Pan, J.; Hou, L.; Wang, Q.; He, Y.; Wu, L.; Mondal, A.N.; Xu, T. Preparation of bipolar membranes by electrospinning. *Mater. Chem. Phys.* **2017**, *186*, 484–491. [CrossRef]

33. Shen, C.; Wycisk, R.; Pintauro, P.N. High performance electrospun bipolar membrane with a 3D junction. *Energy Environ. Sci.* **2017**, *10*, 1435–1442. [CrossRef]

34. Pourcelly, G. Electrodialysis with bipolar membranes: Principles, optimization, and applications. *Russ. J. Electrochem.* **2002**, *38*, 919–926. [CrossRef]

35. Jaroszek, H.; Dydo, P. Ion-exchange membranes in chemical synthesis-a review. *Open Chem.* **2016**, *14*, 1–19. [CrossRef]

36. Huang, C.; Xu, T. Electrodialysis with bipolar membranes for sustainable development. *Environ. Sci. Technol.* **2006**, *40*, 5233–5243. [CrossRef]

37. Mani, K.N.; Chlanda, F.P.; Byszewski, C.H. Aquatech membrane technology for recovery of acid/base values for salt streams. *Desalination* **1988**, *68*, 149–166. [CrossRef]

38. Mani, K.N. Electrodialysis water splitting technology. *J. Membr. Sci.* **1991**, *58*, 117–138. [CrossRef]

39. Zhang, Y.; Paepen, S.; Pinoy, L.; Meesschaert, B.; Van Der Bruggen, B. Selectrodialysis: Fractionation of divalent ions from monovalent ions in a novel electrodialysis stack. *Sep. Purif. Technol.* **2012**, *88*, 191–201. [CrossRef]

40. Alhéritière, C.; Ernst, W.R.; Davis, T.A. Metathesis of magnesium and sodium salt systems by electrodialysis. *Desalination* **1998**, *115*, 189–198. [CrossRef]

41. Chen, Q.-B.; Ren, H.; Tian, Z.; Sun, L.; Wang, J. Conversion and pre-concentration of SWRO reject brine into high solubility liquid salts (HSLS) by using electrodialysis metathesis. *Sep. Purif. Technol.* **2019**, *213*, 587–598. [CrossRef]

42. Alvarado, L.; Chen, A. Electrodeionization: Principles, strategies and applications. *Electrochim. Acta* **2014**, *132*, 583–597. [CrossRef]

43. Hakim, A.N.; Khoiruddin, K.; Ariono, D.; Wenten, I.G. Ionic Separation in Electrodeionization System: Mass Transfer Mechanism and Factor Affecting Separation Performance. *Sep. Purif. Rev.* **2019**, 1–23. [CrossRef]

44. Wood, J.; Gifford, J.; Arba, J.; Shaw, M. Production of ultrapure water by continuous electrodeionization. *Desalination* **2010**, *250*, 973–976. [CrossRef]

45. Dzyazko, Y.S.; Belyakov, V.N. Purification of a diluted nickel solution containing nickel by a process combining ion exchange and electrodialysis. *Desalination* **2004**, *162*, 179–189. [CrossRef]

46. Feng, X.; Wu, Z.; Chen, X. Removal of metal ions from electroplating effluent by EDI process and recycle of purified water. *Sep. Purif. Technol.* **2007**, *57*, 257–263. [CrossRef]

47. Mahmoud, A.; Hoadley, A.F.A. An evaluation of a hybrid ion exchange electrodialysis process in the recovery of heavy metals from simulated dilute industrial wastewater. *Water Res.* **2012**, *46*, 3364–3376. [CrossRef]

48. Souilah, O.; Akretche, D.E.; Amara, M. Water reuse of an industrial effluent by means of electrodeionisation. *Desalination* **2004**, *167*, 49–54. [CrossRef]

49. Spoor, P.B.; Koene, L.; ter Veen, W.R.; Janssen, L.J.J. Continuous deionization of a dilute nickel solution. *Chem. Eng. J.* **2002**, *85*, 127–135. [CrossRef]

50. Spoor, P.B.; Grabovska, L.; Koene, L.; Janssen, L.J.J.; Ter Veen, W.R. Pilot scale deionisation of a galvanic nickel solution using a hybrid ion-exchange/electrodialysis system. *Chem. Eng. J.* **2002**, *89*, 193–202. [CrossRef]

51. Park, S.; Kwak, R. Microscale electrodeionization: In situ concentration profiling and flow visualization. *Water Res.* **2020**, *170*, 115310. [CrossRef] [PubMed]

52. Pan, S.Y.; Snyder, S.W.; Ma, H.W.; Lin, Y.J.; Chiang, P.C. Energy-efficient resin wafer electrodeionization for impaired water reclamation. *J. Clean. Prod.* **2018**, *174*, 1464–1474. [CrossRef]

53. Mei, Y.; Tang, C.Y. Recent developments and future perspectives of reverse electrodialysis technology: A review. *Desalination* **2018**, *425*, 156–174. [CrossRef]

54. Tufa, R.A.; Pawlowski, S.; Veerman, J.; Bouzek, K.; Fontananova, E.; di Profio, G.; Velizarov, S.; Goulão Crespo, J.; Nijmeijer, K.; Curcio, E. Progress and prospects in reverse electrodialysis for salinity gradient energy conversion and storage. *Appl. Energy* **2018**, *225*, 290–331. [CrossRef]

55. Cipollina, A.; Micale, G.; Tamburini, A.; Tedesco, M.; Gurreri, L.; Veerman, J.; Grasman, S. Reverse electrodialysis: Applications. In *Sustainable Energy from Salinity Gradients*; Cipollina, A., Micale, G., Eds.; Woodhead Publishing: Cambridge, UK; Elsevier: Amsterdam, The Netherlands, 2016; pp. 135–180. ISBN 9780081003237.

56. Tamburini, A.; Cipollina, A.; Tedesco, M.; Gurreri, L.; Ciofalo, M.; Micale, G. The REAPower Project: Power Production From Saline Waters and Concentrated Brines. In *Current Trends and Future Developments on (Bio-) Membranes-Membrane Desalination Systems: The Next Generation*; Basile, A., Curcio, E., Inamuddin, I., Eds.; Elsevier: Amsterdam, The Netherlands, 2019; pp. 407–448.

57. Tian, H.; Wang, Y.; Pei, Y.; Crittenden, J.C. Unique applications and improvements of reverse electrodialysis: A review and outlook. *Appl. Energy* **2020**, *262*, 114482. [CrossRef]

58. Avci, A.H.; Tufa, R.A.; Fontananova, E.; Di Profio, G.; Curcio, E. Reverse Electrodialysis for energy production from natural river water and seawater. *Energy* **2018**, *165*, 512–521. [CrossRef]

59. Sata, T. *Ion. Exchange Membranes: Preparation, Characterization, Modification and Application*; Royal Society of Chemistry: Cambridge, UK, 2004; ISBN 978-0-85404-590-7.

60. Luo, T.; Abdu, S.; Wessling, M. Selectivity of ion exchange membranes: A review. *J. Membr. Sci.* **2018**, *555*, 429–454. [CrossRef]

61. Kontturi, K.; Murtomäki, L.; Manzanares, J.A. *Ionic Transport. Processes in Electrochemistry and Membrane Science*; Oxford University Press Inc.: New York, NY, USA, 2008; ISBN 978-0-19-953381-7.

62. Manzanares, J.A.; Vergara, G.; Mafé, S.; Kontturi, K.; Viinikka, P. Potentiometric Determination of Transport Numbers of Ternary Electrolyte Systems in Charged Membranes. *J. Phys. Chem. B* **1998**, *102*, 1301–1307. [CrossRef]

63. Kraaijeveld, G.; Sumberova, V.; Kuindersma, S.; Wesselingh, H. Modelling electrodialysis using the Maxwell-Stefan description. *Chem. Eng. J. Biochem. Eng. J.* **1995**, *57*, 163–176. [CrossRef]

64. Plntauro, P.N.; Bennion, D.N. Mass Transport of Electrolytes in Membranes. 1. Development of Mathematical Transport Model. *Ind. Eng. Chem. Fundam.* **1984**, *23*, 230–234. [CrossRef]

65. Wesselingh, J.A.; Vonk, P.; Kraaijeveld, G. Exploring the Maxwell-Stefan description of ion exchange. *Chem. Eng. J. Biochem. Eng. J.* **1995**, *57*, 75–89. [CrossRef]

66. Sata, T. Studies on ion exchange membranes with permselectivity for specific ions in electrodialysis. *J. Membr. Sci.* **1994**, *93*, 117–135. [CrossRef]

67. Balster, J.; Yildirim, M.H.; Stamatialis, D.F.; Ibanez, R.; Lammertink, R.G.H.; Jordan, V.; Wessling, M. Morphology and microtopology of cation-exchange polymers and the origin of the overlimiting current. *J. Phys. Chem. B* **2007**, *111*, 2152–2165. [CrossRef]

68. Długołecki, P.; Anet, B.; Metz, S.J.; Nijmeijer, K.; Wessling, M. Transport limitations in ion exchange membranes at low salt concentrations. *J. Membr. Sci.* **2010**, *346*, 163–171. [CrossRef]

69. Galama, A.H.; Vermaas, D.A.; Veerman, J.; Saakes, M.; Rijnaarts, H.H.M.; Post, J.W.; Nijmeijer, K. Membrane resistance: The effect of salinity gradients over a cation exchange membrane. *J. Membr. Sci.* **2014**, *467*, 279–291. [CrossRef]

70. Gohil, G.S.; Shahi, V.K.; Rangarajan, R. Comparative studies on electrochemical characterization of homogeneous and heterogeneous type of ion-exchange membranes. *J. Membr. Sci.* **2004**, *240*, 211–219. [CrossRef]

71. Mehdizadeh, S.; Yasukawa, M.; Abo, T.; Kakihana, Y.; Higa, M. Effect of spacer geometry on membrane and solution compartment resistances in reverse electrodialysis. *J. Membr. Sci.* **2019**, *572*, 271–280. [CrossRef]

72. Galama, A.H.; Hoog, N.A.; Yntema, D.R. Method for determining ion exchange membrane resistance for electrodialysis systems. *Desalination* **2016**, *380*, 1–11. [CrossRef]

73. Silva, R.F.; De Francesco, M.; Pozio, A. Tangential and normal conductivities of Nafion® membranes used in polymer electrolyte fuel cells. *J. Power Sources* **2004**, *134*, 18–26. [CrossRef]

74. Kamcev, J.; Sujanani, R.; Jang, E.S.; Yan, N.; Moe, N.; Paul, D.R.; Freeman, B.D. Salt concentration dependence of ionic conductivity in ion exchange membranes. *J. Membr. Sci.* **2018**, *547*, 123–133. [CrossRef]

75. Zhu, S.; Kingsbury, R.S.; Call, D.F.; Coronell, O. Impact of solution composition on the resistance of ion exchange membranes. *J. Membr. Sci.* **2018**, *554*, 39–47. [CrossRef]

76. Larchet, C.; Nouri, S.; Auclair, B.; Dammak, L.; Nikonenko, V. Application of chronopotentiometry to determine the thickness of diffusion layer adjacent to an ion-exchange membrane under natural convection. *Adv. Colloid Interface Sci.* **2008**, *139*, 45–61. [CrossRef]

77. Zabolotsky, V.I.; Nikonenko, V.V. Effect of structural membrane inhomogeneity on transport properties. *J. Membr. Sci.* **1993**, *79*, 181–198. [CrossRef]

78. Veerman, J. The Effect of the NaCl Bulk Concentration on the Resistance of Ion Exchange Membranes—Measuring and Modeling. *Energies* **2020**, *13*, 1946. [CrossRef]

79. Porada, S.; van Egmond, W.J.; Post, J.W.; Saakes, M.; Hamelers, H.V.M. Tailoring ion exchange membranes to enable low osmotic water transport and energy efficient electrodialysis. *J. Membr. Sci.* **2018**, *552*, 22–30. [CrossRef]

80. Kamcev, J.; Doherty, C.M.; Lopez, K.P.; Hill, A.J.; Paul, D.R.; Freeman, B.D. Effect of fixed charge group concentration on salt permeability and diffusion coefficients in ion exchange membranes. *J. Membr. Sci.* **2018**, *566*, 307–316. [CrossRef]

81. Kamcev, J.; Paul, D.R.; Manning, G.S.; Freeman, B.D. Predicting salt permeability coefficients in highly swollen, highly charged ion exchange membranes. *ACS Appl. Mater. Interfaces* **2017**, *9*, 4044–4056. [CrossRef]

82. Kamcev, J.; Paul, D.R.; Manning, G.S.; Freeman, B.D. Ion Diffusion Coefficients in Ion Exchange Membranes: Significance of Counterion Condensation. *Macromolecules* **2018**, *51*, 5519–5529. [CrossRef]

83. Ciofalo, M.; Di Liberto, M.; Gurreri, L.; La Cerva, M.; Scelsi, L.; Micale, G. Mass transfer in ducts with transpiring walls. *Int. J. Heat Mass Transf.* **2019**, *132*, 1074–1086. [CrossRef]

84. Spiegler, K.S. Polarization at ion exchange membrane-solution interfaces. *Desalination* **1971**, *9*, 367–385. [CrossRef]

85. Helfferich, F. *Ion. Exchange*; McGraw-Hill: New York, NY, USA, 1962.

86. Levich, V.G. *Physicochemical Hydrodynamics*; Prentice-Hall: Englewood Cliffs, NJ, USA, 1962.

87. Krol, J.J.; Wessling, M.; Strathmann, H. Concentration polarization with monopolar ion exchange membranes: Current-voltage curves and water dissociation. *J. Membr. Sci.* **1999**, *162*, 145–154. [CrossRef]

88. Kwak, R.; Guan, G.; Peng, W.K.; Han, J. Microscale electrodialysis: Concentration profiling and vortex visualization. *Desalination* **2013**, *308*, 138–146. [CrossRef]

89. Rubinstein, I.; Shtilman, L. Voltage against current curves of cation exchange membranes. *J. Chem. Soc. Faraday Trans. 2 Mol. Chem. Phys.* **1979**, *75*, 231–246. [CrossRef]

90. Lee, H.J.; Strathmann, H.; Moon, S.H. Determination of the limiting current density in electrodialysis desalination as an empirical function of linear velocity. *Desalination* **2006**, *190*, 43–50. [CrossRef]

91. Urtenov, M.K.; Uzdenova, A.M.; Kovalenko, A.V.; Nikonenko, V.V.; Pismenskaya, N.D.; Vasil'eva, V.I.; Sistat, P.; Pourcelly, G. Basic mathematical model of overlimiting transfer enhanced by electroconvection in flow-through electrodialysis membrane cells. *J. Membr. Sci.* **2013**, *447*, 190–202. [CrossRef]

92. La Cerva, M.; Gurreri, L.; Tedesco, M.; Cipollina, A.; Ciofalo, M.; Tamburini, A.; Micale, G. Determination of limiting current density and current efficiency in electrodialysis units. *Desalination* **2018**, *445*, 138–148. [CrossRef]

93. Cowan, D.A.; Brown, J.H. Effect of Turbulence on Limiting Current in Electrodialysis Cells. *Ind. Eng. Chem.* **1959**, *51*, 1445–1448. [CrossRef]

94. Ben Sik Ali, M.; Mnif, A.; Hamrouni, B. Modelling of the limiting current density of an electrodialysis process by response surface methodology. *Ionics (Kiel)* **2018**, *24*, 617–628. [CrossRef]

95. Forgacs, C.; Ishibashi, N.; Leibovitz, J.; Sinkovic, J.; Spiegler, K.S. Polarization at ion-exchange membranes in electrodialysis. *Desalination* **1972**, *10*, 181–214. [CrossRef]

96. Kharkats, Y.I. Mechanism of "supralimiting" currents at ion-exchange mebrane/electrolyte interfaces. *Sov. Electrochem.* **1985**, *21*, 917–920.

97. Simons, R. The origin and elimination of water splitting in ion exchange membranes during water demineralisation by electrodialysis. *Desalination* **1979**, *28*, 41–42. [CrossRef]

98. Simons, R. Water splitting in ion exchange membranes. *Electrochim. Acta* **1985**, *30*, 275–282. [CrossRef]

99. Nikonenko, V.; Urtenov, M.; Mareev, S. Pourcelly Mathematical Modeling of the Effect of Water Splitting on Ion Transfer in the Depleted Diffusion Layer Near an Ion-Exchange Membrane. *Membranes* **2020**, *10*, 22. [CrossRef]

100. Uzdenova, A. 2D Mathematical Modelling of Overlimiting Transfer Enhanced by Electroconvection in Flow-Through Electrodialysis Membrane Cells in Galvanodynamic Mode. *Membranes* **2019**, *9*, 39. [CrossRef]
101. Uzdenova, A.; Urtenov, M. Potentiodynamic and Galvanodynamic Regimes of Mass Transfer in Flow-Through Electrodialysis Membrane Systems: Numerical Simulation of Electroconvection and Current-Voltage Curve. *Membranes* **2020**, *10*, 49. [CrossRef]
102. Nikonenko, V.V.; Pismenskaya, N.D.; Belova, E.I.; Sistat, P.; Huguet, P.; Pourcelly, G.; Larchet, C. Intensive current transfer in membrane systems: Modelling, mechanisms and application in electrodialysis. *Adv. Colloid Interface Sci.* **2010**, *160*, 101–123. [CrossRef]
103. Nikonenko, V.V.; Kovalenko, A.V.; Urtenov, M.K.; Pismenskaya, N.D.; Han, J.; Sistat, P.; Pourcelly, G. Desalination at overlimiting currents: State-of-the-art and perspectives. *Desalination* **2014**, *342*, 85–106. [CrossRef]
104. Mareev, S.A.; Nebavskiy, A.V.; Nichka, V.S.; Urtenov, M.K.; Nikonenko, V.V. The nature of two transition times on chronopotentiograms of heterogeneous ion exchange membranes: 2D modelling. *J. Membr. Sci.* **2019**, *575*, 179–190. [CrossRef]
105. Nikonenko, V.; Nebavsky, A.; Mareev, S.; Kovalenko, A.; Urtenov, M.; Pourcelly, G. Modelling of Ion Transport in Electromembrane Systems: Impacts of Membrane Bulk and Surface Heterogeneity. *Appl. Sci.* **2018**, *9*, 25. [CrossRef]
106. Andersen, M.B.; Wang, K.M.; Schiffbauer, J.; Mani, A. Confinement effects on electroconvective instability. *Electrophoresis* **2017**, *38*, 702–711. [CrossRef] [PubMed]
107. Pham, S.V.; Kwon, H.; Kim, B.; White, J.K.; Lim, G.; Han, J. Helical vortex formation in three-dimensional electrochemical systems with ion-selective membranes. *Phys. Rev. E* **2016**, *93*, 033114. [CrossRef] [PubMed]
108. Karatay, E.; Druzgalski, C.L.; Mani, A. Simulation of chaotic electrokinetic transport: Performance of commercial software versus custom-built direct numerical simulation codes. *J. Colloid Interface Sci.* **2015**, *446*, 67–76. [CrossRef]
109. Zaltzman, B.; Rubinstein, I. Electro-osmotic slip and electroconvective instability. *J. Fluid Mech.* **2007**, *579*, 173–226. [CrossRef]
110. Krol, J.J.; Wessling, M.; Strathmann, H. Chronopotentiometry and overlimiting ion transport through monopolar ion exchange membranes. *J. Membr. Sci.* **1999**, *162*, 155–164. [CrossRef]
111. Rubinstein, I. Electroconvection at an electrically inhomogeneous permselective interface. *Phys. Fluids A* **1991**, *3*, 2301–2309. [CrossRef]
112. Maletzki, F.; Rösler, H.W.; Staude, E. Ion transfer across electrodialysis membranes in the overlimiting current range: Stationary voltage current characteristics and current noise power spectra under different conditions of free convection. *J. Membr. Sci.* **1992**, *71*, 105–116. [CrossRef]
113. Rubinstein, I.; Staude, E.; Kedem, O. Role of the membrane surface in concentration polarization at ion-exchange membrane. *Desalination* **1988**, *69*, 101–114. [CrossRef]
114. Zabolotsky, V.I.; Nikonenko, V.V.; Pismenskaya, N.D.; Laktionov, E.V.; Urtenov, M.K.; Strathmann, H.; Wessling, M.; Koops, G.H. Coupled transport phenomena in overlimiting current electrodialysis. *Sep. Purif. Technol.* **1998**, *14*, 255–267. [CrossRef]
115. Dukhin, S.S. Electrokinetic phenomena of the second kind and their applications. *Adv. Colloid Interface Sci.* **1991**, *35*, 173–196. [CrossRef]
116. Rubinstein, I.; Zaltzman, B.; Kedem, O. Electric fields in and around ion-exchange membranes. *J. Membr. Sci.* **1997**, *125*, 17–21. [CrossRef]
117. Choi, J.-H.; Lee, H.-J.; Moon, S.-H. Effects of Electrolytes on the Transport Phenomena in a Cation-Exchange Membrane. *J. Colloid Interface Sci.* **2001**, *238*, 188–195. [CrossRef] [PubMed]
118. Ibanez, R.; Stamatialis, D.F.; Wessling, M. Role of membrane surface in concentration polarization at cation exchange membranes. *J. Membr. Sci.* **2004**, *239*, 119–128. [CrossRef]
119. Pismenskaia, N.; Sistat, P.; Huguet, P.; Nikonenko, V.; Pourcelly, G. Chronopotentiometry applied to the study of ion transfer through anion exchange membranes. *J. Membr. Sci.* **2004**, *228*, 65–76. [CrossRef]
120. Volodina, E.; Pismenskaya, N.; Nikonenko, V.; Larchet, C.; Pourcelly, G. Ion transfer across ion-exchange membranes with homogeneous and heterogeneous surfaces. *J. Colloid Interface Sci.* **2005**, *285*, 247–258. [CrossRef]

121. Gil, V.V.; Andreeva, M.A.; Jansezian, L.; Han, J.; Pismenskaya, N.D.; Nikonenko, V.V.; Larchet, C.; Dammak, L. Impact of heterogeneous cation-exchange membrane surface modification on chronopotentiometric and current–voltage characteristics in NaCl, CaCl2and MgCl2solutions. *Electrochim. Acta* **2018**, *281*, 472–485. [CrossRef]

122. Nebavskaya, K.A.; Sarapulova, V.V.; Sabbatovskiy, K.G.; Sobolev, V.D.; Pismenskaya, N.D.; Sistat, P.; Cretin, M.; Nikonenko, V.V. Impact of ion exchange membrane surface charge and hydrophobicity on electroconvection at underlimiting and overlimiting currents. *J. Membr. Sci.* **2017**, *523*, 36–44. [CrossRef]

123. Nikonenko, V.V.; Mareev, S.A.; Pis'menskaya, N.D.; Uzdenova, A.M.; Kovalenko, A.V.; Urtenov, M.K.; Pourcelly, G. Effect of electroconvection and its use in intensifying the mass transfer in electrodialysis (Review). *Russ. J. Electrochem.* **2017**, *53*, 1122–1144. [CrossRef]

124. Butylskii, D.; Moroz, I.; Tsygurina, K.; Mareev, S. Effect of Surface Inhomogeneity of Ion-Exchange Membranes on the Mass Transfer Efficiency in Pulsed Electric Field Modes. *Membranes* **2020**, *10*, 40. [CrossRef]

125. Titorova, V.; Sabbatovskiy, K.; Sarapulova, V.; Kirichenko, E.; Sobolev, V.; Kirichenko, K. Characterization of MK-40 membrane modified by layers of cation exchange and anion exchange polyelectrolytes. *Membranes* **2020**, *10*, 20. [CrossRef] [PubMed]

126. Zabolotsky, V.I.; Nikonenko, V.V.; Pismenskaya, N.D. On the role of gravitational convection in the transfer enhancement of salt ions in the course of dilute solution electrodialysis. *J. Membr. Sci.* **1996**, *119*, 171–181. [CrossRef]

127. Larchet, C.; Zabolotsky, V.I.; Pismenskaya, N.; Nikonenko, V.V.; Tskhay, A.; Tastanov, K.; Pourcelly, G. Comparison of different ED stack conceptions when applied for drinking water production from brackish waters. *Desalination* **2008**, *222*, 489–496. [CrossRef]

128. Isaacson, M.S.; Sonin, A.A. Sherwood Number and Friction Factor Correlations for Electrodialysis Systems, with Application to Process Optimization. *Ind. Eng. Chem. Process. Des. Dev.* **1976**, *15*, 313–321. [CrossRef]

129. Sonin, A.A.; Probstein, R.F. A hydrodynamic theory of desalination by electrodialysis. *Desalination* **1968**, *5*, 293–329. [CrossRef]

130. Malek, P.; Ortiz, J.M.; Richards, B.S.; Schäfer, A.I. Electrodialytic removal of NaCl from water: Impacts of using pulsed electric potential on ion transport and water dissociation phenomena. *J. Membr. Sci.* **2013**, *435*, 99–109. [CrossRef]

131. Tadimeti, J.G.D.; Chattopadhyay, S. Uninterrupted swirling motion facilitating ion transport in electrodialysis. *Desalination* **2016**, *392*, 54–62. [CrossRef]

132. Geraldes, V.; Afonso, M.D. Limiting current density in the electrodialysis of multi-ionic solutions. *J. Membr. Sci.* **2010**, *360*, 499–508. [CrossRef]

133. Sonin, A.A.; Isaacson, M.S. Optimization of Flow Design in Forced Flow Electrochemical Systems, with Special Application to Electrodialysis. *Ind. Eng. Chem. Process. Des. Dev.* **1974**, *13*, 241–248. [CrossRef]

134. Fidaleo, M.; Moresi, M. Optimal strategy to model the electrodialytic recovery of a strong electrolyte. *J. Membr. Sci.* **2005**, *260*, 90–111. [CrossRef]

135. Belfort, G.; Guter, G.A. An experimental study of electrodialysis hydrodynamics. *Desalination* **1972**, *10*, 221–262. [CrossRef]

136. Lee, H.J.; Sarfert, F.; Strathmann, H.; Moon, S.H. Designing of an electrodialysis desalination plant. *Desalination* **2002**, *142*, 267–286. [CrossRef]

137. Tanaka, Y. Limiting current density of an ion-exchange membrane and of an electrodialyzer. *J. Membr. Sci.* **2005**, *266*, 6–17. [CrossRef]

138. Gurreri, L.; Tamburini, A.; Cipollina, A.; Micale, G.; Ciofalo, M. Flow and mass transfer in spacer-filled channels for reverse electrodialysis: A CFD parametrical study. *J. Membr. Sci.* **2016**, *497*, 300–317. [CrossRef]

139. La Cerva, M.; Di Liberto, M.; Gurreri, L.; Tamburini, A.; Cipollina, A.; Micale, G.; Ciofalo, M. Coupling CFD with a one-dimensional model to predict the performance of reverse electrodialysis stacks. *J. Membr. Sci.* **2017**, *541*, 595–610. [CrossRef]

140. Gurreri, L.; Tamburini, A.; Cipollina, A.; Micale, G.; Ciofalo, M. Pressure drop at low Reynolds numbers in woven-spacer-filled channels for membrane processes: CFD prediction and experimental validation. *Desalin. Water Treat.* **2017**, *61*, 170–182. [CrossRef]

141. Kuroda, O.; Takahashi, S.; Nomura, M. Characteristics of flow and mass transfer rate in an electrodialyzer compartment including spacer. *Desalination* **1983**, *46*, 225–232. [CrossRef]

142. Winograd, Y.; Solan, A.; Toren, M. Mass transfer in narrow channels in the presence of turbulence promoters. *Desalination* **1973**, *13*, 171–186. [CrossRef]

143. Da Costa, A.R.; Fane, A.G.; Fell, C.J.D.; Franken, A.C.M. Optimal channel spacer design for ultrafiltration. *J. Membr. Sci.* **1991**, *62*, 275–291. [CrossRef]

144. Koutsou, C.P.; Yiantsios, S.G.; Karabelas, A.J. A numerical and experimental study of mass transfer in spacer-filled channels: Effects of spacer geometrical characteristics and Schmidt number. *J. Membr. Sci.* **2009**, *326*, 234–251. [CrossRef]

145. Campione, A.; Cipollina, A.; Bogle, I.D.L.; Gurreri, L.; Tamburini, A.; Tedesco, M.; Micale, G. A hierarchical model for novel schemes of electrodialysis desalination. *Desalination* **2019**, *465*, 79–93. [CrossRef]

146. Culcasi, A.; Gurreri, L.; Zaffora, A.; Cosenza, A.; Tamburini, A.; Cipollina, A.; Micale, G. Ionic shortcut currents via manifolds in reverse electrodialysis stacks. *Desalination* **2020**, *485*, 114450. [CrossRef]

147. Wagholikar, V.V.; Zhuang, H.; Jiao, Y.; Moe, N.E.; Ramanan, H.; Goh, L.M.; Barber, J.; Lee, K.S.; Lee, H.P.; Fuh, J.Y.H. Modeling cell pair resistance and spacer shadow factors in electro-separation processes. *J. Membr. Sci.* **2017**, *543*, 151–162. [CrossRef]

148. Kim, H.K.; Lee, M.S.; Lee, S.Y.; Choi, Y.W.; Jeong, N.J.; Kim, C.S. High power density of reverse electrodialysis with pore-filling ion exchange membranes and a high-open-area spacer. *J. Mater. Chem. A* **2015**, *3*, 16302–16306. [CrossRef]

149. Ciofalo, M.; La Cerva, M.; Di Liberto, M.; Gurreri, L.; Cipollina, A.; Micale, G. Optimization of net power density in Reverse Electrodialysis. *Energy* **2019**, *181*, 576–588. [CrossRef]

150. Barakat, M.A. New trends in removing heavy metals from industrial wastewater. *Arab. J. Chem.* **2011**, *4*, 361–377. [CrossRef]

151. Scarazzato, T.; Panossian, Z.; Tenório, J.A.S.; Pérez-Herranz, V.; Espinosa, D.C.R. A review of cleaner production in electroplating industries using electrodialysis. *J. Clean. Prod.* **2017**, *168*, 1590–1602. [CrossRef]

152. Arar, Ö.; Yüksel, Ü.; Kabay, N.; Yüksel, M. Various applications of electrodeionization (EDI) method for water treatment-A short review. *Desalination* **2014**, *342*, 16–22. [CrossRef]

153. Benvenuti, T.; García-Gabaldón, M.; Ortega, E.M.; Rodrigues, M.A.S.; Bernardes, A.M.; Pérez-Herranz, V.; Zoppas-Ferreira, J. Influence of the co-ions on the transport of sulfate through anion exchange membranes. *J. Membr. Sci.* **2017**, *542*, 320–328. [CrossRef]

154. Martí-Calatayud, M.; García-Gabaldón, M.; Pérez-Herranz, V. Mass Transfer Phenomena during Electrodialysis of Multivalent Ions: Chemical Equilibria and Overlimiting Currents. *Appl. Sci.* **2018**, *8*, 1566. [CrossRef]

155. Nemati, M.; Hosseini, S.M.; Shabanian, M. Novel electrodialysis cation exchange membrane prepared by 2-acrylamido-2-methylpropane sulfonic acid; heavy metal ions removal. *J. Hazard. Mater.* **2017**, *337*, 90–104. [CrossRef]

156. Martí-Calatayud, M.C.; García-Gabaldón, M.; Pérez-Herranz, V. Effect of the equilibria of multivalent metal sulfates on the transport through cation-exchange membranes at different current regimes. *J. Membr. Sci.* **2013**, *443*, 181–192. [CrossRef]

157. Sharma, P.; Shahi, V.K. Assembly of MIL-101(Cr)-sulphonated poly (ether sulfone) membrane matrix for selective electrodialytic separation of Pb2+ from mono-/bi-valent ions. *Chem. Eng. J.* **2020**, *382*, 122688. [CrossRef]

158. Vallois, C.; Sistat, P.; Roualdès, S.; Pourcelly, G. Separation of H+/Cu2+cations by electrodialysis using modified proton conducting membranes. *J. Membr. Sci.* **2003**, *216*, 13–25. [CrossRef]

159. Chang, J.H.; Huang, C.P.; Cheng, S.F.; Shen, S.Y. Transport characteristics and removal efficiency of copper ions in the electrodialysis process under electroconvection operation. *Process. Saf. Environ. Prot.* **2017**, *112*, 235–242. [CrossRef]

160. Barros, K.S.; Scarazzato, T.; Espinosa, D.C.R. Evaluation of the effect of the solution concentration and membrane morphology on the transport properties of Cu (II) through two monopolar cation–exchange membranes. *Sep. Purif. Technol.* **2018**, *193*, 184–192. [CrossRef]

161. Mahmoud, A.; Muhr, L.; Vasiluk, S.; Aleynikoff, A.; Lapicque, F. Investigation of transport phenomena in a hybrid ion exchange-electrodialysis system for the removal of copper ions. *J. Appl. Electrochem.* **2003**, *33*, 875–884. [CrossRef]

162. Scarazzato, T.; Panossian, Z.; García-Gabaldón, M.; Ortega, E.M.; Tenório, J.A.S.; Pérez-Herranz, V.; Espinosa, D.C.R. Evaluation of the transport properties of copper ions through a heterogeneous ion-exchange membrane in etidronic acid solutions by chronopotentiometry. *J. Membr. Sci.* **2017**, *535*, 268–278. [CrossRef]

163. Aouad, F.; Lindheimer, A.; Gavach, C. Transport properties of electrodialysis membranes in the presence of Zn^{2+} complexes with Cl^-. *J. Membr. Sci.* **1997**, *123*, 207–223. [CrossRef]

164. Rodrigues, M.A.S.; Amado, F.D.R.; Bischoff, M.R.; Ferreira, C.A.; Bernardes, A.M.; Ferreira, J.Z. Transport of zinc complexes through an anion exchange membrane. *Desalination* **2008**, *227*, 241–252. [CrossRef]

165. Dalla Costa, R.F.; Antônio Siqueira Rodrigues, M.; Ferreira, J.Z. Transport of Trivalent and Hexavalent Chromium through Different Ion-Selective Membranes in Acidic Aqueous Media. *Sep. Sci. Technol.* **1998**, *33*, 1135–1143. [CrossRef]

166. Vallejo, M.E.; Persin, F.; Innocent, C.; Sistat, P.; Pourcelly, G. Electrotransport of Cr(VI) through an anion exchange membrane. *Sep. Purif. Technol.* **2000**, *21*, 61–69. [CrossRef]

167. Rodrigues, M.A.S.; Dalla Costa, R.F.; Bernardes, A.M.; Zoppas Ferreira, J. Influence of ligand exchange on the treatment of trivalent chromium solutions by electrodialysis. *Electrochim. Acta* **2001**, *47*, 753–758. [CrossRef]

168. Çengeloğlu, Y.; Tor, A.; Kir, E.; Ersöz, M. Transport of hexavalent chromium through anion-exchange membranes. *Desalination* **2003**, *154*, 239–246. [CrossRef]

169. Hosseini, S.M.; Sohrabnejad, S.; Nabiyouni, G.; Jashni, E.; Van der Bruggen, B.; Ahmadi, A. Magnetic cation exchange membrane incorporated with cobalt ferrite nanoparticles for chromium ions removal via electrodialysis. *J. Membr. Sci.* **2019**, *583*, 292–300. [CrossRef]

170. Jashni, E.; Hosseini, S.M. Promoting the electrochemical and separation properties of heterogeneous cation exchange membrane by embedding 8-hydroxyquinoline ligand: Chromium ions removal. *Sep. Purif. Technol.* **2020**, *234*, 116118. [CrossRef]

171. Hosseini, S.M.; Alibakhshi, H.; Jashni, E.; Parvizian, F.; Shen, J.N.; Taheri, M.; Ebrahimi, M.; Rafiei, N. A novel layer-by-layer heterogeneous cation exchange membrane for heavy metal ions removal from water. *J. Hazard. Mater.* **2020**, *381*, 120884. [CrossRef]

172. Barros, K.S.; Espinosa, D.C.R. Chronopotentiometry of an anion-exchange membrane for treating a synthesized free-cyanide effluent from brass electrodeposition with EDTA as chelating agent. *Sep. Purif. Technol.* **2018**, *201*, 244–255. [CrossRef]

173. Mohammadi, T.; Moheb, A.; Sadrzadeh, M.; Razmi, A. Modeling of metal ion removal from wastewater by electrodialysis. *Sep. Purif. Technol.* **2005**, *41*, 73–82. [CrossRef]

174. Sadrzadeh, M.; Razmi, A.; Mohammadi, T. Separation of different ions from wastewater at various operating conditions using electrodialysis. *Sep. Purif. Technol.* **2007**, *54*, 147–156. [CrossRef]

175. Itoi, S.; Nakamura, I.; Kawahara, T. Electrodialytic recovery process of metal finishing waste water. *Desalination* **1980**, *32*, 383–389. [CrossRef]

176. Benvenuti, T.; Krapf, R.S.; Rodrigues, M.A.S.; Bernardes, A.M.; Zoppas-Ferreira, J. Recovery of nickel and water from nickel electroplating wastewater by electrodialysis. *Sep. Purif. Technol.* **2014**, *129*, 106–112. [CrossRef]

177. Benvenuti, T.; Siqueira Rodrigues, M.A.; Bernardes, A.M.; Zoppas-Ferreira, J. Closing the loop in the electroplating industry by electrodialysis. *J. Clean. Prod.* **2017**, *155*, 130–138. [CrossRef]

178. Tzanetakis, N.; Taama, W.M.; Scott, K.; Jachuck, R.J.J.; Slade, R.S.; Varcoe, J. Comparative performance of ion exchange membranes for electrodialysis of nickel and cobalt. *Sep. Purif. Technol.* **2003**, *30*, 113–127. [CrossRef]

179. Li, C.L.; Zhao, H.X.; Tsuru, T.; Zhou, D.; Matsumura, M. Recovery of spent electroless nickel plating bath by electrodialysis. *J. Membr. Sci.* **1999**, *157*, 241–249. [CrossRef]

180. Peng, C.; Jin, R.; Li, G.; Li, F.; Gu, Q. Recovery of nickel and water from wastewater with electrochemical combination process. *Sep. Purif. Technol.* **2014**, *136*, 42–49. [CrossRef]

181. Lu, H.; Wang, Y.; Wang, J. Recovery of Ni2+ and pure water from electroplating rinse wastewater by an integrated two-stage electrodeionization process. *J. Clean. Prod.* **2015**, *92*, 257–266. [CrossRef]

182. Dzyazko, Y.S.; Ponomaryova, L.N.; Rozhdestvenskaya, L.M.; Vasilyuk, S.L.; Belyakov, V.N. Electrodeionization of low-concentrated multicomponent Ni2+-containing solutions using organic-inorganic ion-exchanger. *Desalination* **2014**, *342*, 43–51. [CrossRef]

183. Zhao, C.; Zhang, L.; Ge, R.; Zhang, A.; Zhang, C.; Chen, X. Treatment of low-level Cu(II) wastewater and regeneration through a novel capacitive deionization-electrodeionization (CDI-EDI) technology. *Chemosphere* **2019**, *217*, 763–772. [CrossRef]

184. Dermentzis, K. Removal of nickel from electroplating rinse waters using electrostatic shielding electrodialysis/electrodeionization. *J. Hazard. Mater.* **2010**, *173*, 647–652. [CrossRef] [PubMed]

185. Mohammadi, T.; Moheb, A.; Sadrzadeh, M.; Razmi, A. Separation of copper ions by electrodialysis using Taguchi experimental design. *Desalination* **2004**, *169*, 21–31. [CrossRef]

186. Korngold, E.; Kock, K.; Strathmann, H. Electrodialysis in advanced waste water treatment. *Desalination* **1978**, *24*, 129–139. [CrossRef]

187. Chiapello, J.M.; Gal, J.Y. Recovery by electrodialysis of cyanide electroplating rinse waters. *J. Membr. Sci.* **1992**, *68*, 283–291. [CrossRef]

188. Scarazzato, T.; Buzzi, D.C.; Bernardes, A.M.; Romano Espinosa, D.C. Treatment of wastewaters from cyanide-free plating process by electrodialysis. *J. Clean. Prod.* **2015**, *91*, 241–250. [CrossRef]

189. Scarazzato, T.; Panossian, Z.; Tenório, J.A.S.; Pérez-Herranz, V.; Espinosa, D.C.R. Water reclamation and chemicals recovery from a novel cyanide-free copper plating bath using electrodialysis membrane process. *Desalination* **2018**, *436*, 114–124. [CrossRef]

190. Zhelonkina, E.A.; Shishkina, S.V.; Mikhailova, I.Y.; Ananchenko, B.A. Study of electrodialysis of a copper chloride solution at overlimiting currents. *Pet. Chem.* **2017**, *57*, 947–953. [CrossRef]

191. Peng, C.; Liu, Y.; Bi, J.; Xu, H.; Ahmed, A.S. Recovery of copper and water from copper-electroplating wastewater by the combination process of electrolysis and electrodialysis. *J. Hazard. Mater.* **2011**, *189*, 814–820. [CrossRef]

192. Dong, Y.; Liu, J.; Sui, M.; Qu, Y.; Ambuchi, J.J.; Wang, H.; Feng, Y. A combined microbial desalination cell and electrodialysis system for copper-containing wastewater treatment and high-salinity-water desalination. *J. Hazard. Mater.* **2017**, *321*, 307–315. [CrossRef]

193. Song, Y.; Sun, T.; Cang, L.; Wu, S.; Zhou, D. Migration and transformation of Cu(II)-EDTA during electrodialysis accompanied by an electrochemical process with different compartment designs. *Electrochim. Acta* **2019**, *295*, 605–614. [CrossRef]

194. Nataraj, S.K.; Hosamani, K.M.; Aminabhavi, T.M. Potential application of an electrodialysis pilot plant containing ion-exchange membranes in chromium removal. *Desalination* **2007**, *217*, 181–190. [CrossRef]

195. Chen, S.-S.; Li, C.-W.; Hsu, H.-D.; Lee, P.-C.; Chang, Y.-M.; Yang, C.-H. Concentration and purification of chromate from electroplating wastewater by two-stage electrodialysis processes. *J. Hazard. Mater.* **2009**, *161*, 1075–1080. [CrossRef] [PubMed]

196. Dos Santos, C.S.L.; Miranda Reis, M.H.; Cardoso, V.L.; De Resende, M.M. Electrodialysis for removal of chromium (VI) from effluent: Analysis of concentrated solution saturation. *J. Environ. Chem. Eng.* **2019**, *7*, 103380. [CrossRef]

197. Alvarado, L.; Ramírez, A.; Rodríguez-Torres, I. Cr(VI) removal by continuous electrodeionization: Study of its basic technologies. *Desalination* **2009**, *249*, 423–428. [CrossRef]

198. Alvarado, L.; Torres, I.R.; Chen, A. Integration of ion exchange and electrodeionization as a new approach for the continuous treatment of hexavalent chromium wastewater. *Sep. Purif. Technol.* **2013**, *105*, 55–62. [CrossRef]

199. Xing, Y.; Chen, X.; Wang, D. Electrically regenerated ion exchange for removal and recovery of Cr(VI) from wastewater. *Environ. Sci. Technol.* **2007**, *41*, 1439–1443. [CrossRef]

200. Xing, Y.; Chen, X.; Yao, P.; Wang, D. Continuous electrodeionization for removal and recovery of Cr(VI) from wastewater. *Sep. Purif. Technol.* **2009**, *67*, 123–126. [CrossRef]

201. Xing, Y.; Chen, X.; Wang, D. Variable effects on the performance of continuous electrodeionization for the removal of Cr(VI) from wastewater. *Sep. Purif. Technol.* **2009**, *68*, 357–362. [CrossRef]

202. Jiang, C.; Chen, H.; Zhang, Y.; Feng, H.; Shehzad, M.A.; Wang, Y.; Xu, T. Complexation Electrodialysis as a general method to simultaneously treat wastewaters with metal and organic matter. *Chem. Eng. J.* **2018**, *348*, 952–959. [CrossRef]

203. Wu, X.; Zhu, H.; Liu, Y.; Chen, R.; Qian, Q.; Van der Bruggen, B. Cr(III) recovery in form of Na_2CrO_4 from aqueous solution using improved bipolar membrane electrodialysis. *J. Membr. Sci.* **2020**, *604*, 118097. [CrossRef]

204. Zhang, Z.; Liba, D.; Alvarado, L.; Chen, A. Separation and recovery of Cr(III) and Cr(VI) using electrodeionization as an efficient approach. *Sep. Purif. Technol.* **2014**, *137*, 86–93. [CrossRef]

205. Tor, A.; Büyükerkek, T.; Çengeloğlu, Y.; Ersöz, M. Simultaneous recovery of Cr(III) and Cr(VI) from the aqueous phase with ion-exchange membranes. *Desalination* **2005**, *171*, 233–241. [CrossRef]

206. Raghava Rao, J.; Prasad, B.G.S.; Narasimhan, V.; Ramasami, T.; Shah, P.R.; Khan, A.A. Electrodialysis in the recovery and reuse of chromium from industrial effluents. *J. Membr. Sci.* **1989**, *46*, 215–224.

207. Lambert, J.; Rakib, M.; Durand, G.; Avila-Rodríguez, M. Treatment of solutions containing trivalent chromium by electrodialysis. *Desalination* **2006**, *191*, 100–110. [CrossRef]

208. Lambert, J.; Avila-Rodriguez, M.; Durand, G.; Rakib, M. Separation of sodium ions from trivalent chromium by electrodialysis using monovalent cation selective membranes. *J. Membr. Sci.* **2006**, *280*, 219–225. [CrossRef]

209. Rodrigues, M.A.S.; Amado, F.D.R.; Xavier, J.L.N.; Streit, K.F.; Bernardes, A.M.; Ferreira, J.Z. Application of photoelectrochemical-electrodialysis treatment for the recovery and reuse of water from tannery effluents. *J. Clean. Prod.* **2008**, *16*, 605–611. [CrossRef]

210. Deghles, A.; Kurt, U. Treatment of tannery wastewater by a hybrid electrocoagulation/electrodialysis process. *Chem. Eng. Process. Process. Intensif.* **2016**, *104*, 43–50. [CrossRef]

211. Marder, L.; Sulzbach, G.O.; Bernardes, A.M.; Zoppas Ferreira, J. Removal of cadmium and cyanide from aqueous solutions through electrodialysis. *J. Braz. Chem. Soc.* **2003**, *14*, 610–615. [CrossRef]

212. Marder, L.; Bernardes, A.M.; Zoppas Ferreira, J. Cadmium electroplating wastewater treatment using a laboratory-scale electrodialysis system. *Sep. Purif. Technol.* **2004**, *37*, 247–255. [CrossRef]

213. Mehellou, A.; Delimi, R.; Benredjem, Z.; Saaidia, S.; Allat, L.; Innocent, C. Improving the efficiency and selectivity of Cd 2+ removal using a modified resin in the continuous electropermutation process. *Sep. Sci. Technol.* **2019**, *55*, 2049–2060.

214. Mohammadi, T.; Razmi, A.; Sadrzadeh, M. Effect of operating parameters on Pb^{2+} separation from wastewater using electrodialysis. *Desalination* **2004**, *167*, 379–385. [CrossRef]

215. Sadrzadeh, M.; Mohammadi, T.; Ivakpour, J.; Kasiri, N. Separation of lead ions from wastewater using electrodialysis: Comparing mathematical and neural network modeling. *Chem. Eng. J.* **2008**, *144*, 431–441. [CrossRef]

216. Abou-Shady, A.; Peng, C.; Almeria, O.J.; Xu, H. Effect of pH on separation of Pb (II) and NO^{3-} from aqueous solutions using electrodialysis. *Desalination* **2012**, *285*, 46–53. [CrossRef]

217. Abou-Shady, A.; Peng, C.; Bi, J.; Xu, H.; Almeria, O.J. Recovery of Pb (II) and removal of NO3- from aqueous solutions using integrated electrodialysis, electrolysis, and adsorption process. *Desalination* **2012**, *286*, 304–315. [CrossRef]

218. Gherasim, C.V.; Křivčík, J.; Mikulášek, P. Investigation of batch electrodialysis process for removal of lead ions from aqueous solutions. *Chem. Eng. J.* **2014**, *256*, 324–334. [CrossRef]

219. Barros, K.S.; Ortega, E.M.; Pérez-Herranz, V.; Espinosa, D.C.R. Evaluation of brass electrodeposition at RDE from cyanide-free bath using EDTA as a complexing agent. *J. Electroanal. Chem.* **2020**, *865*, 114129. [CrossRef]

220. Barros, K.S.; Scarazzato, T.; Pérez-Herranz, V.; Espinosa, D.C.R. Treatment of cyanide-free wastewater from brass electrodeposition with edta by electrodialysis: Evaluation of underlimiting and overlimiting operations. *Membranes* **2020**, *10*, 69. [CrossRef]

221. Min, K.J.; Choi, S.Y.; Jang, D.; Lee, J.; Park, K.Y. Separation of metals from electroplating wastewater using electrodialysis. *Energy Sourcespart. A Recover. Util. Environ. Eff.* **2019**, *41*, 2471–2480. [CrossRef]

222. Zuo, W.; Zhang, G.; Meng, Q.; Zhang, H. Characteristics and application of multiple membrane process in plating wastewater reutilization. *Desalination* **2008**, *222*, 187–196. [CrossRef]

223. Peng, C.; Meng, H.; Song, S.; Lu, S.; Lopez-Vaidivieso, A. Elimination of Cr(VI) from electroplating wastewater by electrodialysis following chemical precipitation. *Sep. Sci. Technol.* **2004**, *39*, 1501–1517. [CrossRef]

224. Babilas, D.; Dydo, P. Selective zinc recovery from electroplating wastewaters by electrodialysis enhanced with complex formation. *Sep. Purif. Technol.* **2018**, *192*, 419–428. [CrossRef]

225. Babilas, D.; Dydo, P. Zinc salt recovery from electroplating industry wastes by electrodialysis enhanced with complex formation. *Sep. Sci. Technol.* **2019**, 1–9. [CrossRef]

226. Frioui, S.; Oumeddour, R.; Lacour, S. Highly selective extraction of metal ions from dilute solutions by hybrid electrodialysis technology. *Sep. Purif. Technol.* **2017**, *174*, 264–274. [CrossRef]

227. Cherif, A.T.; Elmidaoui, A.; Gavach, C. Separation of Ag+, Zn2+ and Cu2+ ions by electrodialysis with monovalent cation specific membrane and EDTA. *J. Membr. Sci.* **1993**, *76*, 39–49. [CrossRef]

228. Cifuentes, L.; Crisóstomo, G.; Ibáñez, J.P.; Casas, J.M.; Alvarez, F.; Cifuentes, G. On the electrodialysis of aqueous H2SO4–CuSO4 electrolytes with metallic impurities. *J. Membr. Sci.* **2002**, *207*, 1–16. [CrossRef]

229. Cifuentes, L.; García, I.; Arriagada, P.; Casas, J.M. The use of electrodialysis for metal separation and water recovery from CuSO4-H2SO4-Fe solutions. *Sep. Purif. Technol.* **2009**, *68*, 105–108. [CrossRef]

230. Liu, Y.; Zhu, H.; Zhang, M.; Chen, R.; Chen, X.; Zheng, X.; Jin, Y. Cr(VI) recovery from chromite ore processing residual using an enhanced electrokinetic process by bipolar membranes. *J. Membr. Sci.* **2018**, *566*, 190–196. [CrossRef]

231. Reig, M.; Vecino, X.; Valderrama, C.; Gibert, O.; Cortina, J.L. Application of selectrodialysis for the removal of as from metallurgical process waters: Recovery of Cu and Zn. *Sep. Purif. Technol.* **2018**, *195*, 404–412. [CrossRef]

232. Zheng, Y.; Gao, X.; Wang, X.; Li, Z.; Wang, Y.; Gao, C. Application of electrodialysis to remove copper and cyanide from simulated and real gold mine effluents. *RSC Adv.* **2015**, *5*, 19807–19817. [CrossRef]

233. Yeon, K.H.; Song, J.H.; Moon, S.H. A study on stack configuration of continuous electrodeionization for removal of heavy metal ions from the primary coolant of a nuclear power plant. *Water Res.* **2004**, *38*, 1911–1921. [CrossRef]

234. Yeon, K.H.; Seong, J.H.; Rengaraj, S.; Moon, S.H. Electrochemical characterization of ion-exchange resin beds and removal of cobalt by electrodeionization for high purity water production. *Sep. Sci. Technol.* **2003**, *38*, 443–462. [CrossRef]

235. Zhang, Y.; Wang, L.; Xuan, S.; Lin, X.; Luo, X. Variable effects on electrodeionization for removal of Cs+ ions from simulated wastewater. *Desalination* **2014**, *344*, 212–218. [CrossRef]

236. Jiang, B.; Li, F.; Zhao, X. Removal of trace Cs(I), Sr(II), and Co(II) in aqueous solutions using continuous electrodeionization (CEDI). *Desalin. Water Treat.* **2019**, *155*, 175–182. [CrossRef]

237. Zahakifar, F.; Keshtkar, A.R.; Souderjani, E.Z.; Moosavian, M.A. Use of response surface methodology for optimization of thorium(IV) removal from aqueous solutions by electrodeionization (EDI). *Prog. Nucl. Energy* **2020**, *124*, 103335. [CrossRef]

238. Regel-Rosocka, M. A review on methods of regeneration of spent pickling solutions from steel processing. *J. Hazard. Mater.* **2010**, *177*, 57–69. [CrossRef] [PubMed]

239. Agrawal, A.; Sahu, K.K. An overview of the recovery of acid from spent acidic solutions from steel and electroplating industries. *J. Hazard. Mater.* **2009**, *171*, 61–75. [CrossRef]

240. Urano, K.; Ase, T.; Naito, Y. Recovery of acid from wastewater by electrodialysis. *Desalination* **1984**, *51*, 213–226. [CrossRef]

241. Pourcelly, G.; Tugas, I.; Gavach, C. Electrotransport of sulphuric acid in special anion exchange membranes for the recovery of acids. *J. Membr. Sci.* **1994**, *97*, 99–107. [CrossRef]

242. Jia, Y.X.; Li, F.J.; Chen, X.; Wang, M. Model analysis on electrodialysis for inorganic acid recovery and its experimental validation. *Sep. Purif. Technol.* **2018**, *190*, 261–267. [CrossRef]

243. Wang, L.; Li, Z.; Xu, Z.; Zhang, F.; Efome, J.E.; Li, N. Proton blockage membrane with tertiary amine groups for concentration of sulfonic acid in electrodialysis. *J. Membr. Sci.* **2018**, *555*, 78–87. [CrossRef]

244. Guo, R.Q.; Wang, B.B.; Jia, Y.X.; Wang, M. Development of acid block anion exchange membrane by structure design and its possible application in waste acid recovery. *Sep. Purif. Technol.* **2017**, *186*, 188–196. [CrossRef]

245. Bai, T.; Wang, M.; Zhang, B.; Jia, Y.; Chen, Y. Anion-exchange membrane with ion-nanochannels to beat trade-off between membrane conductivity and acid blocking performance for waste acid reclamation. *J. Membr. Sci.* **2019**, *573*, 657–667. [CrossRef]

246. Zhang, N.; Liu, Y.; Liu, R.; She, Z.; Tan, M.; Mao, D.; Fu, R.; Zhang, Y. Polymer inclusion membrane (PIM) containing ionic liquid as a proton blocker to improve waste acid recovery efficiency in electrodialysis process. *J. Membr. Sci.* **2019**, *581*, 18–27. [CrossRef]

247. Cong, M.Y.; Jia, Y.X.; Wang, H.; Wang, M. Preparation of acid block anion exchange membrane with quaternary ammonium groups by homogeneous amination for electrodialysis-based acid enrichment. *Sep. Purif. Technol.* **2020**, *238*, 116396. [CrossRef]

248. Bai, T.T.; Cong, M.Y.; Jia, Y.X.; Ma, K.K.; Wang, M. Preparation of self-crosslinking anion exchange membrane with acid block performance from side-chain type polysulfone. *J. Membr. Sci.* **2020**, *599*, 117831. [CrossRef]

249. Paquay, E.; Clarinval, A.M.; Delvaux, A.; Degrez, M.; Hurwitz, H.D. Applications of electrodialysis for acid pickling wastewater treatment. *Chem. Eng. J.* **2000**, *79*, 197–201. [CrossRef]

250. Chapotot, A.; Lopez, V.; Lindheimer, A.; Aouad, N.; Gavach, C. Electrodialysis of acid solutions with metallic divalent salts: Cation-exchange membranes with improved permeability to protons. *Desalination* **1995**, *101*, 141–153. [CrossRef]

251. Xu, T. Electrodialysis processes with bipolar membranes (EDBM) in environmental protection—A review. *Resour. Conserv. Recycl.* **2002**, *37*, 1–22.

252. Baltazar, V.; Harris, G.B.; White, C.W. The selective recovery and concentration of sulphuric acid by electrodialysis. *Hydrometallurgy* **1992**, *30*, 463–481. [CrossRef]

253. Tran, A.T.K.; Mondal, P.; Lin, J.; Meesschaert, B.; Pinoy, L.; Van der Bruggen, B. Simultaneous regeneration of inorganic acid and base from a metal washing step wastewater by bipolar membrane electrodialysis after pretreatment by crystallization in a fluidized pellet reactor. *J. Membr. Sci.* **2015**, *473*, 118–127. [CrossRef]

254. Jia, Y.; Chen, X.; Wang, M.; Wang, B. A win-win strategy for the reclamation of waste acid and conversion of organic acid by a modified electrodialysis. *Sep. Purif. Technol.* **2016**, *171*, 11–16. [CrossRef]

255. Song, P.; Wang, M.; Zhang, B.; Jia, Y.; Chen, Y. Fabrication of proton permselective composite membrane for electrodialysis-based waste acid reclamation. *J. Membr. Sci.* **2019**, *592*, 117366. [CrossRef]

256. Boucher, M.; Turcotte, N.; Guillemette, V.; Lantagne, G.; Chapotot, A.; Pourcelly, G.; Sandeaux, R.; Gavach, C. Recovery of spent acid by electrodialysis in the zinc hydrometallurgy industry: Performance study of different cation-exchange membranes. *Hydrometallurgy* **1997**, *45*, 137–160. [CrossRef]

257. Sistat, P.; Pourcelly, G.; Gavach, C.; Turcotte, N.; Boucher, M. Electrodialysis of acid effluents containing metallic divalent salts: Recovery of acid with a cation-exchange membrane modified in situ. *J. Appl. Electrochem.* **1997**, *27*, 65–70. [CrossRef]

258. Wang, M.; Liu, X.; Jia, Y.X.; Wang, X.L. The improvement of comprehensive transport properties to heterogeneous cation exchange membrane by the covalent immobilization of polyethyleneimine. *Sep. Purif. Technol.* **2015**, *140*, 69–76. [CrossRef]

259. He, Y.; Ge, L.; Ge, Z.J.; Zhao, Z.; Sheng, F.; Liu, X.; Ge, X.; Yang, Z.; Fu, R.; Liu, Z.; et al. Monovalent cations permselective membranes with zwitterionic side chains. *J. Membr. Sci.* **2018**, *563*, 320–325. [CrossRef]

260. Sheng, F.; Afsar, N.U.; Zhu, Y.; Ge, L.; Xu, T. PVA-based mixed matrix membranes comprising ZSM-5 for cations separation. *Membranes* **2020**, *10*, 114. [CrossRef] [PubMed]

261. Ge, L.; Wu, B.; Li, Q.; Wang, Y.; Yu, D.; Wu, L.; Pan, J.; Miao, J.; Xu, T. Electrodialysis with nanofiltration membrane (EDNF) for high-efficiency cations fractionation. *J. Membr. Sci.* **2016**, *498*, 192–200. [CrossRef]

262. Liu, Y.; Ke, X.; Zhu, H.; Chen, R.; Chen, X.; Zheng, X.; Jin, Y.; Van der Bruggen, B. Treatment of raffinate generated via copper ore hydrometallurgical processing using a bipolar membrane electrodialysis system. *Chem. Eng. J.* **2020**, *382*, 122956. [CrossRef]

263. Yuzer, B.; Aydin, M.I.; Hasançebi, B.; Selcuk, H. Application of an electrodialysis process to recover nitric acid from aluminum finishing industry waste. *Desalin. Water Treat.* **2019**, *172*, 199–205. [CrossRef]

264. Zhang, X.; Li, C.; Wang, X.; Wang, Y.; Xu, T. Recovery of hydrochloric acid from simulated chemosynthesis aluminum foils wastewater: An integration of diffusion dialysis and conventional electrodialysis. *J. Membr. Sci.* **2012**, *409–410*, 257–263. [CrossRef]

265. Zhuang, J.X.; Chen, Q.; Wang, S.; Zhang, W.M.; Song, W.G.; Wan, L.J.; Ma, K.S.; Zhang, C.N. Zero discharge process for foil industry waste acid reclamation: Coupling of diffusion dialysis and electrodialysis with bipolar membranes. *J. Membr. Sci.* **2013**, *432*, 90–96. [CrossRef]

266. Aydin, M.I.; Yuzer, B.; Hasancebi, B.; Selcuk, H. Application of electrodialysis membrane process to recovery sulfuric acid and wastewater in the chalcopyrite mining industry. *Desalin. Water Treat.* **2019**, *172*, 206–211. [CrossRef]

267. Heinonen, J.; Zhao, Y.; Van der Bruggen, B. A process combination of ion exchange and electrodialysis for the recovery and purification of hydroxy acids from secondary sources. *Sep. Purif. Technol.* **2020**, *240*, 116642. [CrossRef]

268. Li, M.; Sun, M.; Liu, W.; Zhang, X.; Wu, C.; Wu, Y. Quaternized graphene oxide modified PVA-QPEI membranes with excellent selectivity for alkali recovery through electrodialysis. *Chem. Eng. Res. Des.* **2020**, *153*, 875–886. [CrossRef]

269. Davis, J.R.; Chen, Y.; Baygents, J.C.; Farrell, J. Production of Acids and Bases for Ion Exchange Regeneration from Dilute Salt Solutions Using Bipolar Membrane Electrodialysis. *ACS Sustain. Chem. Eng.* **2015**, *3*, 2337–2342. [CrossRef]

270. Graillon, S.; Persin, F.; Pourcelly, G.; Gavach, C. Development of electrodialysis with bipolar membrane for the treatment of concentrated nitrate effluents. *Desalination* **1996**, *107*, 159–169. [CrossRef]

271. Ben Ali, M.A.; Rakib, M.; Laborie, S.; Viers, P.; Durand, G. Coupling of bipolar membrane electrodialysis and ammonia stripping for direct treatment of wastewaters containing ammonium nitrate. *J. Membr. Sci.* **2004**, *244*, 89–96.

272. Cherif, A.T.; Molenat, J.; Elmidaoui, A. Nitric acid and sodium hydroxide generation by electrodialysis using bipolar membranes. *J. Appl. Electrochem.* **1997**, *27*, 1069–1074. [CrossRef]

273. Monat, L.; Chaudhury, S.; Nir, O. Enhancing the Sustainability of Phosphogypsum Recycling by Integrating Electrodialysis with Bipolar Membranes. *ACS Sustain. Chem. Eng.* **2020**, *8*, 2490–2497. [CrossRef]

274. Li, Y.; Shi, S.; Cao, H.; Wu, X.; Zhao, Z.; Wang, L. Bipolar membrane electrodialysis for generation of hydrochloric acid and ammonia from simulated ammonium chloride wastewater. *Water Res.* **2016**, *89*, 201–209. [CrossRef] [PubMed]

275. Lv, Y.; Yan, H.; Yang, B.; Wu, C.; Zhang, X.; Wang, X. Bipolar membrane electrodialysis for the recycling of ammonium chloride wastewater: Membrane selection and process optimization. *Chem. Eng. Res. Des.* **2018**, *138*, 105–115. [CrossRef]

276. Van Linden, N.; Bandinu, G.L.; Vermaas, D.A.; Spanjers, H.; van Lier, J.B. Bipolar membrane electrodialysis for energetically competitive ammonium removal and dissolved ammonia production. *J. Clean. Prod.* **2020**, *259*, 120788. [CrossRef]

277. Trivedi, G.; Shah, B.; Adhikary, S.; Rangarajan, R. Studies on bipolar membranes: Part III: Conversion of sodium phosphate to phosphoric acid and sodium hydroxide. *React. Funct. Polym.* **1999**, *39*, 91–97. [CrossRef]

278. Wei, Y.; Wang, Y.; Zhang, X.; Xu, T. Treatment of simulated brominated butyl rubber wastewater by bipolar membrane electrodialysis. *Sep. Purif. Technol.* **2011**, *80*, 196–201. [CrossRef]

279. Wei, Y.; Wang, Y.; Zhang, X.; Xu, T. Comparative study on the treatment of simulated brominated butyl rubber wastewater by using bipolar membrane electrodialysis (BMED) and conventional electrodialysis (ED). *Sep. Purif. Technol.* **2013**, *110*, 164–169. [CrossRef]

280. Wang, D.; Meng, W.; Lei, Y.; Li, C.; Cheng, J.; Qu, W.; Wang, G.; Zhang, M.; Li, S. The novel strategy for increasing the efficiency and yield of the bipolar membrane electrodialysis by the double conjugate salts stress. *Polymers* **2020**, *12*, 343. [CrossRef]

281. Ghyselbrecht, K.; Huygebaert, M.; Van der Bruggen, B.; Ballet, R.; Meesschaert, B.; Pinoy, L. Desalination of an industrial saline water with conventional and bipolar membrane electrodialysis. *Desalination* **2013**, *318*, 9–18. [CrossRef]

282. Noguchi, M.; Nakamura, Y.; Shoji, T.; Iizuka, A.; Yamasaki, A. Simultaneous removal and recovery of boron from waste water by multi-step bipolar membrane electrodialysis. *J. Water Process. Eng.* **2018**, *23*, 299–305. [CrossRef]

283. Nagasawa, H.; Iizuka, A.; Yamasaki, A.; Yanagisawa, Y. Utilization of bipolar membrane electrodialysis for the removal of boron from aqueous solution. *Ind. Eng. Chem. Res.* **2011**, *50*, 6325–6330. [CrossRef]

284. Sun, M.; Li, M.; Zhang, X.; Wu, C.; Wu, Y. Graphene oxide modified porous P84 co-polyimide membranes for boron recovery by bipolar membrane electrodialysis process. *Sep. Purif. Technol.* **2020**, *232*, 115963. [CrossRef]

285. Reig, M.; Valderrama, C.; Gibert, O.; Cortina, J.L. Selectrodialysis and bipolar membrane electrodialysis combination for industrial process brines treatment: Monovalent-divalent ions separation and acid and base production. *Desalination* **2016**, *399*, 88–95. [CrossRef]

286. Liu, K.J.; Nagasubramanian, K.; Chlanda, F.P. Membrane electrodialysis process for recovery of sulfur dioxide from power plant stack gases. *J. Membr. Sci.* **1978**, *3*, 71–83. [CrossRef]

287. Liu, K.J.; Chlanda, F.P.; Nagasubramanian, K. Application of bipolar membrane technology: A novel process for control of sulfur dioxide from flue gases. *J. Membr. Sci.* **1978**, *3*, 57–70. [CrossRef]

288. Zhang, X.; Ye, C.; Pi, K.; Huang, J.; Xia, M.; Gerson, A.R. Sustainable treatment of desulfurization wastewater by ion exchange and bipolar membrane electrodialysis hybrid technology. *Sep. Purif. Technol.* **2019**, *211*, 330–339. [CrossRef]

289. Tian, W.; Wang, X.; Fan, C.; Cui, Z. Optimal treatment of hypersaline industrial wastewater via bipolar membrane electrodialysis. *ACS Sustain. Chem. Eng.* **2019**, *7*, 12358–12368. [CrossRef]

290. Luo, Z.; Wang, D.; Zhu, D.; Xu, J.; Jiang, H.; Geng, W.; Wei, W.; Lian, Z. Separation of fluoride and chloride ions from ammonia-based flue gas desulfurization slurry using a two-stage electrodialysis. *Chem. Eng. Res. Des.* **2019**, *147*, 73–82. [CrossRef]

291. Wang, Y.; Li, W.; Yan, H.; Xu, T. Removal of heat stable salts (HSS) from spent alkanolamine wastewater using electrodialysis. *J. Ind. Eng. Chem.* **2018**, *57*, 356–362. [CrossRef]

292. Meng, H.; Zhang, S.; Li, C.; Li, L. Removal of heat stable salts from aqueous solutions of N-methyldiethanolamine using a specially designed three-compartment configuration electrodialyzer. *J. Membr. Sci.* **2008**, *322*, 436–440. [CrossRef]

293. Chen, F.; Chi, Y.; Zhang, M.; Liu, Z.; Fei, X.; Yang, K.; Fu, C. Removal of heat stable salts from N-methyldiethanolamine wastewater by anion exchange resin coupled three-compartment electrodialysis. *Sep. Purif. Technol.* **2020**, *242*, 116777. [CrossRef]

294. Bazhenov, S.; Rieder, A.; Schallert, B.; Vasilevsky, V.; Unterberger, S.; Grushevenko, E.; Volkov, V.; Volkov, A. Reclaiming of degraded MEA solutions by electrodialysis: Results of ED pilot campaign at post-combustion CO_2 capture pilot plant. *Int. J. Greenh. Gas. Control.* **2015**, *42*, 593–601. [CrossRef]

295. Grushevenko, E.; Bazhenov, S.; Vasilevsky, V.; Novitsky, E.; Shalygin, M.; Volkov, A. Effect of Carbon Dioxide Loading on Removal of Heat Stable Salts from Amine Solvent by Electrodialysis. *Membranes* **2019**, *9*, 152. [CrossRef] [PubMed]

296. Iizuka, A.; Hashimoto, K.; Nagasawa, H.; Kumagai, K.; Yanagisawa, Y.; Yamasaki, A. Carbon dioxide recovery from carbonate solutions using bipolar membrane electrodialysis. *Sep. Purif. Technol.* **2012**, *101*, 49–59. [CrossRef]

297. Jiang, C.; Li, S.; Zhang, D.; Yang, Z.; Yu, D.; Chen, X.; Wang, Y.; Xu, T. Mathematical modelling and experimental investigation of CO_2 absorber recovery using an electro-acidification method. *Chem. Eng. J.* **2019**, *360*, 654–664. [CrossRef]

298. Wang, Q.; Yang, P.; Cong, W. Cation-exchange membrane fouling and cleaning in bipolar membrane electrodialysis of industrial glutamate production wastewater. *Sep. Purif. Technol.* **2011**, *79*, 103–113. [CrossRef]

299. Shen, J.; Huang, J.; Liu, L.; Ye, W.; Lin, J.; Van der Bruggen, B. The use of BMED for glyphosate recovery from glyphosate neutralization liquor in view of zero discharge. *J. Hazard. Mater.* **2013**, *260*, 660–667. [CrossRef]

300. Ye, W.; Huang, J.; Lin, J.; Zhang, X.; Shen, J.; Luis, P.; Van Der Bruggen, B. Environmental evaluation of bipolar membrane electrodialysis for NaOH production from wastewater: Conditioning NaOH as a CO_2 absorbent. *Sep. Purif. Technol.* **2015**, *144*, 206–214. [CrossRef]

301. Wang, Q.; Jiang, C.; Wang, Y.; Yang, Z.; Xu, T. Reclamation of Aniline Wastewater and CO_2 Capture Using Bipolar Membrane Electrodialysis. *ACS Sustain. Chem. Eng.* **2016**, *4*, 5743–5751. [CrossRef]

302. Loza, N.V.; Loza, S.A.; Romanyuk, N.A.; Kononenko, N.A. Experimental and Theoretical Studies of Electrodialysis of Model Solutions Containing Aniline and Sulfuric Acid. *Russ. J. Electrochem.* **2019**, *55*, 871–877. [CrossRef]

303. Peng, Z.; Sun, Y. Leakage circuit characteristics of a bipolar membrane electrodialyzer with 5 BP-A-C units. *J. Membr. Sci.* **2020**, *597*, 117762. [CrossRef]

304. Schlichter, B.; Mavrov, V.; Erwe, T.; Chmiel, H. Regeneration of bonding agents loaded with heavy metals by electrodialysis with bipolar membranes. *J. Membr. Sci.* **2004**, *232*, 99–105. [CrossRef]

305. Wei, Y.; Li, C.; Wang, Y.; Zhang, X.; Li, Q.; Xu, T. Regenerating sodium hydroxide from the spent caustic by bipolar membrane electrodialysis (BMED). *Sep. Purif. Technol.* **2012**, *86*, 49–54. [CrossRef]

306. Rohman, F.S.; Othman, M.R.; Aziz, N. Modeling of batch electrodialysis for hydrochloric acid recovery. *Chem. Eng. J.* **2010**, *162*, 466–479. [CrossRef]

307. Merkel, A.; Ashrafi, A.M.; Ondrušek, M. The use of electrodialysis for recovery of sodium hydroxide from the high alkaline solution as a model of mercerization wastewater. *J. Water Process. Eng.* **2017**, *20*, 123–129. [CrossRef]

308. Bailly, M. Production of organic acids by bipolar electrodialysis: Realizations and perspectives. *Desalination* **2002**, *144*, 157–162. [CrossRef]

309. Liu, J.; Wu, S.; Lu, Y.; Liu, Q.; Jiao, Q.; Wang, X.; Zhang, H. An integrated electrodialysis-biocatalysis-spray-drying process for efficient recycling of keratin acid hydrolysis industrial wastewater. *Chem. Eng. J.* **2016**, *302*, 146–154. [CrossRef]

310. Vertova, A.; Aricci, G.; Rondinini, S.; Miglio, R.; Carnelli, L.; D'Olimpio, P. Electrodialytic recovery of light carboxylic acids from industrial aqueous wastes. *J. Appl. Electrochem.* **2009**, *39*, 2051–2059. [CrossRef]

311. Wang, Q.; Cheng, G.; Sun, X.; Jin, B. Recovery of lactic acid from kitchen garbage fermentation broth by four-compartment configuration electrodialyzer. *Process. Biochem.* **2006**, *41*, 152–158. [CrossRef]

312. Zhang, Y.; Liu, R.; Lang, Q.; Tan, M.; Zhang, Y. Composite anion exchange membrane made by layer-by-layer method for selective ion separation and water migration control. *Sep. Purif. Technol.* **2018**, *192*, 278–286. [CrossRef]

313. Scoma, A.; Varela-Corredor, F.; Bertin, L.; Gostoli, C.; Bandini, S. Recovery of VFAs from anaerobic digestion of dephenolized Olive Mill Wastewaters by Electrodialysis. *Sep. Purif. Technol.* **2016**, *159*, 81–91. [CrossRef]

314. Pan, X.R.; Li, W.W.; Huang, L.; Liu, H.Q.; Wang, Y.K.; Geng, Y.K.; Kwan-Sing Lam, P.; Yu, H.Q. Recovery of high-concentration volatile fatty acids from wastewater using an acidogenesis-electrodialysis integrated system. *Bioresour. Technol.* **2018**, *260*, 61–67. [CrossRef] [PubMed]

315. Dai, K.; Wen, J.-L.; Wang, Y.-L.; Wu, Z.-G.; Zhao, P.-J.; Zhang, H.-H.; Wang, J.-J.; Zeng, R.J.; Zhang, F. Impacts of medium composition and applied current on recovery of volatile fatty acids during coupling of electrodialysis with an anaerobic digester. *J. Clean. Prod.* **2019**, *207*, 483–489. [CrossRef]

316. Yu, L.; Guo, Q.; Hao, J.; Jiang, W. Recovery of acetic acid from dilute wastewater by means of bipolar membrane electrodialysis. *Desalination* **2000**, *129*, 283–288. [CrossRef]

317. Yu, L.; Lin, T.; Guo, Q.; Hao, J. Relation between mass transfer and operation parameters in the electrodialysis recovery of acetic acid. *Desalination* **2003**, *154*, 147–152. [CrossRef]

318. Zhang, X.; Li, C.; Wang, Y.; Luo, J.; Xu, T. Recovery of acetic acid from simulated acetaldehyde wastewaters: Bipolar membrane electrodialysis processes and membrane selection. *J. Membr. Sci.* **2011**, *379*, 184–190. [CrossRef]

319. Ferrer, J.S.J.; Laborie, S.; Durand, G.; Rakib, M. Formic acid regeneration by electromembrane processes. *J. Membr. Sci.* **2006**, *280*, 509–516. [CrossRef]

320. Lameloise, M.L.; Lewandowski, R. Recovering l-malic acid from a beverage industry waste water: Experimental study of the conversion stage using bipolar membrane electrodialysis. *J. Membr. Sci.* **2012**, *403–404*, 196–202. [CrossRef]

321. Achoh, A.; Zabolotsky, V.; Melnikov, S. Conversion of water-organic solution of sodium naphtenates into naphtenic acids and alkali by electrodialysis with bipolar membranes. *Sep. Purif. Technol.* **2019**, *212*, 929–940. [CrossRef]

322. Fakhru'l-Razi, A.; Pendashteh, A.; Abdullah, L.C.; Biak, D.R.A.; Madaeni, S.S.; Abidin, Z.Z. Review of technologies for oil and gas produced water treatment. *J. Hazard. Mater.* **2009**, *170*, 530–551. [CrossRef]

323. Millar, G.J.; Couperthwaite, S.J.; Moodliar, C.D. Strategies for the management and treatment of coal seam gas associated water. *Renew. Sustain. Energy Rev.* **2016**, *57*, 669–691. [CrossRef]

324. Onishi, V.C.; Reyes-Labarta, J.A.; Caballero, J.A. Membrane Desalination in Shale Gas Industry: Applications and Perspectives. In *Current Trends and Future Developments on (Bio-) Membranes*; Basile, A., Curcio, E., Inamuddin, I., Eds.; Elsevier: Amsterdam, The Netherlands, 2019; pp. 243–267. ISBN 9780128135518.

325. Chang, H.; Li, T.; Liu, B.; Vidic, R.D.; Elimelech, M.; Crittenden, J.C. Potential and implemented membrane-based technologies for the treatment and reuse of flowback and produced water from shale gas and oil plays: A review. *Desalination* **2019**, *455*, 34–57. [CrossRef]

326. Hamawand, I.; Yusaf, T.; Hamawand, S.G. Coal seam gas and associated water: A review paper. *Renew. Sustain. Energy Rev.* **2013**, *22*, 550–560. [CrossRef]

327. Rezakazemi, M.; Khajeh, A.; Mesbah, M. Membrane filtration of wastewater from gas and oil production. *Environ. Chem. Lett.* **2018**, *16*, 367–388. [CrossRef]

328. Arthur, J.D.; Langhus, B.G.; Patel, C. *Technical Summary of Oil & Gas: Produced Water Treatment Technologies*; Tulsa World: Tulsa, OK, USA, 2005.

329. Sirivedhin, T.; McCue, J.; Dallbauman, L. Reclaiming produced water for beneficial use: Salt removal by electrodialysis. *J. Membr. Sci.* **2004**, *243*, 335–343. [CrossRef]

330. Hao, H.; Huang, X.; Gao, C.; Gao, X. Application of an integrated system of coagulation and electrodialysis for treatment of wastewater produced by fracturing. *Desalin. Water Treat.* **2014**, *55*, 2034–2043. [CrossRef]

331. McGovern, R.K.; Weiner, A.M.; Sun, L.; Chambers, C.G.; Zubair, S.M.; Lienhard, V.J.H. On the cost of electrodialysis for the desalination of high salinity feeds. *Appl. Energy* **2014**, *136*, 649–661. [CrossRef]

332. Peraki, M.; Ghazanfari, E.; Pinder, G.F.; Harrington, T.L. Electrodialysis: An application for the environmental protection in shale-gas extraction. *Sep. Purif. Technol.* **2016**, *161*, 96–103. [CrossRef]

333. McGovern, R.K.; Zubair, S.M.; Lienhard, V.J.H. The cost effectiveness of electrodialysis for diverse salinity applications. *Desalination* **2014**, *348*, 57–65. [CrossRef]

334. Hayes, T.D.; Severin, B.F. Electrodialysis of highly concentrated brines: Effects of calcium. *Sep. Purif. Technol.* **2017**, *175*, 443–453. [CrossRef]

335. Severin, B.F.; Hayes, T.D. Electrodialysis of concentrated brines: Effects of multivalent cations. *Sep. Purif. Technol.* **2019**, *218*, 227–241. [CrossRef]

336. Jing, G.L.; Xing, L.J.; Liu, Y; Du, W.T.; Han, C.J. Development of a four-grade and four-segment electrodialysis setup for desalination of polymer-flooding produced water. *Desalination* **2010**, *264*, 214–219. [CrossRef]

337. Guo, H.; You, F.; Yu, S.; Li, L.; Zhao, D. Mechanisms of chemical cleaning of ion exchange membranes: A case study of plant-scale electrodialysis for oily wastewater treatment. *J. Membr. Sci.* **2015**, *496*, 310–317. [CrossRef]

338. Jing, G.L.; Wang, X.Y.; Han, C.J. The effect of oilfield polymer-flooding wastewater on anion-exchange membrane performance. *Desalination* **2008**, *220*, 386–393. [CrossRef]

339. Guo, H.; Xiao, L.; Yu, S.; Yang, H.; Hu, J.; Liu, G.; Tang, Y. Analysis of anion exchange membrane fouling mechanism caused by anion polyacrylamide in electrodialysis. *Desalination* **2014**, *346*, 46–53. [CrossRef]

340. Zuo, X.; Wang, L.; He, J.; Li, Z.; Yu, S. SEM-EDX studies of SiO2/PVDF membranes fouling in electrodialysis of polymer-flooding produced wastewater: Diatomite, APAM and crude oil. *Desalination* **2014**, *347*, 43–51. [CrossRef]

341. Wang, T.; Yu, S.; Hou, L. Impacts of HPAM molecular weights on desalination performance of ion exchange membranes and fouling mechanism. *Desalination* **2017**, *404*, 50–58. [CrossRef]

342. Xia, Q.; Guo, H.; Ye, Y.; Yu, S.; Li, L.; Li, Q.; Zhang, R. Study on the fouling mechanism and cleaning method in the treatment of polymer flooding produced water with ion exchange membranes. *RSC Adv.* **2018**, *8*, 29947–29957. [CrossRef]

343. Sosa-Fernandez, P.A.; Post, J.W.; Bruning, H.; Leermakers, F.A.M.; Rijnaarts, H.H.M. Electrodialysis-based desalination and reuse of sea and brackish polymer-flooding produced water. *Desalination* **2018**, *447*, 120–132. [CrossRef]

344. Sosa-Fernandez, P.A.; Post, J.W.; Leermakers, F.A.M.; Rijnaarts, H.H.M.; Bruning, H. Removal of divalent ions from viscous polymer-flooding produced water and seawater via electrodialysis. *J. Membr. Sci.* **2019**, *589*, 117251. [CrossRef]

345. Sosa-Fernandez, P.A.; Miedema, S.J.; Bruning, H.; Leermakers, F.A.M.; Rijnaarts, H.H.M.; Post, J.W. Influence of solution composition on fouling of anion exchange membranes desalinating polymer-flooding produced water. *J. Colloid Interface Sci.* **2019**, *557*, 381–394. [CrossRef]

346. Sosa-Fernandez, P.A.; Post, J.W.; Ramdlan, M.S.; Leermakers, F.A.M.; Bruning, H.; Rijnaarts, H.H.M. Improving the performance of polymer-flooding produced water electrodialysis through the application of pulsed electric field. *Desalination* **2020**, *484*, 114424. [CrossRef]

347. Lopez, A.M.; Dunsworth, H.; Hestekin, J.A. Reduction of the shadow spacer effect using reverse electrodeionization and its applications in water recycling for hydraulic fracturing operations. *Sep. Purif. Technol.* **2016**, *162*, 84–90. [CrossRef]

348. AECOM Inc. *Petroleum Refining Water/Wastewater Use and Management*; IPIECA: London, UK, 2010.

349. Parkash, S. Refinery Water Systems. *Refin. Process. Handb.* **2007**, 242–269.

350. Mikhak, Y.; Torabi, M.M.A.; Fouladitajar, A. Refinery and petrochemical wastewater treatment. In *Sustainable Water and Wastewater Processing*; Elsevier: Amsterdam, The Netherlands, 2019; pp. 55–91. ISBN 9780128161708.

351. Gioli, P.; Silingardi, G.E.; Ghiglio, G. High quality water from refinery waste. *Desalination* **1987**, *67*, 271–282. [CrossRef]

352. Venzke, C.D.; Giacobbo, A.; Klauck, C.R.; Viegas, C.; Hansen, E.; Monteiro De Aquim, P.; Antônio, M.; Rodrigues, S.; Moura Bernardes, A. Integrated Membrane Processes (EDR-RO) for Water Reuse in the Petrochemical Industry. *J. Membr. Sci. Res.* **2018**, *4*, 218–226.

353. Hughes, M.; Raubenheimer, A.E.; Viljoen, A.J. Electrodialysis reversal at Tutuka power station, RSA—Seven years' design and operating experience. *Water Sci. Technol.* **1992**, *25*, 277–289. [CrossRef]

354. Turek, M.; Dydo, P. Electrodialysis reversal of calcium sulphate and calcium carbonate supersaturated solution. *Desalination* **2003**, *158*, 91–94. [CrossRef]

355. Turek, M. Electrodialytic desalination and concentration of coal-mine brine. *Desalination* **2004**, *162*, 355–359. [CrossRef]

356. Turek, M.; Laskowska, E.; Mitko, K.; Chorążewska, M.; Dydo, P.; Piotrowski, K.; Jakóbik-Kolon, A. Application of nanofiltration and electrodialysis for improved performance of a salt production plant. *Desalin. Water Treat.* **2017**, *64*, 244–250. [CrossRef]

357. Mitko, K.; Podleśny, B.; Jakóbik-Kolon, A.; Turek, M. Electrodialytic utilization of coal mine brines. *Desalin. Water Treat.* **2017**, *75*, 363–367. [CrossRef]

358. Turek, M.; Bandura, B.; Dydo, P. Power production from coal-mine brine utilizing reversed electrodialysis. *Desalination* **2008**, *221*, 462–466. [CrossRef]

359. Buzzi, D.C.; Viegas, L.S.; Rodrigues, M.A.S.; Bernardes, A.M.; Tenório, J.A.S. Water recovery from acid mine drainage by electrodialysis. *Min. Eng.* **2013**, *40*, 82–89. [CrossRef]

360. Martí-Calatayud, M.C.; Buzzi, D.C.; García-Gabaldón, M.; Bernardes, A.M.; Tenório, J.A.S.; Pérez-Herranz, V. Ion transport through homogeneous and heterogeneous ion-exchange membranes in single salt and multicomponent electrolyte solutions. *J. Membr. Sci.* **2014**, *466*, 45–57. [CrossRef]

361. Bisselink, R.; de Schepper, W.; Trampé, J.; van den Broek, W.; Pinel, I.; Krutko, A.; Groot, N. Mild desalination demo pilot: New normalization approach to effectively evaluate electrodialysis reversal technology. *Water Resour. Ind.* **2015**, *14*, 18–25. [CrossRef]

362. Irfan, M.; Ge, L.; Wang, Y.; Yang, Z.; Xu, T. Hydrophobic Side Chains Impart Anion Exchange Membranes with High Monovalent-Divalent Anion Selectivity in Electrodialysis. *ACS Sustain. Chem. Eng.* **2019**, *7*, 4429–4442. [CrossRef]

363. Irfan, M.; Xu, T.; Ge, L.; Wang, Y.; Xu, T. Zwitterion structure membrane provides high monovalent/divalent cation electrodialysis selectivity: Investigating the effect of functional groups and operating parameters. *J. Membr. Sci.* **2019**, *588*, 117211. [CrossRef]

364. Irfan, M.; Wang, Y.; Xu, T. Novel electrodialysis membranes with hydrophobic alkyl spacers and zwitterion structure enable high monovalent/divalent cation selectivity. *Chem. Eng. J.* **2020**, *383*, 123171. [CrossRef]

365. Van Der Bruggen, B.; Koninckx, A.; Vandecasteele, C. Separation of monovalent and divalent ions from aqueous solution by electrodialysis and nanofiltration. *Water Res.* **2004**, *38*, 1347–1353. [CrossRef]

366. Tufa, R.A.; Hnát, J.; Němeček, M.; Kodým, R.; Curcio, E.; Bouzek, K. Hydrogen production from industrial wastewaters: An integrated reverse electrodialysis—Water electrolysis energy system. *J. Clean. Prod.* **2018**, *203*, 418–426. [CrossRef]

367. Luo, K.; Jia, Y.X.; Liu, X.X.; Wang, M. On-line construction of electrodialyzer with mono-cation permselectivity and its application in reclamation of high salinity wastewater. *Desalination* **2019**, *454*, 38–47. [CrossRef]

368. Djouadi Belkada, F.; Kitous, O.; Drouiche, N.; Aoudj, S.; Bouchelaghem, O.; Abdi, N.; Grib, H.; Mameri, N. Electrodialysis for fluoride and nitrate removal from synthesized photovoltaic industry wastewater. *Sep. Purif. Technol.* **2018**, *204*, 108–115. [CrossRef]

369. Kabay, N.; Arar, Ö.; Samatya, S.; Yüksel, Ü.; Yüksel, M. Separation of fluoride from aqueous solution by electrodialysis: Effect of process parameters and other ionic species. *J. Hazard. Mater.* **2008**, *153*, 107–113. [CrossRef]

370. Wang, X.; Li, N.; Li, J.; Feng, J.; Ma, Z.; Xu, Y.; Sun, Y.; Xu, D.; Wang, J.; Gao, X.; et al. Fluoride removal from secondary effluent of the graphite industry using electrodialysis: Optimization with response surface methodology. *Front. Env. Sci. Eng.* **2019**, *13*. [CrossRef]

371. Kabay, N.; Arar, O.; Acar, F.; Ghazal, A.; Yuksel, U.; Yuksel, M. Removal of boron from water by electrodialysis: Effect of feed characteristics and interfering ions. *Desalination* **2008**, *223*, 63–72. [CrossRef]

372. Melnyk, L.; Goncharuk, V.; Butnyk, I.; Tsapiuk, E. Boron removal from natural and wastewaters using combined sorption/ membrane process. *Desalination* **2005**, *185*, 147–157. [CrossRef]

373. Turek, M.; Dydo, P.; Ciba, J.; Trojanowska, J.; Kluczka, J.; Palka-Kupczak, B. Electrodialytic treatment of boron-containing wastewater with univalent permselective membranes. *Desalination* **2005**, *185*, 139–145. [CrossRef]

374. Turek, M.; Dydo, P.; Trojanowska, J.; Bandura, B. Electrodialytic treatment of boron-containing wastewater. *Desalination* **2007**, *205*, 185–191. [CrossRef]

375. Jiang, B.; Zhang, X.; Zhao, X.; Li, F. Removal of high level boron in aqueous solutions using continuous electrodeionization (CEDI). *Sep. Purif. Technol.* **2018**, *192*, 297–301. [CrossRef]

376. Kozaderova, O.A.; Kim, K.B.; Gadzhiyeva, C.S.; Niftaliev, S.I. Electrochemical characteristics of thin heterogeneous ion exchange membranes. *J. Membr. Sci.* **2020**, *604*, 118081. [CrossRef]

377. Melnikov, S.S.; Mugtamov, O.A.; Zabolotsky, V.I. Study of electrodialysis concentration process of inorganic acids and salts for the two-stage conversion of salts into acids utilizing bipolar electrodialysis. *Sep. Purif. Technol.* **2020**, *235*, 116198. [CrossRef]

378. Zhang, Y.; Shi, Q.; Luo, M.; Wang, H.; Qi, X.; Hou, C.H.; Li, F.; Ai, Z.; Junior, J.T.A. Improved bauxite residue dealkalization by combination of aerated washing and electrodialysis. *J. Hazard. Mater.* **2019**, *364*, 682–690. [CrossRef] [PubMed]

379. Koivisto, E.; Zevenhoven, R. Energy use of flux salt recovery using bipolar membrane electrodialysis for a CO_2 mineralisation process. *Entropy* **2019**, *21*, 395. [CrossRef]

380. Iizuka, A.; Yamashita, Y.; Nagasawa, H.; Yamasaki, A.; Yanagisawa, Y. Separation of lithium and cobalt from waste lithium-ion batteries via bipolar membrane electrodialysis coupled with chelation. *Sep. Purif. Technol.* **2013**, *113*, 33–41. [CrossRef]

381. Lindstrand, V.; Jönsson, A.S.; Sundström, G. Organic fouling of electrodialysis membranes with and without applied voltage. *Desalination* **2000**, *130*, 73–84. [CrossRef]

382. Vanoppen, M.; Bakelants, A.F.A.M.; Gaublomme, D.; Schoutteten, K.V.K.M.; Van Den Bussche, J.; Vanhaecke, L.; Verliefde, A.R.D. Properties governing the transport of trace organic contaminants through ion-exchange membranes. *Environ. Sci. Technol.* **2015**, *49*, 489–497. [CrossRef]

383. Han, L.; Galier, S.; Roux-de Balmann, H. Transfer of neutral organic solutes during desalination by electrodialysis: Influence of the salt composition. *J. Membr. Sci.* **2016**, *511*, 207–218. [CrossRef]

384. Han, L.; Galier, S.; Roux-de Balmann, H. A phenomenological model to evaluate the performances of electrodialysis for the desalination of saline water containing organic solutes. *Desalination* **2017**, *422*, 17–24. [CrossRef]

385. Wang, Q.; Gao, X.; Zhang, Y.; He, Z.; Ji, Z.; Wang, X.; Gao, C. Hybrid RED/ED system: Simultaneous osmotic energy recovery and desalination of high-salinity wastewater. *Desalination* **2017**, *405*, 59–67. [CrossRef]

386. De Schepper, W.; Moraru, M.D.; Jacobs, B.; Oudshoorn, M.; Helsen, J. Electrodialysis of aqueous NaCl-glycerol solutions: A phenomenological comparison of various ion exchange membranes. *Sep. Purif. Technol.* **2019**, *217*, 274–283. [CrossRef]

387. Paltrinieri, L.; Huerta, E.; Puts, T.; Van Baak, W.; Verver, A.B.; Sudhölter, E.J.R.; De Smet, L.C.P.M. Functionalized Anion-Exchange Membranes Facilitate Electrodialysis of Citrate and Phosphate from Model Dairy Wastewater. *Environ. Sci. Technol.* **2019**, *53*, 2396–2404. [CrossRef] [PubMed]

388. Selvaraj, H.; Aravind, P.; Sundaram, M. Four compartment mono selective electrodialysis for separation of sodium formate from industry wastewater. *Chem. Eng. J.* **2018**, *333*, 162–169. [CrossRef]

389. Chao, Y.M.; Liang, T.M. A feasibility study of industrial wastewater recovery using electrodialysis reversal. *Desalination* **2008**, *221*, 433–439. [CrossRef]

390. Yen, F.C.; You, S.J.; Chang, T.C. Performance of electrodialysis reversal and reverse osmosis for reclaiming wastewater from high-tech industrial parks in Taiwan: A pilot-scale study. *J. Environ. Manag.* **2017**, *187*, 393–400. [CrossRef] [PubMed]

391. Rapp, H.J.; Pfromm, P.H. Electrodialysis for chloride removal from the chemical recovery cycle of a Kraft pulp mill. *J. Membr. Sci.* **1998**, *146*, 249–261. [CrossRef]

392. Nataraj, S.K.; Sridhar, S.; Shaikha, I.N.; Reddy, D.S.; Aminabhavi, T.M. Membrane-based microfiltration/electrodialysis hybrid process for the treatment of paper industry wastewater. *Sep. Purif. Technol.* **2007**, *57*, 185–192. [CrossRef]

393. Lafi, R.; Gzara, L.; Lajimi, R.H.; Hafiane, A. Treatment of textile wastewater by a hybrid ultrafiltration/electrodialysis process. *Chem. Eng. Process. Process Intensif.* **2018**, *132*, 105–113. [CrossRef]

394. Lin, J.; Lin, F.; Chen, X.; Ye, W.; Li, X.; Zeng, H.; Van Der Bruggen, B. Sustainable Management of Textile Wastewater: A Hybrid Tight Ultrafiltration/Bipolar-Membrane Electrodialysis Process for Resource Recovery and Zero Liquid Discharge. *Ind. Eng. Chem. Res.* **2019**, *58*, 11003–11012. [CrossRef]

395. Berkessa, Y.W.; Lang, Q.; Yan, B.; Kuang, S.; Mao, D.; Shu, L.; Zhang, Y. Anion exchange membrane organic fouling and mitigation in salt valorization process from high salinity textile wastewater by bipolar membrane electrodialysis. *Desalination* **2019**, *465*, 94–103. [CrossRef]

396. Tamersit, S.; Bouhidel, K.E.; Zidani, Z. Investigation of electrodialysis anti-fouling configuration for desalting and treating tannery unhairing wastewater: Feasibility of by-products recovery and water recycling. *J. Environ. Manag.* **2018**, *207*, 334–340. [CrossRef]

397. Valero, D.; García-García, V.; Expósito, E.; Aldaz, A.; Montiel, V. Application of electrodialysis for the treatment of almond industry wastewater. *J. Membr. Sci.* **2015**, *476*, 580–589. [CrossRef]

398. Kingsbury, R.S.; Liu, F.; Zhu, S.; Boggs, C.; Armstrong, M.D.; Call, D.F.; Coronell, O. Impact of natural organic matter and inorganic solutes on energy recovery from five real salinity gradients using reverse electrodialysis. *J. Membr. Sci.* **2017**, *541*, 621–632. [CrossRef]

399. Lee, H.J.; Oh, S.J.; Moon, S.H. Recovery of ammonium sulfate from fermentation waste by electrodialysis. *Water Res.* **2003**, *37*, 1091–1099. [CrossRef]

400. Luiz, A.; McClure, D.D.; Lim, K.; Leslie, G.; Coster, H.G.L.; Barton, G.W.; Kavanagh, J.M. Potential upgrading of bio-refinery streams by electrodialysis. *Desalination* **2017**, *415*, 20–28. [CrossRef]

401. Luiz, A.; Mcclure, D.D.; Lim, K.; Coster, H.G.L.; Barton, G.W.; Kavanagh, J.M. Towards a model for the electrodialysis of bio-refinery streams. *J. Membr. Sci.* **2019**, *573*, 320–332. [CrossRef]

402. Barros, L.B.M.; Andrade, L.H.; Drewes, J.E.; Amaral, M.C.S. Investigation of electrodialysis configurations for vinasse desalting and potassium recovery. *Sep. Purif. Technol.* **2019**, *229*, 115797. [CrossRef]

403. Milewski, A.; Czechowicz, D.; Jakóbik-Kolon, A.; Dydo, P. Preparation of Glycidol via Dehydrohalogenation of 3-Chloro-1,2-popanediol Using Bipolar Membrane Electrodialysis. *ACS Sustain. Chem. Eng.* **2019**, *7*, 18640–18646. [CrossRef]

404. Lu, H.; Zou, W.; Chai, P.; Wang, J.; Bazinet, L. Feasibility of antibiotic and sulfate ions separation from wastewater using electrodialysis with ultrafiltration membrane. *J. Clean. Prod.* **2016**, *112*, 3097–3105. [CrossRef]

405. Silva, V.; Poiesz, E.; Van Der Heijden, P. Industrial wastewater desalination using electrodialysis: Evaluation and plant design. *J. Appl. Electrochem.* **2013**, *43*, 1057–1067. [CrossRef]

406. Wang, X.; Wang, Y.; Zhang, X.; Feng, H.; Li, C.; Xu, T. Phosphate recovery from excess sludge by conventional electrodialysis (CED) and electrodialysis with bipolar membranes (EDBM). *Ind. Eng. Chem. Res.* **2013**, *52*, 15896–15904. [CrossRef]

407. Zhang, Y.; Desmidt, E.; Van Looveren, A.; Pinoy, L.; Meesschaert, B.; Van Der Bruggen, B. Phosphate separation and recovery from wastewater by novel electrodialysis. *Environ. Sci. Technol.* **2013**, *47*, 5888–5895. [CrossRef]

408. Abou-Shady, A. Recycling of polluted wastewater for agriculture purpose using electrodialysis: Perspective for large scale application. *Chem. Eng. J.* **2017**, *323*, 1–18. [CrossRef]

409. Besseris, G.J. Concurrent multiresponse multifactorial screening of an electrodialysis process of polluted wastewater using robust non-linear Taguchi profiling. *Chemom. Intell. Lab. Syst.* **2020**, *200*, 103997. [CrossRef]

410. Allison, R.P. Electrodialysis reversal in water reuse applications. *Desalination* **1995**, *103*, 11–18. [CrossRef]

411. Del Pino, M.P.; Durham, B. Wastewater reuse through dual-membrane processes: Opportunities for sustainable water resources. *Desalination* **1999**, *124*, 271–277. [CrossRef]

412. Gotor, A.G.; Pérez Baez, S.O.; Espinoza, C.A.; Bachir, S.I. Membrane processes for the recovery and reuse of wastewater in agriculture. *Desalination* **2001**, *137*, 187–192. [CrossRef]

413. Kim, J.O.; Jung, J.T.; Chung, J. Treatment performance of metal membrane microfiltration and electrodialysis integrated system for wastewater reclamation. *Desalination* **2007**, *202*, 343–350. [CrossRef]

414. Goodman, N.B.; Taylor, R.J.; Xie, Z.; Gozukara, Y.; Clements, A. A feasibility study of municipal wastewater desalination using electrodialysis reversal to provide recycled water for horticultural irrigation. *Desalination* **2013**, *317*, 77–83. [CrossRef]

415. Llanos, J.; Cotillas, S.; Cañizares, P.; Rodrigo, M.A. Novel electrodialysis-electrochlorination integrated process for the reclamation of treated wastewaters. *Sep. Purif. Technol.* **2014**, *132*, 362–369. [CrossRef]

416. Snyder, S.A.; Redding, A.M.; Yoon, Y.; Adham, S.; Wert, E.C.; Cannon, F.S.; Oppenheimer, J.; DeCarolis, J. Role of membranes and activated carbon in the removal of endocrine disruptors and pharmaceuticals. *Desalination* **2006**, *202*, 156–181. [CrossRef]

417. Guedes-Alonso, R.; Montesdeoca-Esponda, S.; Pacheco-Juárez, J.; Sosa-Ferrera, Z.; Santana-Rodríguez, J.J. A survey of the presence of pharmaceutical residues in wastewaters. Evaluation of their removal using conventional and natural treatment procedures. *Molecules* **2020**, *25*, 1639. [CrossRef]

418. Gally, C.R.; Benvenuti, T.; Da Trindade, C.D.M.; Rodrigues, M.A.S.; Zoppas-Ferreira, J.; Pérez-Herranz, V.; Bernardes, A.M. Electrodialysis for the tertiary treatment of municipal wastewater: Efficiency of ion removal and ageing of ion exchange membranes. *J. Environ. Chem. Eng.* **2018**, *6*, 5855–5869. [CrossRef]

419. Albornoz, L.L.; Marder, L.; Benvenuti, T.; Bernardes, A.M. Electrodialysis applied to the treatment of an university sewage for water recovery. *J. Environ. Chem. Eng.* **2019**, *7*, 102982. [CrossRef]

420. Zhang, Y.; Pinoy, L.; Meesschaert, B.; Van Der Bruggen, B. A natural driven membrane process for brackish and wastewater treatment: Photovoltaic powered ED and FO hybrid system. *Environ. Sci. Technol.* **2013**, *47*, 10548–10555. [CrossRef] [PubMed]

421. Lu, Y.; He, Z. Mitigation of Salinity Buildup and Recovery of Wasted Salts in a Hybrid Osmotic Membrane Bioreactor-Electrodialysis System. *Environ. Sci. Technol.* **2015**, *49*, 10529–10535. [CrossRef] [PubMed]

422. Zou, S.; He, Z. Electrodialysis recovery of reverse-fluxed fertilizer draw solute during forward osmosis water treatment. *Chem. Eng. J.* **2017**, *330*, 550–558. [CrossRef]

423. Mei, Y.; Tang, C.Y. Co-locating reverse electrodialysis with reverse osmosis desalination: Synergies and implications. *J. Membr. Sci.* **2017**, *539*, 305–312. [CrossRef]

424. Li, W.; Krantz, W.B.; Cornelissen, E.R.; Post, J.W.; Verliefde, A.R.D.; Tang, C.Y. A novel hybrid process of reverse electrodialysis and reverse osmosis for low energy seawater desalination and brine management. *Appl. Energy* **2013**, *104*, 592–602. [CrossRef]

425. Vanoppen, M.; Criel, E.; Walpot, G.; Vermaas, D.A.; Verliefde, A. Assisted reverse electrodialysis—principles, mechanisms, and potential. *npj Clean Water* **2018**, *1*, 9. [CrossRef]

426. La Cerva, M.; Gurreri, L.; Cipollina, A.; Tamburini, A.; Ciofalo, M.; Micale, G. Modelling and cost analysis of hybrid systems for seawater desalination: Electromembrane pre-treatments for Reverse Osmosis. *Desalination* **2019**, *467*, 175–195. [CrossRef]

427. Roman, M.; Van Dijk, L.H.; Gutierrez, L.; Vanoppen, M.; Post, J.W.; Wols, B.A.; Cornelissen, E.R.; Verliefde, A.R.D. Key physicochemical characteristics governing organic micropollutant adsorption and transport in ion-exchange membranes during reverse electrodialysis. *Desalination* **2019**, *468*, 114084. [CrossRef]

428. Nam, J.Y.; Hwang, K.S.; Kim, H.C.; Jeong, H.; Kim, H.; Jwa, E.; Yang, S.C.; Choi, J.; Kim, C.S.; Han, J.H.; et al. Assessing the behavior of the feed-water constituents of a pilot-scale 1000-cell-pair reverse electrodialysis with seawater and municipal wastewater effluent. *Water Res.* **2019**, *148*, 261–271. [CrossRef] [PubMed]

429. Higa, M.; Watanabe, T.; Yasukawa, M.; Endo, N.; Kakihana, Y.; Futamura, H.; Inoue, K.; Miyake, H.; Usui, J.; Hayashi, A.; et al. Sustainable hydrogen production from seawater and sewage treated water using reverse electrodialysis technology. *Water Pract. Technol.* **2019**, *14*, 645–651. [CrossRef]

430. Vanoppen, M.; van Vooren, T.; Gutierrez, L.; Roman, M.; Croué, L.-P.; Verbeken, K.; Philips, J.; Verliefde, A.R.D. Secondary treated domestic wastewater in reverse electrodialysis: What is the best pre-treatment? *Sep. Purif. Technol.* **2019**, *218*, 25–42. [CrossRef]

431. Mehdizadeh, S.; Yasukawa, M.; Suzuki, T.; Higa, M. Reverse electrodialysis for power generation using seawater/municipal wastewater: Effect of coagulation pretreatment. *Desalination* **2020**, *481*, 114356. [CrossRef]

432. Gómez-Coma, L.; Ortiz-Martínez, V.M.; Fallanza, M.; Ortiz, A.; Ibañez, R.; Ortiz, I. Blue energy for sustainable water reclamation in WWTPs. *J. Water Process. Eng.* **2020**, *33*, 101020. [CrossRef]

433. Luque Di Salvo, J.; Cosenza, A.; Tamburini, A.; Micale, G.; Cipollina, A. Long-run operation of a reverse electrodialysis system fed with wastewaters. *J. Environ. Manag.* **2018**, *217*, 871–887. [CrossRef] [PubMed]

434. Mercer, E.; Davey, C.J.; Azzini, D.; Eusebi, A.L.; Tierney, R.; Williams, L.; Jiang, Y.; Parker, A.; Tyrrel, S.; Cartmell, E.; et al. Hybrid membrane distillation reverse electrodialysis configuration for water and energy recovery from human urine: An opportunity for off-grid decentralised sanitation. *J. Membr. Sci.* **2019**, *584*, 343–352. [CrossRef] [PubMed]

435. Desmidt, E.; Ghyselbrecht, K.; Zhang, Y.; Pinoy, L.; Van der Bruggen, B.; Verstraete, W.; Rabaey, K.; Meesschaert, B. Global Phosphorus Scarcity and Full-Scale P-Recovery Techniques: A Review. *Crit. Rev. Environ. Sci. Technol.* **2015**, *45*, 336–384. [CrossRef]

436. Tran, A.T.K.; Zhang, Y.; De Corte, D.; Hannes, J.-B.; Ye, W.; Mondal, P.; Jullok, N.; Meesschaert, B.; Pinoy, L.; Van der Bruggen, B. P-recovery as calcium phosphate from wastewater using an integrated selectrodialysis/crystallization process. *J. Clean. Prod.* **2014**, *77*, 140–151. [CrossRef]

437. Tran, A.T.K.; Zhang, Y.; Lin, J.; Mondal, P.; Ye, W.; Meesschaert, B.; Pinoy, L.; Van Der Bruggen, B. Phosphate pre-concentration from municipal wastewater by selectrodialysis: Effect of competing components. *Sep. Purif. Technol.* **2015**, *141*, 38–47. [CrossRef]

438. Liu, R.; Wang, Y.; Wu, G.; Luo, J.; Wang, S. Development of a selective electrodialysis for nutrient recovery and desalination during secondary effluent treatment. *Chem. Eng. J.* **2017**, *322*, 224–233. [CrossRef]

439. Nur, H.M.; Yüzer, B.; Aydin, M.İ.; Aydin, S.; Öngen, A.; Selçuk, H. Desalination and fate of nutrient transport in domestic wastewater using electrodialysis membrane process. *Desalin. Water Treat.* **2019**, *172*, 323–329. [CrossRef]

440. Rotta, E.H.; Bitencourt, C.S.; Marder, L.; Bernardes, A.M. Phosphorus recovery from low phosphate-containing solution by electrodialysis. *J. Membr. Sci.* **2019**, *573*, 293–300. [CrossRef]

441. Rybalkina, O.; Tsygurina, K.; Melnikova, E.; Mareev, S.; Moroz, I.; Nikonenko, V.; Pismenskaya, N. Partial Fluxes of Phosphoric Acid Anions through Anion-Exchange Membranes in the Course of NaH_2PO_4 Solution Electrodialysis. *Int. J. Mol. Sci.* **2019**, *20*, 3593. [CrossRef] [PubMed]

442. Pismenskaya, N.; Sarapulova, V.; Nevakshenova, E.; Kononenko, N.; Fomenko, M.; Nikonenko, V. Concentration dependencies of diffusion permeability of anion-exchange membranes in sodium hydrogen carbonate, monosodium phosphate, and potassium hydrogen tartrate solutions. *Membranes* **2019**, *9*, 170. [CrossRef] [PubMed]

443. Cai, Y.; Han, Z.; Lin, X.; Duan, Y.; Du, J.; Ye, Z.; Zhu, J. Study on removal of phosphorus as struvite from synthetic wastewater using a pilot-scale electrodialysis system with magnesium anode. *Sci. Total Environ.* **2020**, *726*, 138221. [CrossRef]

444. Vineyard, D.; Hicks, A.; Karthikeyan, K.G.; Barak, P. Economic analysis of electrodialysis, denitrification, and anammox for nitrogen removal in municipal wastewater treatment. *J. Clean. Prod.* **2020**, *262*, 121145. [CrossRef]

445. Wang, X.; Zhang, X.; Wang, Y.; Du, Y.; Feng, H.; Xu, T. Simultaneous recovery of ammonium and phosphorus via the integration of electrodialysis with struvite reactor. *J. Membr. Sci.* **2015**, *490*, 65–71. [CrossRef]

446. Thompson Brewster, E.; Ward, A.J.; Mehta, C.M.; Radjenovic, J.; Batstone, D.J. Predicting scale formation during electrodialytic nutrient recovery. *Water Res.* **2017**, *110*, 202–210. [CrossRef]

447. Ward, A.J.; Arola, K.; Thompson Brewster, E.; Mehta, C.M.; Batstone, D.J. Nutrient recovery from wastewater through pilot scale electrodialysis. *Water Res.* **2018**, *135*, 57–65. [CrossRef]

448. Arola, K.; Ward, A.; Mänttäri, M.; Kallioinen, M.; Batstone, D. Transport of pharmaceuticals during electrodialysis treatment of wastewater. *Water Res.* **2019**, *161*, 496–504. [CrossRef] [PubMed]

449. Van Linden, N.; Spanjers, H.; van Lier, J.B. Application of dynamic current density for increased concentration factors and reduced energy consumption for concentrating ammonium by electrodialysis. *Water Res.* **2019**, *163*, 114856. [CrossRef]

450. Vecino, X.; Reig, M.; Gibert, O.; Valderrama, C.; Cortina, J.L. Integration of liquid-liquid membrane contactors and electrodialysis for ammonium recovery and concentration as a liquid fertilizer. *Chemosphere* **2020**, *245*, 125606. [CrossRef]

451. Gao, F.; Wang, L.; Wang, J.; Zhang, H.; Lin, S. Nutrient recovery from treated wastewater by a hybrid electrochemical sequence integrating bipolar membrane electrodialysis and membrane capacitive deionization. *Environ. Sci. Water Res. Technol.* **2020**, *6*, 383–391. [CrossRef]

452. Tao, B.; Passanha, P.; Kumi, P.; Wilson, V.; Jones, D.; Esteves, S. Recovery and concentration of thermally hydrolysed waste activated sludge derived volatile fatty acids and nutrients by microfiltration, electrodialysis and struvite precipitation for polyhydroxyalkanoates production. *Chem. Eng. J.* **2016**, *295*, 11–19. [CrossRef]

453. López-Gómez, J.P.; Alexandri, M.; Schneider, R.; Latorre-Sánchez, M.; Coll Lozano, C.; Venus, J. Organic fraction of municipal solid waste for the production of L-lactic acid with high optical purity. *J. Clean. Prod.* **2020**, *247*, 119165. [CrossRef]

454. Rose, C.; Parker, A.; Jefferson, B.; Cartmell, E. The characterization of feces and urine: A review of the literature to inform advanced treatment technology. *Crit. Rev. Environ. Sci. Technol.* **2015**, *45*, 1827–1879. [CrossRef]

455. Escher, B.I.; Pronk, W.; Suter, M.J.F.; Maurer, M. Monitoring the removal efficiency of pharmaceuticals and hormones in different treatment processes of source-separated urine with bioassays. *Environ. Sci. Technol.* **2006**, *40*, 5095–5101. [CrossRef] [PubMed]

456. Pronk, W.; Biebow, M.; Boller, M. Electrodialysis for recovering salts from a urine solution containing micropollutants. *Environ. Sci. Technol.* **2006**, *40*, 2414–2420. [CrossRef]

457. Pronk, W.; Zuleeg, S.; Lienert, J.; Escher, B.; Koller, M.; Berner, A.; Koch, G.; Boller, M. Pilot experiments with electrodialysis and ozonation for the production of a fertiliser from urine. *Water Sci. Technol.* **2007**, *56*, 219–227. [CrossRef]

458. Pronk, W.; Koné, D. Options for urine treatment in developing countries. *Desalination* **2009**, *248*, 360–368. [CrossRef]

459. De Paepe, J.; Lindeboom, R.E.F.; Vanoppen, M.; De Paepe, K.; Demey, D.; Coessens, W.; Lamaze, B.; Verliefde, A.R.D.; Clauwaert, P.; Vlaeminck, S.E. Refinery and concentration of nutrients from urine with electrodialysis enabled by upstream precipitation and nitrification. *Water Res.* **2018**, *144*, 76–86. [CrossRef]

460. Hjorth, M.; Christensen, K.V.; Christensen, M.L.; Sommer, S.G. Solid-liquid separation of animal slurry in theory and practice. *Sustain. Agric.* **2009**, *2*, 953–986.

461. Nasir, I.M.; Mohd Ghazi, T.I.; Omar, R. Anaerobic digestion technology in livestock manure treatment for biogas production: A review. *Eng. Life Sci.* **2012**, *12*, 258–269. [CrossRef]

462. FAO. *Food Outlook, Biannual Report on Global Food Markets*; FAO: Rome, Italy, 2017.

463. Mondor, M.; Masse, L.; Ippersiel, D.; Lamarche, F.; Massé, D.I. Use of electrodialysis and reverse osmosis for the recovery and concentration of ammonia from swine manure. *Bioresour. Technol.* **2008**, *99*, 7363–7368. [CrossRef]

464. Mondor, M.; Ippersiel, D.; Lamarche, F.; Masse, L. Fouling characterization of electrodialysis membranes used for the recovery and concentration of ammonia from swine manure. *Bioresour. Technol.* **2009**, *100*, 566–571. [CrossRef]

465. Ippersiel, D.; Mondor, M.; Lamarche, F.; Tremblay, F.; Dubreuil, J.; Masse, L. Nitrogen potential recovery and concentration of ammonia from swine manure using electrodialysis coupled with air stripping. *J. Environ. Manag.* **2012**, *95*, S165–S169. [CrossRef]

466. Shi, L.; Xie, S.; Hu, Z.; Wu, G.; Morrison, L.; Croot, P.; Hu, H.; Zhan, X. Nutrient recovery from pig manure digestate using electrodialysis reversal: Membrane fouling and feasibility of long-term operation. *J. Membr. Sci.* **2019**, *573*, 560–569. [CrossRef]

467. Shi, L.; Hu, Z.; Simplicio, W.S.; Qiu, S.; Xiao, L.; Harhen, B.; Zhan, X. Antibiotics in nutrient recovery from pig manure via electrodialysis reversal: Sorption and migration associated with membrane fouling. *J. Membr. Sci.* **2020**, *597*, 117633. [CrossRef]

468. Ye, Z.L.; Ghyselbrecht, K.; Monballiu, A.; Pinoy, L.; Meesschaert, B. Fractionating various nutrient ions for resource recovery from swine wastewater using simultaneous anionic and cationic selective-electrodialysis. *Water Res.* **2019**, *160*, 424–434. [CrossRef]

469. Shi, L.; Hu, Y.; Xie, S.; Wu, G.; Hu, Z.; Zhan, X. Recovery of nutrients and volatile fatty acids from pig manure hydrolysate using two-stage bipolar membrane electrodialysis. *Chem. Eng. J.* **2018**, *334*, 134–142. [CrossRef]

470. Duong, H.C.; Ansari, A.J.; Nghiem, L.D.; Cao, H.T.; Vu, T.D.; Nguyen, T.P. Membrane Processes for the Regeneration of Liquid Desiccant Solution for Air Conditioning. *Curr. Pollut. Rep.* **2019**, *5*, 308–318. [CrossRef]

471. Li, X.W.; Zhang, X.S. Photovoltaic-electrodialysis regeneration method for liquid desiccant cooling system. *Sol. Energy* **2009**, *83*, 2195–2204. [CrossRef]

472. Li, X.W.; Zhang, X.S.; Quan, S. Single-stage and double-stage photovoltaic driven regeneration for liquid desiccant cooling system. *Appl. Energy* **2011**, *88*, 4908–4917. [CrossRef]

473. Cheng, Q.; Zhang, X.S.; Li, X.W. Double-stage photovoltaic/thermal ED regeneration for liquid desiccant cooling system. *Energy Build.* **2012**, *51*, 64–72. [CrossRef]

474. Cheng, Q.; Xu, Y.; Zhang, X.S. Experimental investigation of an electrodialysis regenerator for liquid desiccant. *Energy Build.* **2013**, *67*, 419–425. [CrossRef]

475. Cheng, Q.; Zhang, X.; Jiao, S. Influence of concentration difference between dilute cells and regenerate cells on the performance of electrodialysis regenerator. *Energy* **2017**, *140*, 646–655. [CrossRef]

476. Cheng, Q.; Jiao, S. Experimental and theoretical research on the current efficiency of the electrodialysis regenerator for liquid desiccant air-conditioning system using LiCl solution. *Int. J. Refrig.* **2018**, *96*, 1–9. [CrossRef]

477. Guo, Y.; Ma, Z.; Al-Jubainawi, A.; Cooper, P.; Nghiem, L.D. Using electrodialysis for regeneration of aqueous lithium chloride solution in liquid desiccant air conditioning systems. *Energy Build.* **2016**, *116*, 285–295. [CrossRef]

478. Al-Jubainawi, A.; Ma, Z.; Guo, Y.; Nghiem, L.D.; Cooper, P.; Li, W. Factors governing mass transfer during membrane electrodialysis regeneration of LiCl solution for liquid desiccant dehumidification systems. *Sustain. Cities Soc.* **2017**, *28*, 30–41. [CrossRef]

479. Guo, Y.; Al-Jubainawi, A.; Ma, Z. Performance investigation and optimisation of electrodialysis regeneration for LiCl liquid desiccant cooling systems. *Appl. Therm. Eng.* **2019**, *149*, 1023–1034. [CrossRef]

480. Guo, Y.; Al-Jubainawi, A.; Peng, X. Modelling and the feasibility study of a hybrid electrodialysis and thermal regeneration method for LiCl liquid desiccant dehumidification. *Appl. Energy* **2019**, *239*, 1014–1036. [CrossRef]

481. Al-Jubainawi, A.; Ma, Z.; Guo, Y.; Nghiem, L.D. Effect of regulating main governing factors on the selectivity membranes of electrodialysis used for LiCl liquid desiccant regeneration. *J. Build. Eng.* **2020**, *28*, 101022. [CrossRef]

482. Pei, W.; Cheng, Q.; Jiao, S.; Liu, L. Performance evaluation of the electrodialysis regenerator for the lithium bromide solution with high concentration in the liquid desiccant air-conditioning system. *Energy* **2019**, *187*, 115928. [CrossRef]

483. Sun, B.; Zhang, M.; Huang, S.; Su, W.; Zhou, J.; Zhang, X. Performance evaluation on regeneration of high-salt solutions used in air conditioning systems by electrodialysis. *J. Membr. Sci.* **2019**, *582*, 224–235. [CrossRef]

484. International Desalination Association (IDA). *Desalination Yearbook 2016–2017*; International Desalination Association (IDA): Topsfield, MA, USA, 2017.

485. Jones, E.; Qadir, M.; van Vliet, M.T.H.; Smakhtin, V.; Kang, S. The state of desalination and brine production: A global outlook. *Sci. Total Environ.* **2019**, *657*, 1343–1356. [CrossRef]

486. Gude, V.G. Desalination and sustainability–An appraisal and current perspective. *Water Res.* **2016**, *89*, 87–106. [CrossRef]

487. Baker, R.W. *Membrane Technology and Applications*; John Wiley & Sons, Ltd.: Chichester, UK, 2012; ISBN 9781118359686.

488. Giwa, A.; Dufour, V.; Al Marzooqi, F.; Al Kaabi, M.; Hasan, S.W. Brine management methods: Recent innovations and current status. *Desalination* **2017**, *407*, 1–23. [CrossRef]

489. Mavukkandy, M.O.; Chabib, C.M.; Mustafa, I.; Al Ghaferi, A.; AlMarzooqi, F. Brine management in desalination industry: From waste to resources generation. *Desalination* **2019**, *472*, 114187. [CrossRef]

490. Panagopoulos, A.; Haralambous, K.J.; Loizidou, M. Desalination brine disposal methods and treatment technologies—A review. *Sci. Total Environ.* **2019**, *693*, 133545. [CrossRef]

491. Korngold, E.; Aronov, L.; Daltrophe, N. Electrodialysis of brine solutions discharged from an RO plant. *Desalination* **2009**, *242*, 215–227. [CrossRef]

492. Oren, Y.; Korngold, E.; Daltrophe, N.; Messalem, R.; Volkman, Y.; Aronov, L.; Weismann, M.; Bouriakov, N.; Glueckstern, P.; Gilron, J. Pilot studies on high recovery BWRO-EDR for near zero liquid discharge approach. *Desalination* **2010**, *261*, 321–330. [CrossRef]

493. Xu, X.; Lin, L.; Ma, G.; Wang, H.; Jiang, W.; He, Q.; Nirmalakhandan, N.; Xu, P. Study of polyethyleneimine coating on membrane permselectivity and desalination performance during pilot-scale electrodialysis of reverse osmosis concentrate. *Sep. Purif. Technol.* **2018**, *207*, 396–405. [CrossRef]

494. Yan, H.; Wang, Y.; Wu, L.; Shehzad, M.A.; Jiang, C.; Fu, R.; Liu, Z.; Xu, T. Multistage-batch electrodialysis to concentrate high-salinity solutions: Process optimisation, water transport, and energy consumption. *J. Membr. Sci.* **2019**, *570–571*, 245–257. [CrossRef]

495. Jiang, C.; Wang, Y.; Zhang, Z.; Xu, T. Electrodialysis of concentrated brine from RO plant to produce coarse salt and freshwater. *J. Membr. Sci.* **2014**, *450*, 323–330. [CrossRef]

496. Reig, M.; Casas, S.; Aladjem, C.; Valderrama, C.; Gibert, O.; Valero, F.; Centeno, C.M.; Larrotcha, E.; Cortina, J.L. Concentration of NaCl from seawater reverse osmosis brines for the chlor-alkali industry by electrodialysis. *Desalination* **2014**, *342*, 107–117. [CrossRef]

497. Tanaka, Y.; Reig, M.; Casas, S.; Aladjem, C.; Cortina, J.L. Computer simulation of ion-exchange membrane electrodialysis for salt concentration and reduction of RO discharged brine for salt production and marine environment conservation. *Desalination* **2015**, *367*, 76–89. [CrossRef]

498. Reig, M.; Farrokhzad, H.; Van der Bruggen, B.; Gibert, O.; Cortina, J.L. Synthesis of a monovalent selective cation exchange membrane to concentrate reverse osmosis brines by electrodialysis. *Desalination* **2015**, *375*, 1–9. [CrossRef]

499. Zhang, W.; Miao, M.; Pan, J.; Sotto, A.; Shen, J.; Gao, C.; Van der Bruggen, B. Separation of divalent ions from seawater concentrate to enhance the purity of coarse salt by electrodialysis with monovalent-selective membranes. *Desalination* **2017**, *411*, 28–37. [CrossRef]

500. Liu, J.; Yuan, J.; Ji, Z.; Wang, B.; Hao, Y.; Guo, X. Concentrating brine from seawater desalination process by nanofiltration-electrodialysis integrated membrane technology. *Desalination* **2016**, *390*, 53–61. [CrossRef]

501. McGovern, R.K.; Zubair, S.M.; Lienhard, V.J.H. The benefits of hybridising electrodialysis with reverse osmosis. *J. Membr. Sci.* **2014**, *469*, 326–335. [CrossRef]

502. Chung, H.W.; Nayar, K.G.; Swaminathan, J.; Chehayeb, K.M.; Lienhard, V.J.H. Thermodynamic analysis of brine management methods: Zero-discharge desalination and salinity-gradient power production. *Desalination* **2017**, *404*, 291–303. [CrossRef]

503. Chehayeb, K.M.; Farhat, D.M.; Nayar, K.G.; Lienhard, J.H. Optimal design and operation of electrodialysis for brackish-water desalination and for high-salinity brine concentration. *Desalination* **2017**, *420*, 167–182. [CrossRef]

504. Chehayeb, K.M.; Nayar, K.G.; Lienhard, J.H. On the merits of using multi-stage and counterflow electrodialysis for reduced energy consumption. *Desalination* **2018**, *439*, 1–16. [CrossRef]

505. Nayar, K.G.; Fernandes, J.; McGovern, R.K.; Dominguez, K.P.; McCance, A.; Al-Anzi, B.S.; Lienhard, J.H. Cost and energy requirements of hybrid RO and ED brine concentration systems for salt production. *Desalination* **2019**, *456*, 97–120. [CrossRef]

506. Nayar, K.G.; Fernandes, J.; McGovern, R.K.; Al-Anzi, B.S.; Lienhard, J.H. Cost and energy needs of RO-ED-crystallizer systems for zero brine discharge seawater desalination. *Desalination* **2019**, *457*, 115–132. [CrossRef]

507. Cappelle, M.; Walker, W.S.; Davis, T.A. Improving Desalination Recovery Using Zero Discharge Desalination (ZDD): A Process Model for Evaluating Technical Feasibility. *Ind. Eng. Chem. Res.* **2017**, *56*, 10448–10460. [CrossRef]

508. Reahl, E.R. Reclaiming reverse osmosis blowdown with electrodialysis reversal. *Desalination* **1990**, *78*, 77–89. [CrossRef]

509. Zhang, Y.; Ghyselbrecht, K.; Meesschaert, B.; Pinoy, L.; Van der Bruggen, B. Electrodialysis on RO concentrate to improve water recovery in wastewater reclamation. *J. Membr. Sci.* **2011**, *378*, 101–110. [CrossRef]

510. Zhang, Y.; Ghyselbrecht, K.; Vanherpe, R.; Meesschaert, B.; Pinoy, L.; Van der Bruggen, B. RO concentrate minimization by electrodialysis: Techno-economic analysis and environmental concerns. *J. Environ. Manag.* **2012**, *107*, 28–36. [CrossRef]

511. Tran, A.T.K.; Zhang, Y.; Jullok, N.; Meesschaert, B.; Pinoy, L.; Van der Bruggen, B. RO concentrate treatment by a hybrid system consisting of a pellet reactor and electrodialysis. *Chem. Eng. Sci.* **2012**, *79*, 228–238. [CrossRef]

512. Ghyselbrecht, K.; Van Houtte, E.; Pinoy, L.; Verbauwhede, J.; Van Der Bruggen, B.; Meesschaert, B. Treatment of RO concentrate by means of a combination of a willow field and electrodialysis. *Resour. Conserv. Recycl.* **2012**, *65*, 116–123. [CrossRef]

513. Ribera-Pi, J.; Badia-Fabregat, M.; Espí, J.; Clarens, F.; Jubany, I.; Martínez-Lladó, X. Decreasing environmental impact of landfill leachate treatment by MBR, RO and EDR hybrid treatment. *Environ. Technol. (UK)* **2020**. [CrossRef]

514. Zhang, Y.; Van der Bruggen, B.; Pinoy, L.; Meesschaert, B. Separation of nutrient ions and organic compounds from salts in RO concentrates by standard and monovalent selective ion-exchange membranes used in electrodialysis. *J. Membr. Sci.* **2009**, *332*, 104–112. [CrossRef]

515. Banasiak, L.J.; Van der Bruggen, B.; Schäfer, A.I. Sorption of pesticide endosulfan by electrodialysis membranes. *Chem. Eng. J.* **2011**, *166*, 233–239. [CrossRef]

516. Venzke, C.D.; Giacobbo, A.; Ferreira, J.Z.; Bernardes, A.M.; Rodrigues, M.A.S. Increasing water recovery rate of membrane hybrid process on the petrochemical wastewater treatment. *Process. Saf. Environ. Prot.* **2018**, *117*, 152–158. [CrossRef]

517. Zhao, D.; Lee, L.Y.; Ong, S.L.; Chowdhury, P.; Siah, K.B.; Ng, H.Y. Electrodialysis reversal for industrial reverse osmosis brine treatment. *Sep. Purif. Technol.* **2019**, *213*, 339–347. [CrossRef]

518. Van Linden, N.; Shang, R.; Stockinger, G.; Heijman, B.; Spanjers, H. Separation of natural organic matter and sodium chloride for salt recovery purposes in zero liquid discharge. *Water Resour. Ind.* **2020**, *23*, 100117. [CrossRef]

519. Xu, Y.; Sun, Y.; Ma, Z.; Wang, R.; Wang, X.; Wang, J.; Wang, L.; Gao, X.; Gao, J. Response surface modeling and optimization of electrodialysis for reclamation of RO concentrates in coal-fired power plants. *Sep. Sci. Technol.* **2019**, 1–11. [CrossRef]

520. Qiu, Y.; Ruan, H.; Tang, C.; Yao, L.; Shen, J.; Sotto, A. Study on Recovering High-Concentration Lithium Salt from Lithium-Containing Wastewater Using a Hybrid Reverse Osmosis (RO)–Electrodialysis (ED) Process. *ACS Sustain. Chem. Eng.* **2019**, *7*, 13481–13490. [CrossRef]

521. Vaudevire, E.; Radmanesh, F.; Kolkman, A.; Vughs, D.; Cornelissen, E.; Post, J.; van der Meer, W. Fate and removal of trace pollutants from an anion exchange spent brine during the recovery process of natural organic matter and salts. *Water Res.* **2019**, *154*, 34–44. [CrossRef]

522. Haddad, M.; Bazinet, L.; Barbeau, B. Eco-efficient treatment of ion exchange spent brine via electrodialysis to recover NaCl and minimize waste disposal. *Sci. Total Environ.* **2019**, *690*, 400–409. [CrossRef] [PubMed]

523. Ibáñez, R.; Pérez-González, A.; Gómez, P.; Urtiaga, A.M.; Ortiz, I. Acid and base recovery from softened reverse osmosis (RO) brines. Experimental assessment using model concentrates. *Desalination* **2013**, *309*, 165–170. [CrossRef]

524. Fernandez-Gonzalez, C.; Dominguez-Ramos, A.; Ibañez, R.; Chen, Y.; Irabien, A. Valorization of desalination brines by electrodialysis with bipolar membranes using nanocomposite anion exchange membranes. *Desalination* **2017**, *406*, 16–24. [CrossRef]

525. Herrero-Gonzalez, M.; Diaz-Guridi, P.; Dominguez-Ramos, A.; Ibañez, R.; Irabien, A. Photovoltaic solar electrodialysis with bipolar membranes. *Desalination* **2018**, *433*, 155–163. [CrossRef]

526. Herrero-Gonzalez, M.; Diaz-Guridi, P.; Dominguez-Ramos, A.; Irabien, A.; Ibañez, R. Highly concentrated HCl and NaOH from brines using electrodialysis with bipolar membranes. *Sep. Purif. Technol.* **2020**, *242*, 116785. [CrossRef]

527. Herrero-Gonzalez, M.; Admon, N.; Dominguez-Ramos, A.; Ibañez, R.; Wolfson, A.; Irabien, A. Environmental sustainability assessment of seawater reverse osmosis brine valorization by means of electrodialysis with bipolar membranes. *Environ. Sci. Pollut. Res.* **2020**, *27*, 1256–1266. [CrossRef]

528. Yang, Y.; Gao, X.; Fan, A.; Fu, L.; Gao, C. An innovative beneficial reuse of seawater concentrate using bipolar membrane electrodialysis. *J. Membr. Sci.* **2014**, *449*, 119–126. [CrossRef]

529. Reig, M.; Casas, S.; Gibert, O.; Valderrama, C.; Cortina, J.L. Integration of nanofiltration and bipolar electrodialysis for valorization of seawater desalination brines: Production of drinking and waste water treatment chemicals. *Desalination* **2016**, *382*, 13–20. [CrossRef]

530. Reig, M.; Casas, S.; Valderrama, C.; Gibert, O.; Cortina, J.L. Integration of monopolar and bipolar electrodialysis for valorization of seawater reverse osmosis desalination brines: Production of strong acid and base. *Desalination* **2016**, *398*, 87–97. [CrossRef]

531. Chen, B.; Jiang, C.; Wang, Y.; Fu, R.; Liu, Z.; Xu, T. Selectrodialysis with bipolar membrane for the reclamation of concentrated brine from RO plant. *Desalination* **2018**, *442*, 8–15. [CrossRef]

532. Badruzzaman, M.; Oppenheimer, J.; Adham, S.; Kumar, M. Innovative beneficial reuse of reverse osmosis concentrate using bipolar membrane electrodialysis and electrochlorination processes. *J. Membr. Sci.* **2009**, *326*, 392–399. [CrossRef]

533. Wang, M.; Wang, K.; Jia, Y.; Ren, Q. The reclamation of brine generated from desalination process by bipolar membrane electrodialysis. *J. Membr. Sci.* **2014**, *452*, 54–61. [CrossRef]

534. Yip, N.Y.; Brogioli, D.; Hamelers, H.V.M.; Nijmeijer, K. Salinity gradients for sustainable energy: Primer, progress, and prospects. *Environ. Sci. Technol.* **2016**, *50*, 12072–12094. [CrossRef]

535. Lee, S.; Choi, J.; Park, Y.G.; Shon, H.; Ahn, C.H.; Kim, S.H. Hybrid desalination processes for beneficial use of reverse osmosis brine: Current status and future prospects. *Desalination* **2019**, *454*, 104–111. [CrossRef]

536. Tufa, R.A.; Noviello, Y.; Di Profio, G.; Macedonio, F.; Ali, A.; Drioli, E.; Fontananova, E.; Bouzek, K.; Curcio, E. Integrated membrane distillation-reverse electrodialysis system for energy-efficient seawater desalination. *Appl. Energy* **2019**, *253*, 113551. [CrossRef]

537. Choi, J.; Oh, Y.; Chae, S.; Hong, S. Membrane capacitive deionization-reverse electrodialysis hybrid system for improving energy efficiency of reverse osmosis seawater desalination. *Desalination* **2019**, *462*, 19–28. [CrossRef]

538. Luo, F.; Wang, Y.; Jiang, C.; Wu, B.; Feng, H.; Xu, T. A power free electrodialysis (PFED) for desalination. *Desalination* **2017**, *404*, 138–146. [CrossRef]

539. Mei, Y.; Li, X.; Yao, Z.; Qing, W.; Fane, A.G.; Tang, C.Y. Simulation of an energy self-sufficient electrodialysis desalination stack for salt removal efficiency and fresh water recovery. *J. Membr. Sci.* **2020**, *598*, 117771. [CrossRef]

540. Kwon, K.; Han, J.; Park, B.H.; Shin, Y.; Kim, D. Brine recovery using reverse electrodialysis in membrane-based desalination processes. *Desalination* **2015**, *362*, 1–10. [CrossRef]

541. Tufa, R.A.; Curcio, E.; Brauns, E.; van Baak, W.; Fontananova, E.; Di Profio, G. Membrane Distillation and Reverse Electrodialysis for Near-Zero Liquid Discharge and low energy seawater desalination. *J. Membr. Sci.* **2015**, *496*, 325–333. [CrossRef]

542. Tufa, R.A.; Rugiero, E.; Chanda, D.; Hnàt, J.; van Baak, W.; Veerman, J.; Fontananova, E.; Di Profio, G.; Drioli, E.; Bouzek, K.; et al. Salinity gradient power-reverse electrodialysis and alkaline polymer electrolyte water electrolysis for hydrogen production. *J. Membr. Sci.* **2016**, *514*, 155–164. [CrossRef]

543. Mehdizadeh, S.; Yasukawa, M.; Kuno, M.; Kawabata, Y.; Higa, M. Evaluation of energy harvesting from discharge solutions in a salt production plant by reverse electrodialysis (RED). *Desalination* **2019**, *467*, 95–102. [CrossRef]

544. Tian, H.; Wang, Y.; Pei, Y. Energy capture from thermolytic solutions and simulated sunlight coupled with hydrogen peroxide production and wastewater remediation. *Water Res.* **2020**, *170*, 115318. [CrossRef]

545. Xu, P.; Zheng, D.; Xu, H. The feasibility and mechanism of reverse electrodialysis enhanced photocatalytic fuel cell-Fenton system on advanced treatment of coal gasification wastewater. *Sep. Purif. Technol.* **2019**, *220*, 183–188. [CrossRef]

546. D'Angelo, A.; Tedesco, M.; Cipollina, A.; Galia, A.; Micale, G.; Scialdone, O. Reverse electrodialysis performed at pilot plant scale: Evaluation of redox processes and simultaneous generation of electric energy and treatment of wastewater. *Water Res.* **2017**, *125*, 123–131. [CrossRef]

547. Ma, P.; Hao, X.; Galia, A.; Scialdone, O. Development of a process for the treatment of synthetic wastewater without energy inputs using the salinity gradient of wastewaters and a reverse electrodialysis stack. *Chemosphere* **2020**, *248*, 125994. [CrossRef]

548. Van Egmond, W.J.; Starke, U.K.; Saakes, M.; Buisman, C.J.N.; Hamelers, H.V.M. Energy efficiency of a concentration gradient flow battery at elevated temperatures. *J. Power Sources* **2017**, *340*, 71–79. [CrossRef]

549. Krakhella, K.W.; Morales, M.; Bock, R.; Seland, F.; Burheim, O.S.; Einarsrud, K.E. Electrodialytic energy storage system: Permselectivity, stack measurements and life-cycle analysis. *Energies* **2020**, *13*, 1247. [CrossRef]

550. Xia, J.; Eigenberger, G.; Strathmann, H.; Nieken, U. Acid-base flow battery, based on reverse electrodialysis with bi-polar membranes: Stack experiments. *Processes* **2020**, *8*, 99. [CrossRef]

551. Martí-Calatayud, M.C.; Buzzi, D.C.; García-Gabaldón, M.; Ortega, E.; Bernardes, A.M.; Tenório, J.A.S.; Pérez-Herranz, V. Sulfuric acid recovery from acid mine drainage by means of electrodialysis. *Desalination* **2014**, *343*, 120–127. [CrossRef]

552. Dara, S.; Bonakdarpour, A.; Ho, M.; Govindarajan, R.; Wilkinson, D.P. Conversion of saline waste-water and gaseous carbon dioxide to (bi)carbonate salts, hydrochloric acid and desalinated water for on-site industrial utilization. *React. Chem. Eng.* **2019**, *4*, 141–150. [CrossRef]

553. Wu, D.; Chen, G.Q.; Hu, B.; Deng, H. Feasibility and energy consumption analysis of phenol removal from salty wastewater by electro-electrodialysis. *Sep. Purif. Technol.* **2019**, *215*, 44–50. [CrossRef]

554. Magro, C.; Paz-Garcia, J.M.; Ottosen, L.M.; Mateus, E.P.; Ribeiro, A.B. Sustainability of construction materials: Electrodialytic technology as a tool for mortars production. *J. Hazard. Mater.* **2019**, *363*, 421–427. [CrossRef]

555. Liu, Z.; Zhang, L.; Li, L.; Zhang, S. Separation of olefin/paraffin by electrodialysis. *Sep. Purif. Technol.* **2019**, *218*, 20–24. [CrossRef]

556. Xia, A.; Wei, P.; Sun, C.; Show, P.L.; Huang, Y.; Fu, Q. Hydrogen fermentation of organic wastewater with high ammonium concentration via electrodialysis system. *Bioresour. Technol.* **2019**, *288*, 121560. [CrossRef]

557. Raschitor, A.; Llanos, J.; Rodrigo, M.A.; Cañizares, P. Is it worth using the coupled electrodialysis/electro-oxidation system for the removal of pesticides? Process modelling and role of the pollutant. *Chemosphere* **2020**, *246*, 125781. [CrossRef]

558. Liu, M.J.; Neo, B.S.; Tarpeh, W.A. Building an operational framework for selective nitrogen recovery via electrochemical stripping. *Water Res.* **2020**, *169*, 115226. [CrossRef]

559. Hara, K.; Kishimoto, N.; Kato, M.; Otsu, H. Efficacy of a two-compartment electrochemical flow cell introduced into a reagent-free UV/chlorine advanced oxidation process. *Chem. Eng. J.* **2020**, *388*, 124385. [CrossRef]

560. Cerrillo-Gonzalez, M.M.; Villen-Guzman, M.; Vereda-Alonso, C.; Gomez-Lahoz, C.; Rodriguez-Maroto, J.M.; Paz-Garcia, J.M. Recovery of Li and Co from LiCoO2 via hydrometallurgical-electrodialytic treatment. *Appl. Sci.* **2020**, *10*, 2367. [CrossRef]

561. Sadyrbaeva, T.Z. Removal of chromium(VI) from aqueous solutions using a novel hybrid liquid membrane—electrodialysis process. *Chem. Eng. Process. Process. Intensif.* **2016**, *99*, 183–191. [CrossRef]

562. Wang, B.; Li, Z.; Lang, Q.; Tan, M.; Ratanatamskul, C.; Lee, M.; Liu, Y.; Zhang, Y. A comprehensive investigation on the components in ionic liquid-based polymer inclusion membrane for Cr(VI) transport during electrodialysis. *J. Membr. Sci.* **2020**, *604*, 118016. [CrossRef]

563. Meng, X.; Li, J.; Lv, Y.; Feng, Y.; Zhong, Y. Electro-membrane extraction of cadmium(II) by bis(2-ethylhexyl) phosphate/kerosene/polyvinyl chloride polymer inclusion membrane. *J. Hazard. Mater.* **2020**, *386*, 121990. [CrossRef] [PubMed]

564. Pan, Y.; Zhu, T.; He, Z. Minimizing effects of chloride and calcium towards enhanced nutrient recovery from sidestream centrate in a decoupled electrodialysis driven by solar energy. *J. Clean. Prod.* **2020**, *263*, 121419. [CrossRef]

565. Alkhadra, M.A.; Conforti, K.M.; Gao, T.; Tian, H.; Bazant, M.Z. Continuous Separation of Radionuclides from Contaminated Water by Shock Electrodialysis. *Environ. Sci. Technol.* **2020**, *45*, 527–536. [CrossRef]

566. Lee, J.B.; Park, K.K.; Eum, H.M.; Lee, C.W. Desalination of a thermal power plant wastewater by membrane capacitive deionization. *Desalination* **2006**, *196*, 125–134. [CrossRef]

567. Xu, P.; Drewes, J.E.; Heil, D.; Wang, G. Treatment of brackish produced water using carbon aerogel-based capacitive deionization technology. *Water Res.* **2008**, *42*, 2605–2617. [CrossRef]

568. Liang, P.; Yuan, L.; Yang, X.; Zhou, S.; Huang, X. Coupling ion-exchangers with inexpensive activated carbon fiber electrodes to enhance the performance of capacitive deionization cells for domestic wastewater desalination. *Water Res.* **2013**, *47*, 2523–2530. [CrossRef]

569. Oren, Y. Capacitive deionization (CDI) for desalination and water treatment—Past, present and future (a review). *Desalination* **2008**, *228*, 10–29. [CrossRef]

570. Anderson, M.A.; Cudero, A.L.; Palma, J. Capacitive deionization as an electrochemical means of saving energy and delivering clean water. Comparison to present desalination practices: Will it compete? *Electrochim. Acta* **2010**, *55*, 3845–3856. [CrossRef]

571. Liu, P.; Yan, T.; Zhang, J.; Shi, L.; Zhang, D. Separation and recovery of heavy metal ions and salt ions from wastewater by 3D graphene-based asymmetric electrodes: Via capacitive deionization. *J. Mater. Chem. A* **2017**, *5*, 14748–14757. [CrossRef]

572. Choi, J.; Dorji, P.; Shon, H.K.; Hong, S. Applications of capacitive deionization: Desalination, softening, selective removal, and energy efficiency. *Desalination* **2019**, *449*, 118–130. [CrossRef]

573. Pawlowski, S.; Huertas, R.M.; Galinha, C.F.; Crespo, J.G.; Velizarov, S. On operation of reverse electrodialysis (RED) and membrane capacitive deionisation (MCDI) with natural saline streams: A critical review. *Desalination* **2020**, *476*, 114183. [CrossRef]

574. Folaranmi, G.; Bechelany, M.; Sistat, P.; Cretin, M.; Zaviska, F. Towards Electrochemical Water Desalination Techniques: A Review on Capacitive Deionization, Membrane Capacitive Deionization and Flow Capacitive Deionization. *Membranes* **2020**, *10*, 96. [CrossRef] [PubMed]

575. Zhou, Y.; Hu, C.; Liu, H.; Qu, J. Potassium-Ion Recovery with a Polypyrrole Membrane Electrode in Novel Redox Transistor Electrodialysis. *Environ. Sci. Technol.* **2020**, *54*, 4592–4600. [CrossRef]

576. Yang, E.; Chae, K.-J.; Choi, M.-J.; He, Z.; Kim, I.S. Critical review of bioelectrochemical systems integrated with membrane-based technologies for desalination, energy self-sufficiency, and high-efficiency water and wastewater treatment. *Desalination* **2019**, *452*, 40–67. [CrossRef]

577. Logan, B.E.; Rabaey, K. Conversion of wastes into bioelectricity and chemicals by using microbial electrochemical technologies. *Science* **2012**, *337*, 686–690. [CrossRef]

578. Srikanth, S.; Kumar, M.; Puri, S.K. Bio-electrochemical system (BES) as an innovative approach for sustainable waste management in petroleum industry. *Bioresour. Technol.* **2018**, *265*, 506–518. [CrossRef] [PubMed]

579. Wang, H.; Ren, Z.J. A comprehensive review of microbial electrochemical systems as a platform technology. *Biotechnol. Adv.* **2013**, *31*, 1796–1807. [CrossRef] [PubMed]

580. Cao, X.; Huang, X.; Liang, P.; Xiao, K.; Zhou, Y.; Zhang, X.; Logan, B.E. A new method for water desalination using microbial desalination cells. *Environ. Sci. Technol.* **2009**, *43*, 7148–7152. [CrossRef] [PubMed]

581. Saeed, H.M.; Husseini, G.A.; Yousef, S.; Saif, J.; Al-Asheh, S.; Abu Fara, A.; Azzam, S.; Khawaga, R.; Aidan, A. Microbial desalination cell technology: A review and a case study. *Desalination* **2015**, *359*, 1–13. [CrossRef]

582. Wang, Y.; Xu, A.; Cui, T.; Zhang, J.; Yu, H.; Han, W.; Shen, J.; Li, J.; Sun, X.; Wang, L. Construction and application of a 1-liter upflow-stacked microbial desalination cell. *Chemosphere* **2020**, *248*, 126028. [CrossRef] [PubMed]

583. Jafary, T.; Al-Mamun, A.; Alhimali, H.; Baawain, M.S.; Rahman, M.S.; Rahman, S.; Dhar, B.R.; Aghbashlo, M.; Tabatabaei, M. Enhanced power generation and desalination rate in a novel quadruple microbial desalination cell with a single desalination chamber. *Renew. Sustain. Energy Rev.* **2020**, *127*, 109855. [CrossRef]

584. Lan, J.; Ren, Y.; Lu, Y.; Liu, G.; Luo, H.; Zhang, R. Combined microbial desalination and chemical-production cell with Fenton process for treatment of electroplating wastewater nanofiltration concentrate. *Chem. Eng. J.* **2019**, *359*, 1139–1149. [CrossRef]

585. Li, X.; Jin, X.; Zhao, N.; Angelidaki, I.; Zhang, Y. Novel bio-electro-Fenton technology for azo dye wastewater treatment using microbial reverse-electrodialysis electrolysis cell. *Bioresour. Technol.* **2017**, *228*, 322–329. [CrossRef] [PubMed]

586. Kim, Y.; Logan, B.E. Microbial reverse electrodialysis cells for synergistically enhanced power production. *Environ. Sci. Technol.* **2011**, *45*, 5834–5839. [CrossRef]

587. Cusick, R.D.; Kim, Y.; Logan, B.E. Energy capture from thermolytic solutions in microbial reverse- electrodialysis cells. *Science* **2012**, *335*, 1474–1477. [CrossRef]

 *membranes*

Article

PVA-Based Mixed Matrix Membranes Comprising ZSM-5 for Cations Separation

Fangmeng Sheng [1], Noor Ul Afsar [1], Yanran Zhu [1], Liang Ge [1,2,*] and Tongwen Xu [1,*]

[1] CAS Key Laboratory of Soft Matter Chemistry, iCHEM (Collaborative Innovation Center of Chemistry for Energy Materials), Department of Applied Chemistry, School of Chemistry and Materials Science, University of Science and Technology of China, Hefei 230026, China; sa052@mail.ustc.edu.cn (F.S.); noor@mail.ustc.edu.cn (N.U.A.); zhuyr@mail.ustc.edu.cn (Y.Z.)

[2] Applied Engineering Technology Research Center for Functional Membranes, Institute of Advanced Technology, University of Science and Technology of China, Hefei 230088, China

* Correspondence: twxu@ustc.edu.cn (T.X.); geliang@ustc.edu.cn (L.G.); Tel.: +86-551-63601581 (T.X.); Fax: +86-551-63602171 (T.X.)

Received: 15 May 2020; Accepted: 28 May 2020; Published: 30 May 2020

Abstract: The traditional ion-exchange membranes face the trade-off effect between the ion flux and perm-selectivity, which limits their application for selective ion separation. Herein, we amalgamated various amounts of the ZSM-5 with the polyvinyl alcohol as ions transport pathways to improve the permeability of monovalent cations and exclusively reject the divalent cations. The highest contents of ZSM-5 in the mixed matrix membranes (MMMs) can be extended up to 60 wt% while the MMMs with optimized content (50 wt%) achieved high perm-selectivity of 34.4 and 3.7 for H^+/Zn^{2+} and Li^+/Mg^{2+} systems, respectively. The obtained results are high in comparison with the commercial CSO membrane. The presence of cationic exchange sites in the ZSM-5 initiated the fast transport of proton, while the microporous crystalline morphology restricted the active transport of larger hydrated cations from the solutions. Moreover, the participating sites and porosity of ZSM-5 granted continuous channels for ions electromigration in order to give high limiting current density to the MMMs. The SEM analysis further exhibited that using ZSM-5 as conventional fillers, gave a uniform and homogenous formation to the membranes. However, the optimized amount of fillers and the assortment of a proper dispersion phase are two critical aspects and must be considered to avoid defects and agglomeration of these enhancers during the formation of membranes.

Keywords: ZSM-5 zeolite; electrodialysis; monovalent cation separation; mixed matrix membrane

---

1. Introduction

Electrodialysis (ED) based on ion-exchange membranes (IEMs) is an important separation technology, which has been widely used for seawater/brackish water desalination, wastewater treatment, acid-base recovery, selective ion separation due to the low energy consumption, no phase transition, and high productivity [1–7]. With the development of innovative IEMs, the membrane-based separation techniques have expanded further and played a crucial role in energy-saving and clean production [8–12]. The conventional IEMs can only separate oppositely charged ions (Donnan effect). However, it may not separate ions with the same charge [5]. The best example can be observed for the selective removal of ions from the industrial waste-acid solution containing Cu^{2+}, Zn^{2+}, Ni^{2+}, and other heavy metal ions that require further acid recovery and extraction of heavy metal ions to prevent environmental pollution and to realize the utilization of resources trapped in the wastewaters [13,14]. Similarly, the enrichment of Li^+ ions and the production of edible salt (NaCl) from brines [15,16] are also critical challenges for the traditional IEMs to overcome. Monovalent cations are always accompanied by the divalent cations such as Mg^{2+} and must be removed selectively to circumvent membranes scaling in the ED process. The scaling may decrease the ion flux and the process efficiency [17–20]. Therefore, it is urgently needed to develop IEMs with unique features to separate monovalent ions from the divalent ions for desired applications. Recently, various researchers employed monovalent ion selective membranes to overwhelm the problems mentioned earlier. For example, Zhang et al. proposed the idea of selectrodialysis to separate monovalent ions from divalent ions [21]. A few studies also show the selective separation of Li^+ from Mg^{2+} by ED with the monovalent ion selective membranes [22,23]. In fact, the commercial monovalent ion selective membranes are expensive and show low permselectivity. In addition, the high area resistance and low limiting current density restrict their industrial applications. Therefore, we focused on developing a facile way to prepare selective membranes for monovalent/divalent cations separation.

The permselectivity of the traditional IEMs could be improved in multiple approaches: (i) crosslinking to get high density of the membranes' structure, [24] (ii) membranes surface coating with a positively charged layer to facilitate the exclusion of multivalent cations via electrostatic repulsion [23,25], and (iii) blending of polymers with either other polymers or microporous crystalline materials such as zeolites to systematize the permselectivity of mixed matrix membranes (MMMs) [26,27]. Zeolites are crystalline aluminosilicates with systematic pores and cavities of molecular dimensions and are famous for their excellent ion-exchange capacity [28]. Zeolites are formed by interlinked tetrahedral of SiO_4 and AlO_4^- sharing an oxygen atom with a large number of exchangeable cationic sites at AlO_4^-. These sites could be easily exchanged by Li^+ and H^+ and afford excellent ion conductivity [29–31]. Ample surface area, strong adsorption capacity, and molecular separation make zeolites conducive for catalysis, heavy metals ions adsorption, and other separation applications [32–34].

Due to the tunable porosity of zeolites, it has been used for ion separation from aqueous solution. For example, Dong et al. [35] introduced NaY zeolite nanoparticles in the polyamide interfacial polymerization to increase the salt rejection up to 98.8%. Similarly, Fathizadeh et al. [36] doped NaX zeolite into the polyamide surface layer to prepare MMMs with enhanced surface properties, such as contact angle, surface roughness, and solid-liquid interface free energy. The reported MMMs showed high water flux, which was 1.8 times higher as compared with the polyamide membrane without zeolite. In addition, ZSM-35 zeolite was also used to construct a thin layer on the poly (ether sulfone) (PES) porous membrane using Nafion solution as a crosslinking agent [31]. With the help of a 0.5 nm pore size, the resultant membrane showed excellent permselectivity for hydrated proton (<0.24 nm) and vanadium ions (0.6 nm). Polysulfone (PS) and zeolite-based membranes have been reported for efficient adsorption of Cu^{2+} ions from aqueous solution [37]. Except for ions permselectivity, various MMMs have been recently investigated for additional separation applications [38,39]. In summary, due to the high permselectivity, dimensional stability, low cost, and easy preparation of membranes, MMMs originated as an exciting theme for modern research. Inspired by the well-crystalline and microporous structure of the zeolite, we prefer to amalgamate this nano stuff inside the PVA polymer for selective ions separation.

In the present study, commercial ZSM-5 zeolite of low price with an appropriate pore size (0.5–0.6 nm) was used to prepare ZSM-5/PVA-based MMMs by simple mixing and a casting procedure. Due to the sieving effect and cationic exchange sites of the ZSM-5 zeolite, we developed low-cost MMMs with various contents of zeolite in the PVA backbone for selective separation of monovalent cations via the ED process. Additionally, the influence of zeolite contents was evaluated to optimize the dosage of zeolite for the membrane's synthesis. The physicochemical properties, such as microstructures analysis, water uptake (WU), area swelling, and membrane's resistance, were examined and explained in detail. The prepared MMMs were investigated for the H^+/Zn^{2+} and Li^+/Mg^{2+} systems, respectively.

2. Experimental

2.1. Materials

ZSM-5 zeolite was purchased from Shentan Environmental Protection New Materials Co. Ltd. (Shanghai, China). Polyvinyl alcohol (PVA) was obtained from Shanghai Macklin Biochemical Co. Ltd. (Shanghai, China). Commercial anion exchange membranes AMX (Neosepta, Tokuyama Co., Tokyo, Japan) were used as auxiliary membranes in the ED experiments. Other reagents, including $LiCl$, $MgCl_2$, $ZnCl_2$, HCl, $NaOH$, and dimethyl sulfoxide (DMSO), were supplied by China National Pharmaceutical Group Industry Co. Ltd. (Beijing, China). All these reagents were of an analytical grade and used without any further purification. Deionized (DI) water was used in all experiments.

2.2. Preparation of ZSM-5/PVA-Based MMMs

PVA (10 wt%) transparent solution was prepared in the DMSO at 100 °C for 1.5 h. Thereafter, the PVA solution was cooled down to room temperature and stabilized for a certain time. The membrane casting solutions were prepared by mixing various amounts of the ZSM-5 in the PVA solution, and the corresponding membranes were named x-ZSM-5 (x = 20, 30, 40, 50, and 60 in weight percent). The mixed solutions were then magnetically stirred and sonicated for 30 mins to get homogenous solutions. Lastly, the resultant mixed solutions were cast on a clean glass plate at 60 °C for 12 h. The membranes were peeled off and hydrated with 0.1 mol L^{-1} solution of HCl for further use.

2.3. Characterization of ZSM-5/PVA-Based MMMs

The surface and cross-sectional morphologies of the prepared membranes were analyzed by scanning electron microscopy (SEM-mask-Hitachi 8220). The adsorption property of the ZSM-5 was measured as described in this study. We made three solutions of different concentrations, i.e., 0.5 mmol L^{-1}, 1 mmol L^{-1}, and 10 mmol L^{-1}, containing equimolar of $LiCl$ and $MgCl_2$, respectively. Then, 2 g of dried ZSM-5 (H-form) was added to each solution (30 mL) and stirred for a certain time (one day and six days) at room temperature. The solutions were centrifuged at 10,000 rpm for 5 min, and the supernatant (2 mL) was collected from each solution and tested by inductively coupled plasma atomic emission spectrometer (ICP-AES, Optima 7300 DV, Waltham, MA, USA) for Li^+ and Mg^{2+} ions, respectively. The crystallinity of the ZSM-5 zeolite powder was analyzed by X-ray diffractions (XRD). The surface aperture analyzer was used to obtain the surface area and pore size distribution of the commercial ZSM-5 zeolite.

Water uptake (WU) and area swelling were measured according to the reported literature [40]. First, the membrane samples (1.5 × 1.5 cm^2) were dipped in DI water for 24 h to fully hydrate and weigh as W_{wet}. The samples were then dried in an oven at 60 °C and reweighed as W_{dry}. The WU was calculated according to the difference in weight using Equation (1). Similarly, the area of the dried (A_{dried}) and wet (A_{wet}) membrane samples were measured and the area swelling can be calculated using Equation (2), as given below.

$$Water\ uptake\ (\%) = \frac{W_{wet} - W_{dry}}{W_{dry}} \times 100 \tag{1}$$

$$Area\ swelling\ (\%) = \frac{A_{wet} - A_{dry}}{A_{dry}} \times 100 \tag{2}$$

2.4. Current-Voltage (I-V) Curves

The I-V curves of ZSM-5/PVA-based MMMs were obtained by using a device with a four-compartment cell as in Figure 1 with an exposed membrane's area of 7.07 cm^2. The anode and cathode chambers (filled with 0.3 mol L^{-1} Na_2SO_4 solution) are separated from the middle two chambers by a couple of membranes, i.e., AMX. The testing membrane was placed in the middle of AMX membranes separating diluted and concentrated chambers. The diluted chamber and the concentrated chamber were filled with the same mixed solution of 0.1 mol L^{-1} LiCl and 0.1 mol L^{-1} MgCl$_2$, respectively. A direct current (DC) power supply (WYL1703, Hangzhou Siling Electrical Instrument Ltd., Hangzhou, China) was connected with a pair of Pt electrodes on either side, and the current density was increased steadily from 0 to 100 mA cm^{-2}, while the difference in potential was recorded using a multimeter (Victor Hi-Tech Co., Ltd., VC890C$^+$, Shenzhen, China) attached with a couple of Ag-AgCl reference electrodes near the membrane surface. Before the test, the membranes were equilibrated in the mixed solution of 0.1 mol L^{-1} LiCl and 0.1 mol L^{-1} MgCl$_2$ for 24 h. Peristaltic pumps (YZ15, Baoding Lead Fluid Co., Ltd., Baoding, China) were used to circulate the corresponding solutions in their respective chambers.

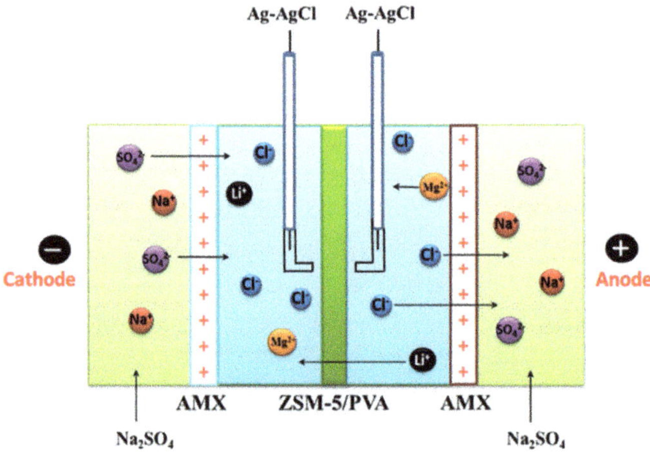

Figure 1. Schematic diagram of the testing device for the I–V curve.

2.5. Evaluation of the Cations Permselectivity

The permselectivity of the prepared membranes with an effective area of 7.07 cm^2 was investigated using a similar device, as in Figure 1. The diluted chamber was filled with a 100 mL solutions of 0.1 mol L^{-1} LiCl/0.1 mol L^{-1} MgCl$_2$ (for Li$^+$/Mg^{2+} system) or 0.5 mol L^{-1} H$_2$SO$_4$/0.23 mol L^{-1} ZnSO$_4$ (for H$^+$/Zn^{2+} system) mixtures while the concentrated chamber was filled with 200 mL of 0.01 mol L^{-1} KCl solution, respectively. An electrodes' rinse solution of 0.3 mol L^{-1} Na$_2$SO$_4$ was circulated in the electrode's chambers. The solutions were circulated with a pair of peristaltic pumps at the flow rate of 5.2 L h^{-1} to avoid the concentration polarization. Two auxiliary membranes (AMX) were used on either side of the testing membrane to complete the ED setup. During the ED experiments, a DC power was used to provide a constant current density of 30 mA cm^{-2}. The samples collected (from the concentrated section) after 1 h, were tested for Li$^+$ and Mg^{2+} ions by ICP-AES, while the H$^+$ ion centration was analyzed by acid-base titration using phenolphthalein as an indicator. The membrane's permselectivity was calculated using Equation (3) as follows.

$$J = \frac{(C_t - C_0).V}{A_m.t} \tag{3}$$

where J (mol cm^{-2} s^{-1}) is the cationic flux while c_t (mol L^{-1}) and c_0 (mol L^{-1}) represent the cation concentrations in the concentrated chamber at time t and 0, respectively. V (dm^3) is the volume of the concentrated solution. A_m means the effective area of the tested membrane.

The perm-selectivity $\left(P_{M^{2+}}^{N^+}\right)$ between monovalent and divalent cations was calculated as reported [41].

$$P_{M^{2+}}^{N^+} = \frac{J_{N^+}}{J_{M^{2+}}}\frac{C_{M^{+2}}}{C_{N^+}} \tag{4}$$

In Equation (4), J_{N^+} and $J_{M^{2+}}$ are the fluxes of monovalent and divalent cations, whereas c_{N^+} (mol L^{-1}) and $c_{M^{2+}}$ (mol L^{-1}) are the average molar concentrations of the monovalent and divalent cations in the diluted chamber during the experiment, respectively.

3. Results and Discussion

3.1. Surface Area and Pore Size Distribution Analysis of the ZSM-5

To characterize the microporous structure of the ZSM-5 zeolite, the nitrogen adsorption-desorption isotherm was measured at 77 K. As shown in Figure 2, the result of isotherm sorption profiles showed a typical type I curve (Figure 2b), which is following the IUPAC classifications and indicates the microporous characteristics of the ZSM-5 material. The Brunauer–Emmett–Teller (BET) surface area was calculated to be 288 m^2 g^{-1}, and the pore size distribution mainly focused on a 0.6 nm (Figure 2c) in agreement with that of the skeletal diagram of ZSM-5 zeolite (Figure 2a). The pore size is suitable for ions separation based on size-selective sieving.

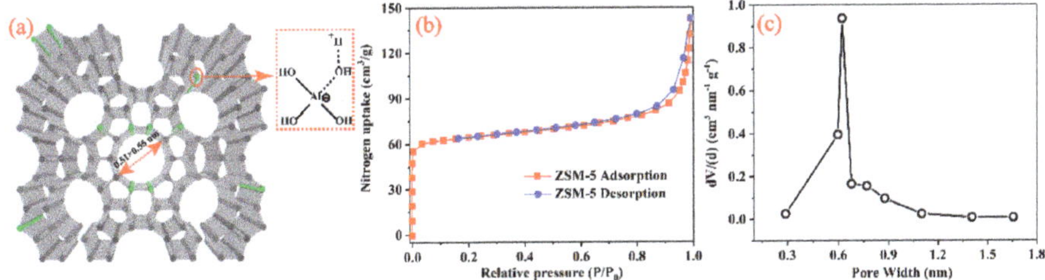

Figure 2. (a) Skeletal diagram, (b) nitrogen adsorption-desorption isotherms, and (c) pore size distribution profiles of ZSM-5 zeolite.

3.2. Adsorption Capacity of ZSM-5

The adsorption capability of the pristine ZSM-5 was evaluated for various concentrations of the binary solution of LiCl/MgCl$_2$ at 25 °C for a different interval of time, as given in Figure 3. The acquired results revealed that ZSM-5 is very selective for the monovalent cation as compared with the divalent cation. For instance, when a high concentration (10 mmol L^{-1} LiCl/MgCl$_2$) solution was used, the adsorption capacity of ZSM-5 for Li$^+$ ion was extraordinary, while the Mg^{2+} did not show noticeable adsorption and we did not include it in Figure 3. When the solution concentration was decreased to 1 and 0.5 mmol L^{-1}, the adsorption capacity further decreased. The adsorption ability of the ZSM-5 can be explained by considering the ion-exchange reaction between the Li$^+$ in the solution and H$^+$ of the ZSM-5 material, i.e., H$^+$ is replaced with Li$^+$ more quickly as compared with Mg^{2+} ions due to size exclusion and less solvation effect. For a different interval of time (Figure 3b), the initial adsorption takes place very promptly, and then the equilibrium is established. Appropriate Li$^+$ ions are maintained inside the ZSM-5 framework for charge compensation and no further exchange takes place. The adsorption results indicate that ZSM-5 are suitable to prepare MMMs for monovalent cations separation.

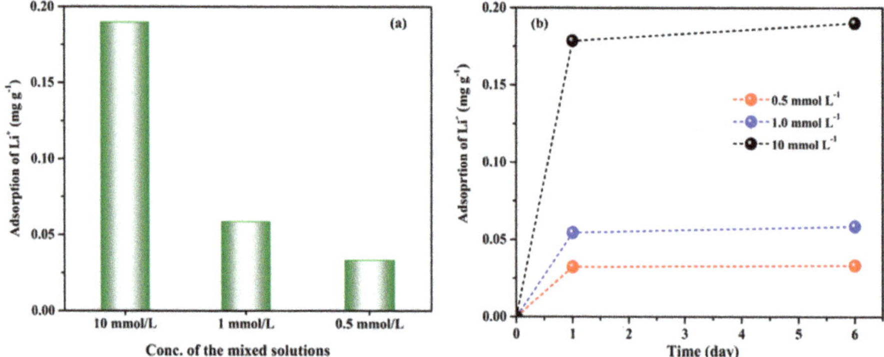

Figure 3. The adsorption property of ZSM-5 for LiCl/MgCl$_2$ mixed solutions: (**a**) Li$^+$ adsorption vs. concentration of solution, (**b**) Li$^+$ adsorption vs. time (day), respectively.

3.3. Morphology

The uniform dispersion of the ZSM-5 was examined using SEM for the representative membranes, as given in Figure 4. The surfaces were assigned as A1, B1, C1, D1, and E1, while the cross-sections were nominated as A2, B2, C2, D2, and E2, respectively. The SEM micrographs reveal that, when the concentration of ZSM-5 was 20 wt%, particles were smaller in dimensions with a consistently disseminated PVA profile. For a minute quantity of ZSM-5, the PVA substrate quickly established physical interaction with the ZSM-5 particles via hydrogen bonding between the –OH group of the PVA and oxygen of the doping material [42]. However, when the quantity of ZSM-5 expanded to a maximum of 60%, the particles lump into agglomerates and are hard to scatter uniformly in the PVA solution, as can be seen in Figure 4E. These lumps are more visible in the cross-section analysis of the membranes. The disturbance of the PVA crystal region (explained in the XRD section) and relaxation of the PVA chains may improve ionic fluxes during the ED process. Thus, it is indispensable to choose the optimal quantity of the ZSM-5 to thoroughly adjust it in the membrane's configuration for desired homogenous membranes.

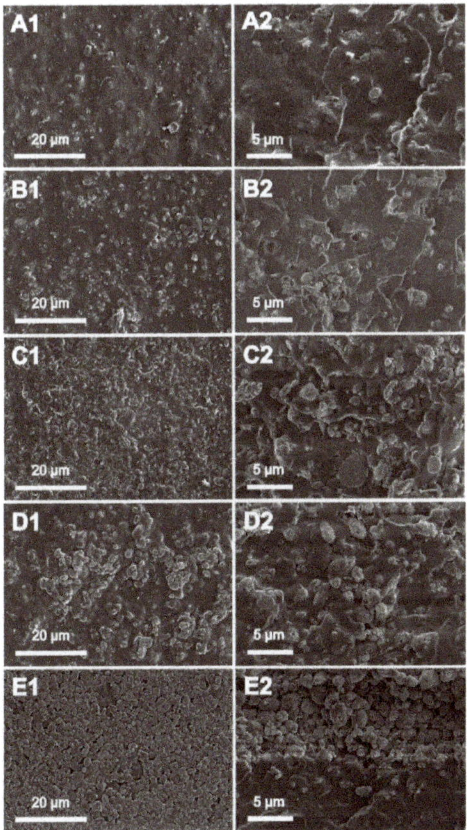

Figure 4. Surface (**A1–E1**) and cross-section (**A2–E2**) morphologies of ZSM-5/PVA-based mixed matrix membranes (MMMs): (**A**), (**B**), (**C**), (**D**) and (**E**) are 20 wt%, 30 wt%, 40 wt%, 50 wt%, and 60 wt% zeolite doping content, respectively.

3.4. XRD Analysis

The ZSM-5 is a microporous crystalline material, as discussed in the earlier section. It was assumed that, during membranes' synthesis, the fabrication process might disrupt its indigenous characteristics, which are crucial for selective cations separation. The powder X-ray diffraction (PXRD) pattern of the pristine ZSM-5 was compared to the ZSM-5 modified membranes, which are given in Figure 5. The modified ZSM-5 membranes demonstrate the identical characteristic peaks as in pristine ZSM-5 at 2θ at 6°–10° (doublet), and 22°–25° (triplet), respectively, which indicated that the crystalline structure of ZSM-5 was not altered after incorporation in the PVA backbone [43]. Moreover, the crystallinity of the PVA (at 2θ of 19.4°) decreased when the dosage of ZSM-5 improved from 20% to 60% [44]. It might be due to the breaking of hydrogen bonds between PVA chains and the insertion of ZSM-5 as an intervening material. Additionally, when the ZSM-5 loading was adequately high (60%), the aggregate of the ZSM-5 nanoparticles may form (SEM section) and, consequently, lower the crystalline behavior of the PVA profile. In conclusion, the XRD pattern of the ZSM-5 doped membranes have no noticeable difference with the original ZSM-5, which justify the synthesis protocol for the high permselective membranes.

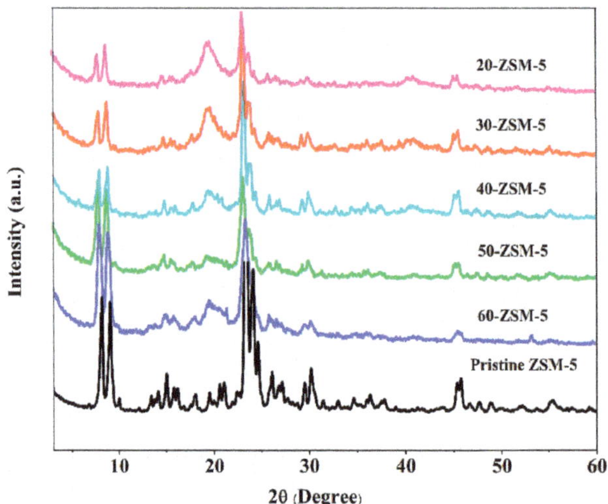

Figure 5. XRD pattern of ZSM-5/PVA-based mixed matrix membranes (MMMs) in comparison with the ZSM-5 zeolite powder.

3.5. Water Uptake (WU) and Area Swelling

The physical characteristics such as water uptake (WU) and area swelling of membranes are essential parameters that appraise the membrane applicability for the purpose applications. This primarily occurs when the PVA substrate is employed for membrane synthesis. In the present case, the modification of PVA with ZSM-5 has acknowledged in depreciating its swelling behavior in the water-based application. Comprehensive articles have been reported in this regard [45–48]. The WU and area swelling results are assessed as given in Figure 6. After alteration with ZSM-5, the WU and area swelling were reduced to 62.8% and 33.7%, respectively. Further raising the amount of ZSM-5 to 60%, both parameters were significantly lowered to 35% and 6.5%, respectively. The decrease in WU and area swelling could be attributed to the engagement of the -OH of the PVA with oxygen sites in the zeolite via hydrogen bonding. For high permselectivity, high membranes WU and area swelling are discouraged to evade the leakage of more hydrophilic cations such as Mg^{2+}, Zn^{2+}, and structural deformation during water-based applications. From Figure 6, it declares that ZSM-5 modified membranes exhibited lower WU and area swellings since we increased the quantity of ZSM-5. It may presume that the interfacial hindrance interaction was lessened and chain motion of the base membrane was restricted, which was the principal theme of ZSM-5 modified membranes. Hence, the developed membranes, particularly 60%, demonstrated considerably lower WU and area swelling than the rest of the membranes due to the more potent inhibition. The membrane's surface inhibition is beneficial to segregate ions on the base of their hydrophilic nature. Additionally, it can be realized that the reduced area swelling may also improve the structural stability of ZSM-5/PVA-based membranes.

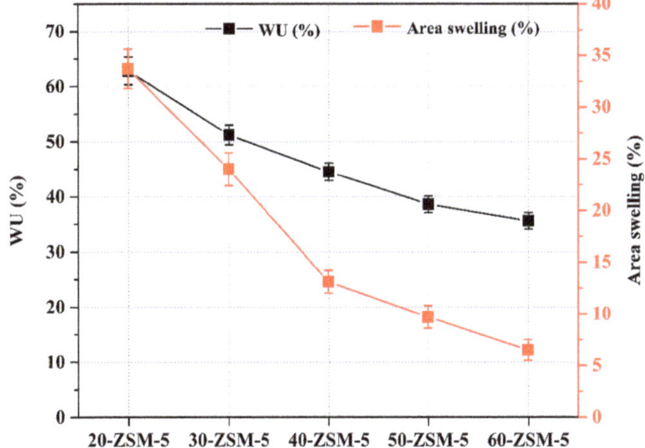

Figure 6. Water uptake (WU) and area swelling of the prepared MMMs.

3.6. Current-Voltage (I–V) Curves

The current–voltage investigation is an essential tool to explain the behavior of IEMs over a range of current applications. Figure 7 exhibited the I-V curves for the ZSM-5 modified membranes. The membrane resistance (R_M) was obtained by plotting the ratio of dE/di vs. current density, as given in Figure 7b, which demonstrated the decrease of membranes' resistance by inserting ZSM-5 in the PVA backbone. This behavior was expected since the cation exchange carriers were introduced with the ZSM-5. For instance, when the ZSM-5 content was increased from 20 to 60 wt%, the R_M decreased from 1.64 $\Omega \cdot cm^2$ to 0.77 $\Omega \cdot cm^2$. This decrease in resistance can be associated with the structure of the ZSM-5 in the membranes. The 20-ZSM-5 membrane with the lowest content does not have adequate cations ways, and the substrate provides a high impedance to the incoming cations. However, when the dosage of ZSM-5 was further increased, the R_M was significantly lowered due to the generation of more channels to predominate cations' transportation. The optimized content plus compactness of the membranes are the critical factors conferring monovalent cations' perm-selectivity to these membranes (see ED section).

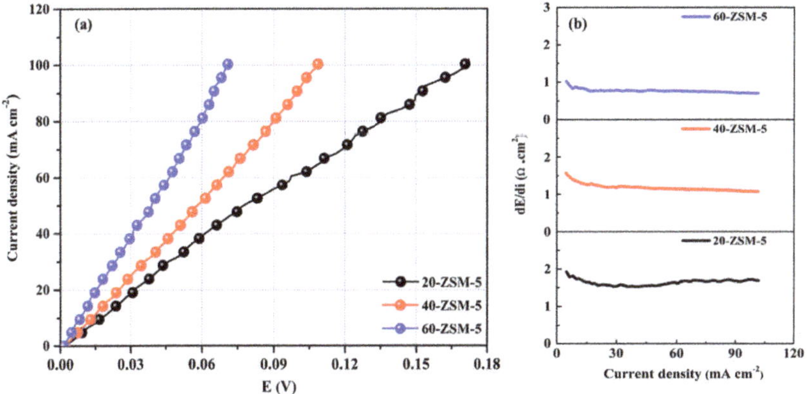

Figure 7. (**a**) Current-voltage curves, and (**b**) membrane resistance of representative's membranes.

Notably, the developed membranes unveil high limiting current density beyond 100 mA cm^{-2}. The high limiting current density is affiliated with the foundation of nanochannels (nano porosity was confirmed in the BET). Moreover, the lowest R_M value calculated for the modified membranes (0.77 $\Omega \cdot cm^2$ for 60-ZSM-5 membrane) is even lower as compared with the reported membranes [49,50]. As asserted earlier, the reduction in the R_M with the dosage of ZSM-5 can be considered a supplementary sign of the successful establishment of ion transport channels in the membrane profile. From the perspective of I-V curves, the MMMs have high limiting current density and low R_M, which indicated more applicability and flexibility of the developed membranes for practical applications.

3.7. Electrodialysis (ED) Experiments

The membranes' performance with various concentrations of the ZSM-5 was inquired for a pair of mixed feed solutions such as 0.1 mol L^{-1} LiCl/0.1 mol L^{-1} $MgCl_2$ and 0.5 mol L^{-1} H_2SO_4/ 0.23 mol L^{-1} $ZnSO_4$, respectively. The ion fluxes and perm-selectivity for both systems were conducted at 30 mA cm^{-2} current density and 25 °C, as represented in Figures 8 and 9, respectively.

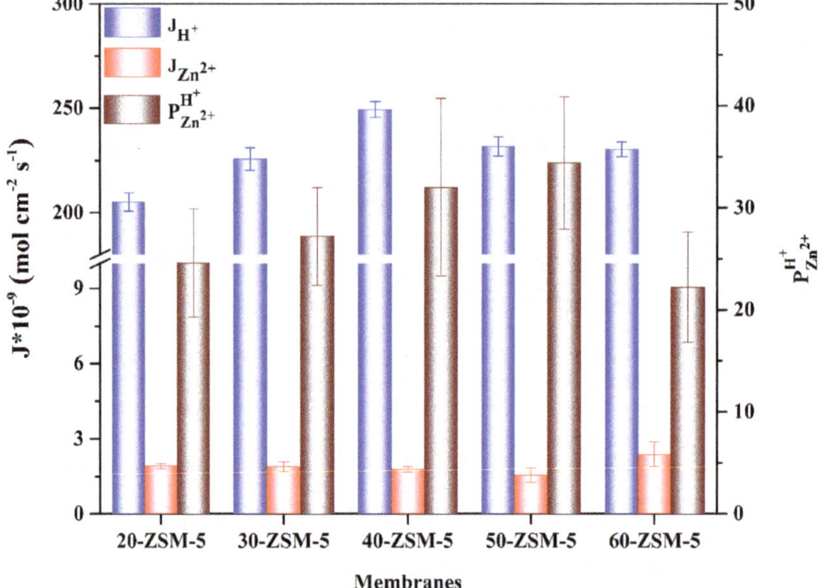

Figure 8. The ion flux and perm-selectivity of ZSM-5/PVA-based MMMs for H^+/Zn^{2+}.

Figure 8 depicts the ions fluxes and perm-selectivity of the modified membranes in a mixture solution (H^+/Zn^{2+} system). The low applied current density (30 mA cm^{-2}) was chosen to avoid the limiting current density's complications. In these experiments, we did not test the PVA base membrane due to the high membrane resistance and no noticeable perm-selectivity. Supplementing the membrane pattern with 20% of ZSM-5, the H^+ flux was approached to 2.05×10^{-7} mol cm^{-2} s^{-1}, while the Zn^{2+} flux was marked to 1.92×10^{-9} mol cm^{-2} s^{-1} and the perm-selectivity was recorded up to 24.6. The high ion flux and perm-selectivity of the ZSM-5 modified membranes could be entirely ascribed to the ZSM-5, which contributed ion-exchange localities to the membrane profile. When the ZSM-5 amount was increased from 20% to 40%, the H^+ flux was increased, i.e., 2.49×10^{-7} mol cm^{-2} s^{-1}, while the Zn^{2+} flux was dropped from 1.92×10^{-9} mol cm^{-2} s^{-1} to 1.72×10^{-9} mol cm^{-2} s^{-1} and the perm-selectivity was further improved to 32.

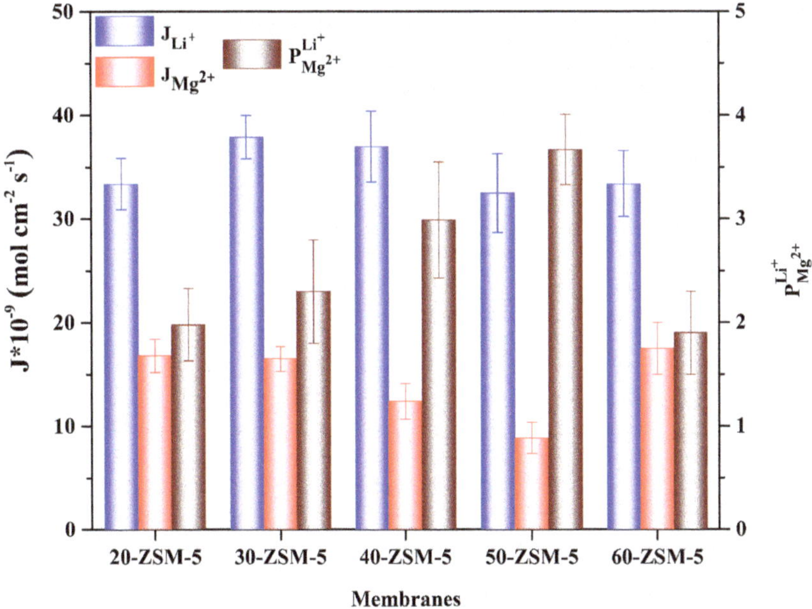

Figure 9. The ion flux and perm-selectivity of ZSM-5/PVA-based MMMs for Li$^+$/Mg^{2+}.

The H$^+$ with the lowest hydrated radius of less than 0.24 nm can efficiently progress through the porous ZSM-5 materials with a pore diameter of about 0.6 nm. Zn^{2+} has high charge density and sustains a larger hydrated radius of 0.6 nm and were screened-out due to the size exclusion (SE) effect. When the ZSM-5 was further increased to 50%, both cations fluxes (H$^+$ flux is less decreased) were decreased because of the more compact structure of membranes. However, the perm-selectivity was increased to 34.4. By, subsequently, increasing the ZSM-5 contents up to 60%, the perm-selectivity was dropped to 22 due to the ineffective adhesion among the PVA and ZSM-5, which produced non-selective voids at the ZSM-5/polymer interface (SEM results) and decreased the membrane perm-selectivity.

We also explored the performance of membranes for the Li$^+$/Mg^{2+} system using the same operational conditions (30 mA cm^{-2}). The adsorption characteristics of the ZSM-5 material exhibited high adsorption selectivity to the monovalent cations. Hence, it is essential to screen-out the divalent cations using the ZSM-5 in the membrane materials for particular applications. As shown in Figure 9, the Li$^+$ ion flux progressed to 3.34 × 10^{-8} mol cm^{-2} s^{-1}, while the lower flux of Mg^{2+}, i.e., 1.68 × 10^{-8} mol cm^{-2} s^{-1}, rationalized the rejection property of membranes (perm-selectivity = 1.98). Likewise, by increasing the quantity of ZSM-5 up to 50% (50-ZSM-5), one can notice insignificant shrinkage in the flux of the monovalent cation, while the flux of divalent cations drastically descended to 8 × 10^{-9} mol cm^{-2} s^{-1} and delivered high perm-selectivity of 3.7. However, further expanding the value of ZSM-5 was not effective, and may create membranes with defects and degrade the perm-selectivity. The high flux of monovalent cations can be characterized by ion size exclusion and the selective replacement of H$^+$ with Li$^+$ ions in comparison with the Mg^{2+} ion, as demonstrated earlier in the adsorption part. The H$^+$ can quickly get through the pores due to the smaller hydrated Stokes radius to obtain high flux. However, the Li$^+$ possesses a lower hydration number and hydration energy (ΔG_{hyd}^0), i.e., 5.2 and −475 kJ mol^{-1} as compared with the Mg^{2+} (10.0, -1830 kJ mol^{-1}) [51]. Therefore, the Li$^+$ can easily de-hydrate to permeate across the pores of ZSM-5.

In view of the ED results, ZSM-5/PVA membranes performed well in both acid recovery and Li+ recovery, respectively. The achieved results are encouraging when compared with the commercial (CSO) and literature reported membranes, as listed in Table 1. The satisfactory performance of prepared

MMMs can be accredited to the cationic exchange sites and crystalline structure. Our membranes exhibited high limiting current density, which is profitable for industrial applications.

Table 1. Characteristics' properties of recently reported membranes and ZSM-5/PVA-based MMMs.

Membranes	Morphology	Systems	Perm-Selectivity	[Ref.]
CSO	dense	H^+/Zn^{2+}, Li^+/Mg^{2+}	3.5, 1.6	[40,52]
SBQAPPO	dense	H^+/Zn^{2+}, Na^+/Mg^{2+}	23.5, 7.4	[52]
Neosepta CMX	dense	Na^+/Mg^{2+}	1.6	[41]
Asymmetric porous	porous	Na^+/Mg^{2+}	3.3	[2]
DL-2540 NF	porous	Li^+/Mg^{2+}	3.3	[20]
ZSM-5/PVA-based MMMs	porous	H^+/Zn^{2+}, Li^+/Mg^{2+}	34.4, 3.7	This work

4. Conclusions

In the present research, the impact of the ZSM-5 zeolite-based membranes on the selective removal of monovalent cations was investigated. For this purpose, we preferred a PVA polymer as a substrate and then consolidated various amounts of ZSM-5 using an economical design to prepare permselective membranes. Initially, the adsorption for the monovalent cation, porosity by BET, and crystalline feature by XRD were exclusively scrutinized, which validated the usage of the ZSM-5 for membranes application and ions separation. The choice of the PVA polymer was productive to establish a strong interaction with the ZSM-5 and to provide excellent dispersion media. Moreover, the SEM micrographs revealed an excellent dispersion of the ZSM-5 by creating permeation pathways without any aggregation for a controlled quantity. The controlled amount of ZSM-5 (40-ZSM-5 and 50-ZSM-5) produced a significant enhancement in both monovalent cation permeability and perm-selectivity. The physiochemical characteristics such as WU, area swelling, and membrane resistance results revealed that the ZSM-5 doping has a substantial impact on desired properties. The MMMs exhibited very high limiting current density, which indicates more applicability and flexibility in the practical application. The conferred strategy is very adaptable for the synthesis of MMMs for a wide variety of zeolites and polymer combinations for envisioned purposes.

Author Contributions: Conceptualization, F.S., L.G. and T.X.; Data curation, L.G.; Funding acquisition, L.G. and T.X.; Investigation, F.S., N.U.A. and Y.Z.; Methodology, Y.Z. and T.X.; Project administration, L.G. and T.X.; Resources, T.X.; Supervision, L.G. and T.X.; Writing—original draft, F.S.; Writing—review & editing, N.U.A., L.G. and T.X. All authors have read and agreed to the published version of the manuscript.

Funding: This research was funded by National Natural Science Foundation of China, grant number 21878282; Key Technologies R & D Program of Anhui Province, grant number 17030901079.

Acknowledgments: This work was partially carried out at the USTC center for Micro- and Nanoscale Research and Fabrication.

Conflicts of Interest: The authors declare no conflict of interest.

References

1. Luo, T.; Abdu, S.; Wessling, M. Selectivity of ion exchange membranes: A review. *J. Membr. Sci.* **2018**, *555*, 429–454. [CrossRef]
2. Hou, L.; Wu, B.; Yu, D.; Wang, S.; Shehzad, M.A.; Fu, R.; Liu, Z.; Li, Q.; He, Y.; Afsar, N.U.; et al. Asymmetric porous monovalent cation perm-selective membranes with an ultrathin polyamide selective layer for cations separation. *J. Membr. Sci.* **2018**, *557*, 49–57. [CrossRef]
3. Amor, Z.; Bariou, B.; Mameri, N.; Taky, M.; Nicolas, S.; Elmidaoui, A. Fluoride removal from brackish water by electrodialysis. *Desalination* **2001**, *133*, 215–223. [CrossRef]
4. Lee, H.-J.; Sarfert, F.; Strathmann, H.; Moon, S.-H. Designing of an electrodialysis desalination plant. *Desalination* **2002**, *142*, 267–286. [CrossRef]
5. Strathmann, H. Electrodialysis, a mature technology with a multitude of new applications. *Desalination* **2010**, *264*, 268–288. [CrossRef]

6. Ahdab, Y.D.; Rehman, D. Brackish water desalination for greenhouses: Improving groundwater quality for irrigation using monovalent selective electrodialysis reversal. *J. Membr. Sci.* **2020**, 118072. [CrossRef]

7. Davis, J.R.; Chen, Y.; Baygents, J.C.; Farrell, J. Production of Acids and Bases for Ion Exchange Regeneration from Dilute Salt Solutions Using Bipolar Membrane Electrodialysis. *ACS Sustain. Chem. Eng.* **2015**, *3*, 2337–2342. [CrossRef]

8. Ghaloussi, R.; Garcia-Vasquez, W.; Chaabane, L.; Dammak, L.; Larchet, C.; Deabatea, S.; Nevakshenova, E.; Nikonenko, V.; Grande, D. Ageing of ion-exchange membranes in electrodialysis: A structural and physicochemical investigation. *J. Membr. Sci.* **2013**, *436*, 68–78. [CrossRef]

9. Scarazzato, T.; Panossian, Z.; Tenório, J.; Pérez-Herranz, V.; Espinosa, D. A review of cleaner production in electroplating industries using electrodialysis. *J. Clean. Prod.* **2017**, *168*, 1590–1602. [CrossRef]

10. Giacalone, F.; Catrini, P.; Tamburini, A.; Cipollina, A.; Piacentino, A.; Micale, G. Exergy analysis of reverse electrodialysis. *Energy Convers. Manag.* **2018**, *164*, 588–602. [CrossRef]

11. Mei, Y.; Tang, C.Y. Recent developments and future perspectives of reverse electrodialysis technology: A review. *Desalination* **2018**, *425*, 156–174. [CrossRef]

12. Bazinet, L.; Lamarche, F.; Ippersiel, D. Bipolar-membrane electrodialysis: Applications of electrodialysis in the food industry. *Trends Food Sci. Technol.* **1998**, *9*, 107–113. [CrossRef]

13. Nemati, M.; Hosseini, S.M.; Shabanian, M. Novel electrodialysis cation exchange membrane prepared by 2-acrylamido-2-methylpropane sulfonic acid; heavy metal ions removal. *J. Hazard. Mater.* **2017**, *337*, 90–104. [CrossRef] [PubMed]

14. López, J.; Reig, M.; Gibert, O.; Cortina, J. Increasing sustainability on the metallurgical industry by integration of membrane nanofiltration processes: Acid recovery. *Sep. Purif. Technol.* **2019**, *226*, 267–277. [CrossRef]

15. Ji, P.-Y.; Ji, Z.; Chen, Q.-B.; Liu, J.; Zhao, Y.-Y.; Wang, S.-Z.; Li, F.; Yuan, J. Effect of coexisting ions on recovering lithium from high Mg2+/Li+ ratio brines by selective-electrodialysis. *Sep. Purif. Technol.* **2018**, *207*, 1–11. [CrossRef]

16. Nayar, K.G.; Fernandes, J.; McGovern, R.K.; Dominguez, K.P.; McCance, A.; Al-Anzi, B. Cost and energy requirements of hybrid RO and ED brine concentration systems for salt production. *Desalination* **2019**, *456*, 97–120. [CrossRef]

17. Kim, S.; Joo, H.; Moon, T.; Kim, S.-H.; Yoon, J. Rapid and selective lithium recovery from desalination brine using an electrochemical system. *Environ. Sci. Process. Impacts* **2019**, *21*, 667–676. [CrossRef]

18. Liu, X.; Chen, X.; He, L.; Zhao, Z.-W. Study on extraction of lithium from salt lake brine by membrane electrolysis. *Desalination* **2015**, *376*, 35–40. [CrossRef]

19. Wen, X.; Ma, P.; Zhu, C.; He, Q.; Deng, X. Preliminary study on recovering lithium chloride from lithium-containing waters by nanofiltration. *Sep. Purif. Technol.* **2006**, *49*, 230–236. [CrossRef]

20. Sun, S.-Y.; Cai, L.-J.; Nie, X.-Y.; Song, X.; Yu, J. Separation of magnesium and lithium from brine using a Desal nanofiltration membrane. *J. Water Process. Eng.* **2015**, *7*, 210–217. [CrossRef]

21. Zhang, Y.; Paepen, S.; Pinoy, L.; Meesschaert, B.; Van Der Bruggen, B. Selectrodialysis: Fractionation of divalent ions from monovalent ions in a novel electrodialysis stack. *Sep. Purif. Technol.* **2012**, *88*, 191–201. [CrossRef]

22. Nie, X.-Y.; Sun, S.-Y.; Sun, Z.; Song, X.; Yu, J. Ion-fractionation of lithium ions from magnesium ions by electrodialysis using monovalent selective ion-exchange membranes. *Desalination* **2017**, *403*, 128–135. [CrossRef]

23. Afsar, N.U.; Ji, W.; Wu, B.; Shehzad, M.A.; Ge, L.; Xu, T. SPPO-based cation exchange membranes with a positively charged layer for cation fractionation. *Desalination* **2019**, *472*, 114145. [CrossRef]

24. Safronova, E.Y.; Golubenko, D.; Shevlyakova, N.; D'Yakova, M.; Tverskoi, V.; Dammak, L.; Grande, D.; Yaroslavtsev, A. New cation-exchange membranes based on cross-linked sulfonated polystyrene and polyethylene for power generation systems. *J. Membr. Sci.* **2016**, *515*, 196–203. [CrossRef]

25. Roghmans, F.; Evdochenko, E.; Martí-Calatayud, M.; Garthe, M.; Tiwari, R.; Walther, A.; Wessling, M. On the permselectivity of cation-exchange membranes bearing an ion selective coating. *J. Membr. Sci.* **2020**, *600*, 117854. [CrossRef]

26. Sulaiman, K.O.; Sajid, M.; Alhooshani, K. Application of porous membrane bag enclosed alkaline treated Y-Zeolite for removal of heavy metal ions from water. *Microchem. J.* **2020**, *152*, 104289. [CrossRef]

27. Ji, W.; Afsar, N.U.; Wu, B.; Sheng, F.; Shehzad, M.A.; Ge, L.; Xu, T. In-situ crosslinked SPPO/PVA composite membranes for alkali recovery via diffusion dialysis. *J. Membr. Sci.* **2019**, *590*, 117267. [CrossRef]

28. Wang, Z.; Yu, J.; Xu, R. Needs and trends in rational synthesis of zeolitic materials. *Chem. Soc. Rev.* **2012**, *41*, 1729–1741. [CrossRef]

29. Chen, Z.; Holmberg, B.; Li, W.; Wang, X.; Deng, W.; Munoz, R.; Yan, Y. Nafion/Zeolite Nanocomposite Membrane by in Situ Crystallization for a Direct Methanol Fuel Cell. *Chem. Mater.* **2006**, *18*, 5669–5675. [CrossRef]

30. Csicsery, S.M. Shape-selective catalysis in zeolites. *Zeolites* **1984**, *4*, 202–213. [CrossRef]

31. Yuan, Z.; Zhu, X.; Li, M.; Lu, W.; Li, X.; Zhang, H. A Highly Ion-Selective Zeolite Flake Layer on Porous Membranes for Flow Battery Applications. *Angew. Chem. Int. Ed.* **2016**, *55*, 3058–3062. [CrossRef] [PubMed]

32. Moliner, M.; Martínez, C.; Corma, A. Multipore Zeolites: Synthesis and Catalytic Applications. *Angew. Chem. Int. Ed.* **2015**, *54*, 3560–3579. [CrossRef] [PubMed]

33. Nur, H.; Kee, G.L.; Hamdan, H.; Mahlia, T.M.I.; Efendi, J.; Metselaar, H.S.C. Organosulfonic acid functionalized zeolite ZSM-5 as temperature tolerant proton conducting material. *Int. J. Hydrog. Energy* **2012**, *37*, 12513–12521. [CrossRef]

34. Qiao, X.; Chung, T.-S.; Rajagopalan, R.; Chung, T.-S. Zeolite filled P84 co-polyimide membranes for dehydration of isopropanol through pervaporation process. *Chem. Eng. Sci.* **2006**, *61*, 6816–6825. [CrossRef]

35. Dong, H.; Zhao, L.; Zhang, L.; Chen, H.; Gao, C.; Ho, W.W. High-flux reverse osmosis membranes incorporated with NaY zeolite nanoparticles for brackish water desalination. *J. Membr. Sci.* **2015**, *476*, 373–383. [CrossRef]

36. Fathizadeh, M.; Aroujalian, A.; Raisi, A. Effect of added NaX nano-zeolite into polyamide as a top thin layer of membrane on water flux and salt rejection in a reverse osmosis process. *J. Membr. Sci.* **2011**, *375*, 88–95. [CrossRef]

37. Hamid, S.A.; Azha, S.F.; Sellaoui, L.; Bonilla-Petriciolet, A. Suzylawati Adsorption of copper (II) cation on polysulfone/zeolite blend sheet membrane: Synthesis, characterization, experiments and adsorption modelling. *Colloids Surfaces A: Physicochem. Eng. Asp.* **2020**, 124980. [CrossRef]

38. Raso, R.; Tovar, M.; Lasobras, J.; Herguido, J.; Kumakiri, I.; Araki, S.; Menéndez, M. Zeolite membranes: Comparison in the separation of H2O/H2/CO2 mixtures and test of a reactor for CO2 hydrogenation to methanol. *Catal. Today* **2020**. [CrossRef]

39. Liu, B.; Kita, H.; Yogo, K. Preparation of Si-rich LTA zeolite membrane using organic template-free solution for methanol dehydration. *Sep. Purif. Technol.* **2020**, *239*, 116533. [CrossRef]

40. Sheng, F.; Hou, L.; Wang, X.; Irfan, M.; Shehzad, M.A.; Wu, B.; Ren, X.; Ge, L.; Xu, T. Electro-nanofiltration membranes with positively charged polyamide layer for cations separation. *J. Membr. Sci.* **2020**, *594*, 117453. [CrossRef]

41. Abdu, S.; Martí-Calatayud, M.-C.; Wong, J.E.; García-Gabaldón, M.; Wessling, M. Layer-by-Layer Modification of Cation Exchange Membranes Controls Ion Selectivity and Water Splitting. *ACS Appl. Mater. Interfaces* **2014**, *6*, 1843–1854. [CrossRef] [PubMed]

42. Li, P.; Chen, H.; Schott, J.A.; Li, B.; Zheng, Y.-P.; Mahurin, S.M.; Jiang, D.-E.; Cui, G.; Hu, X.; Wang, Y.; et al. Porous liquid zeolites: Hydrogen bonding-stabilized H-ZSM-5 in branched ionic liquids. *Nanoscale* **2019**, *11*, 1515–1519. [CrossRef] [PubMed]

43. Jambulingam, R.; Jambulingam, R. Waste crab shell derived CaO impregnated Na-ZSM-5 as a solid base catalyst for the transesterification of neem oil into biodiesel. *Sustain. Environ. Res.* **2017**, *27*, 273–278. [CrossRef]

44. Ali, F.; Maiz, F. Structural, optical and AFM characterization of PVA:La^{3+} polymer films. *Phys. B Condens. Matter* **2018**, *530*, 19–23. [CrossRef]

45. Puguan, J.M.C.; Kim, H.-S.; Lee, K.-J.; Kim, H. Low internal concentration polarization in forward osmosis membranes with hydrophilic crosslinked PVA nanofibers as porous support layer. *Desalination* **2014**, *336*, 24–31. [CrossRef]

46. Park, M.J.; Gonzales, R.R.; Abdel-Wahab, A.; Phuntsho, S.; Shon, H.K. Hydrophilic polyvinyl alcohol coating on hydrophobic electrospun nanofiber membrane for high performance thin film composite forward osmosis membrane. *Desalination* **2018**, *426*, 50–59. [CrossRef]

47. Zhang, Q.; Li, B.; Sun, D.; Yang, P. Preparation and characterization of PVA membrane modified by water-soluble hyperbranched polyester (WHBP) for the dehydration of n-butanol. *J. Appl. Polym. Sci.* **2016**, *133*, 133. [CrossRef]

48. Wu, Y.; Luo, J.; Zhao, L.; Zhang, G.; Wu, C.; Xu, T. QPPO/PVA anion exchange hybrid membranes from double crosslinking agents for acid recovery. *J. Membr. Sci.* **2013**, *428*, 95–103. [CrossRef]

49. Li, J.; Yuan, S.; Wang, J.; Zhu, J.; Shen, J.; Van Der Bruggen, B. Mussel-inspired modification of ion exchange membrane for monovalent separation. *J. Membr. Sci.* **2018**, *553*, 139–150. [CrossRef]
50. Liu, H.; Ruan, H.; Zhao, Y.; Pan, J.; Sotto, A.; Gao, C.; Van Der Bruggen, B.; Shen, J. A facile avenue to modify polyelectrolyte multilayers on anion exchange membranes to enhance monovalent selectivity and durability simultaneously. *J. Membr. Sci.* **2017**, *543*, 310–318. [CrossRef]
51. Marcus, Y. Thermodynamics of solvation of ions. Part 5.—Gibbs free energy of hydration at 298.15 K. *J. Chem. Soc. Faraday Trans.* **1991**, *87*, 2995–2999. [CrossRef]
52. He, Y.; Ge, L.; Ge, Z.; Zhao, Z.; Sheng, F.; Liu, X.; Ge, X.; Yang, Z.; Fu, R.; Liu, Z.; et al. Monovalent cations permselective membranes with zwitterionic side chains. *J. Membr. Sci.* **2018**, *563*, 320–325. [CrossRef]

 *membranes*

Article

Potentiodynamic and Galvanodynamic Regimes of Mass Transfer in Flow-Through Electrodialysis Membrane Systems: Numerical Simulation of Electroconvection and Current-Voltage Curve

Aminat Uzdenova [1,*] and Makhamet Urtenov [2]

[1] Department of Computer Science and Computational Mathematics, Federal State Budgetary Educational Institution of Higher Education "Umar Aliev Karachai-Cherkess State University", 369202 Karachaevsk, Russia
[2] Department of Applied Mathematics, Federal State Budgetary Educational Institution of Higher Education "Kuban State University", 350040 Krasnodar, Russia; urtenovmax@mail.ru
* Correspondence: uzd_am@mail.ru

Received: 12 February 2020; Accepted: 19 March 2020; Published: 20 March 2020

Abstract: Electromembrane devices are usually operated in two electrical regimes: potentiodynamic (PD), when a potential drop in the system is set, and galvanodynamic (GD), when the current density is set. This article theoretically investigates the current-voltage curves (CVCs) of flow-through electrodialysis membrane systems calculated in the PD and GD regimes and compares the parameters of the electroconvective vortex layer for these regimes. The study is based on numerical modelling using a basic model of overlimiting transfer enhanced by electroconvection with a modification of the boundary conditions. The Dankwerts' boundary condition is used for the ion concentration at the inlet boundary of the membrane channel. The Dankwerts' condition allows one to increase the accuracy of the numerical implementation of the boundary condition at the channel inlet. On the CVCs calculated for PD and DG regimes, four main current modes can be distinguished: underlimiting, limiting, overlimiting, and chaotic overlimiting. The effect of the electric field regime is manifested in overlimiting current modes, when a significant electroconvection vortex layer develops in the channel.

Keywords: ion-exchange membrane; electrodialysis; current-voltage curve; electroconvection; potentiodynamic regime; galvanodynamic regime; numerical simulation

1. Introduction

Flow-through electrodialysis (ED) membrane cells are widely used in water purification and the processing of agricultural products (milk, wine, etc.) [1–4]. Electromembrane systems are described by a nonlinear current-voltage curve (CVC), owing largely to the phenomena of concentration polarization, current-induced convection, and water dissociation [5,6]. For dilute electrolyte solutions considered in this article, the main mechanism of overlimiting transfer is electroconvection, as shown by experimental [7–13] and theoretical studies [14–19]. It is customary to distinguish three modes on the CVC of membrane system (Figure 3):

(1) The underlimiting current (ohmic behavior) is the initial linear region of the CVC, which is characterized by a rather high concentration of ions in the region near the membrane. When an electric current flows through the ion-exchange membrane, the ion concentration decreases on one side of the membrane and increases on the other due to the selective transfer of counterions in the membrane (ion concentration polarization). With the increase in the potential drop,

almost complete depletion of ions in the region at the membrane surface in the channel of desalination and the transition of the system to the limiting state are observed [20,21].

(2)   The limiting current is a section of the CVC with a small slope (plateau), which describes the saturation of the current corresponding to the almost complete depletion of ions at the membrane surface [22,23].

(3)   The overlimiting current is the region of secondary current growth: with a further increase in the applied potential drop, the current takes on values greater than the limiting. The increase in the electric current essentially indicates an increase in the conductivity of the depleted region. For dilute electrolyte solutions, electroconvection is the main process that partially destroys the depleted region [7–19]. Electroconvection is the entrainment of liquid molecules by ions that form a space charge at the ion-selective surface under the influence of the electric force [24]. The intensity of electroconvection increases significantly with the passage of the overlimiting current when an extended macroscopic space charge region (SCR) is formed at the interface due to the polarization of the electric double layer (EDL) (Figures 1 and 2).

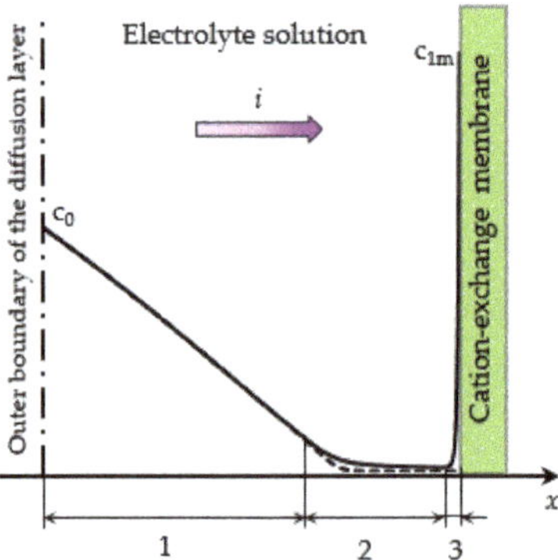

Figure 1. Schematic concentration profiles of cations (c_1, the solid line) and anions (c_2, the dashed line) in the diffusion layer adjacent to the surface of a cation-exchange membrane (CEM) [25]. The current density i is flowing across the system; the electrolyte concentration in the bulk solution, c_0; the cation concentration at the solution/CEM boundary, c_{1m}; different diffusion layer regions are shown: the electroneutral region (**1**), the extended SCR (**2**) and the quasi-equilibrium electric double layer (**3**), respectively.

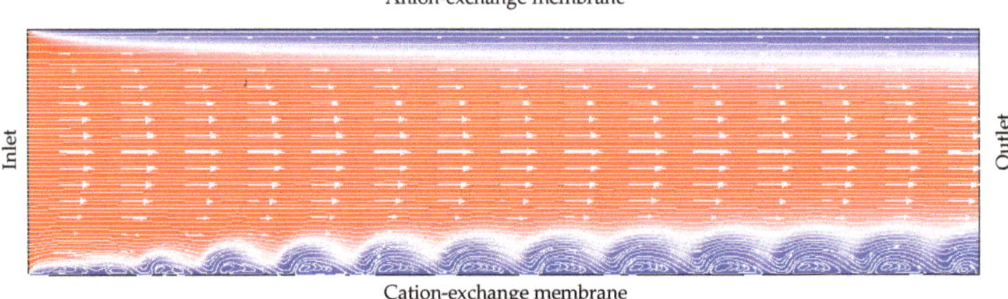

Figure 2. Scheme of the flow of an electrolyte solution in the channel between the anion-exchange and cation-exchange membranes, with taking into account the forced flow (shown by arrows) and the development of an electroconvective vortex layer (at the CEM surface). Ion depletion zones are shown in blue. Based on [19].

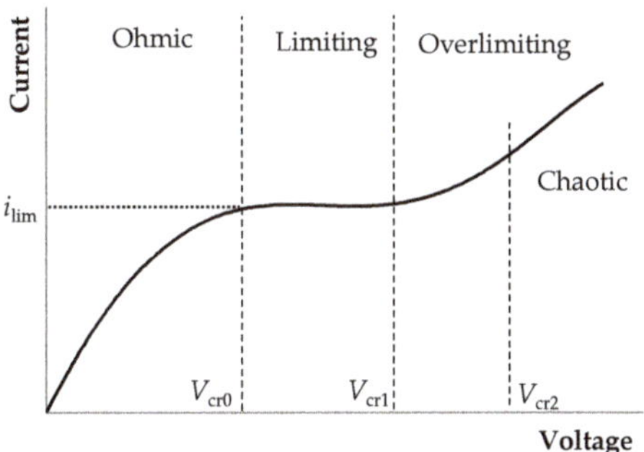

Figure 3. Sketch of a typical current-voltage curve (CVC) of an ion-exchange membrane. The dashed lines V_{cr0}, V_{cr1}, V_{cr2} indicate changes in the CVC regions: underlimiting current, plateau of the limiting current ($i_{\lim}$), overlimiting, overlimiting with chaotic oscillations.

The existence of the extended SCR at the electrolyte solution/membrane interface, which is much larger than the region of the equilibrium EDL when a sufficiently high voltage is applied, was first shown by I. Rubinstein and L. Shtilman on the basis of a numerical solution of the Nernst-Planck and Poisson equations for the potential of the electric field [25].

Later, I. Rubinstein and B. Zaltzman [14] developed a model for describing mass transfer in a diffusion layer at a homogeneous ion-exchange membrane. They found a numerical solution for the Nernst-Planck-Poisson and Navier-Stokes equations under the assumption of local electroneutrality in the solution outside the SCR and using a special condition of electroosmotic slip at the interface of the electroneutral region with the SCR. It was shown that the heterogeneity of the surface is not a necessary condition for the emergence of electroconvection. A characteristic feature of this system is its hydrodynamic instability at sufficiently high potential drop. Several threshold potential drops were established, which separate different phases in the development of electrokinetic instability.

Approaches to electroconvection modelling using the slip condition at the boundary with the SCR were applied by V. Dydek et al. [26], R. Abu-Rjal et al. [27]. Models based on the Nernst-Planck-Poisson and Navier-Stokes equations that directly take into account the formation of the extended SCR were considered in the works of E.A. Demekhin et al. [15,28,29], S.V. Pham et al. [16,30], and K. Druzgalski, E. Karatay et al. [18,31], P. Magnico [32,33]. Numerical studies of electroconvection flows generated at an electrically heterogeneous membrane surface were carried out by S. Davidson et al. [34], M. Andersen et al. [35], V.A. Kirii et al. [36].

The difference in the CVCs of the ion-exchange membrane without a forced flow during the transition between the limiting and overlimiting current regimes at the increasing and decreasing potential drop was theoretically described in [11,16,28,32]. S.V. Pham et al. examined a wavy membrane and explained hysteretic behavior by the fact that in the decreasing regime the existing depletion zone creates a lateral gradient, which creates a high lateral electric field. Thus, an additional lateral volumetric force is created to maintain the vortex flow. E.A. Demekhin et al. [28] investigated the hysteresis behavior of ideal smooth ion-exchange membranes and showed the dependence of the hysteresis amplitude on the coupling coefficient between the hydrodynamics and the electrostatics. Hysteretic amplitude calculations observed by Demekhin et al. has been confirmed by P. Magnico [32].

Two electroconvection kinds can be distinguished in overlimiting current modes in membrane systems. In the case of a curved or electrically heterogeneous surface, the tangential electric field causes a stable electroosmotic transfer, described in the works of S.S. Dukhin and N.A. Mishchuk [37–39]. In the case of homogeneous membranes in the absence of forced fluid flow, such a kind is not realized: electroconvection appears as a result of hydrodynamic instability, as shown by I. Rubinstein and B. Zaltzman [14]. These two kinds are sometimes termed as electrokinetic modes, respectively, of Dukhin and Rubinstein [40].

Electroconvection in ED channels with forced fluid flow was investigated by M.Kh. Urtenov et al. [19,41] and R. Kwak et al. [17,42], R. Abu-Rjal et al. [27], P. Magnico [33]. In such channels, the concentration is distributed unevenly along the length of the channel: as the solution moves between the membranes, the electrolyte concentration decreases and the thickness of the diffusion layer increases. In this case, a tangential bulk electric force is formed, that acts on the SCR at the depleted surface of the membrane, even if the membrane is homogeneous. This force causes stationary electroconvection even at underlimiting current densities. According to the terminology of S.S. Dukhin and N.A. Mishchuk [37–39], this type of electroconvection can be considered as an electroosmosis of the first kind. The bulk force is localized at a relatively small distance from the membrane, where viscous forces play an important role due to the adhesion condition. The contribution of electroconvection to the increase of current becomes significant only at the potential drop corresponding to overlimiting currents. In this case, the SCR thickness increases sharply in comparison with the thickness of the equilibrium double layer. At such distances, the role of viscous forces decreases. Therefore, the main contribution to the development of overlimiting transport belongs to the electroosmosis of the second kind. This mode is similar to the Dukhin-Mishchuk mode described above. Nevertheless, it differs in that in the presence of the forced flow, the tangential force necessary for the occurrence of electroconvection arises due to the inhomogeneity of the longitudinal distribution of concentration, and not due to the electrical inhomogeneity of the surface.

For systems with forced flow at the threshold potential drop, V_{cr1} (Figure 3), single electroconvective vortices rotating in the same direction are formed in the region near the membrane surface (Figure 2) [17,19]. Vortices mix the electrolyte solution in the area near the membrane, which partially destroys the depletion layer and provides the regime of overlimiting current. Due to the forced flow, the vortices move along the solution/membrane interface towards the channel outlet. This movement of the vortices causes current density fluctuations on the CVC [19]. As the potential drop increases, the size of the vortices increases; at a certain potential drop (V_{cr2} on Figure 3) single vortices transform into large vortex complexes consisting of several vortices rotating in opposite directions [19,33]. As a result, the amplitude of the current density oscillations (or potential drop) increases and oscillations become chaotic [19]. P. Magnico investigated the role of electroconvective vortices in the fluid motion using the Lagrangian approach [33]. In this way, trajectories were constructed that reflect the ejection from the mixing layer, trapping by a growing vortex or merging vortices.

The electrical regime in membrane devices as a rule is determined in two ways: potentiodynamic (PD), when a potential drop in the system is set (constant, linearly increasing, periodically changing in time, etc.), and galvanodynamic (GD) when the current density is set (constant, linearly increasing, periodically changing in time, etc.).

Theoretical studies of transport processes taking into account the formation of the extended SCR and the development of electroconvection in membrane systems were mainly carried out for the PD regime using the equations of Navier-Stokes, Nernst-Planck and Poisson for the electric field potential [14–19,28,32,33]. The description of the GD regime caused difficulties associated with the absence of a differential equation for the current density. One approach to describing the ion transport in the membrane system in the GD regime is the decomposition of the system of Nernst-Planck and Poisson equations based on the assumption of local electroneutrality of the electrolyte solution [43,44]. In this approach, the distribution of a current density in the system is obtained using the electric current stream function. However, approaches based on the local electroneutrality assumption do not allow taking explicitly into account the effect of the SCR, which is formed at the solution/membrane boundary. Recently, these difficulties have been overcome using an approach involving the solution of the Poisson equation with a boundary condition determining the potential gradient through the current density [45,46].

This article presents numerical calculations of the CVCs and the hydrodynamic response of the electrolyte solution in flow-through membrane systems in the PD and GD regimes of the electric field.

The structure of the electroconvective vortex layer is compared for these modes. For the first time, the hysteresis amplitude is calculated for flow-through systems in the PD and GD regimes.

2. Mathematical Models

CVCs were calculated:

(1) for the PD regime, when the potential drop, $\Delta\varphi$, is set to increase from 0 to a certain value, then to decrease from this value to 0:

$$\Delta\varphi = \begin{cases} \alpha t, & t \leq t_1, \\ 2\alpha t_1 - \alpha t, & t > t_1, \end{cases} \qquad (1)$$

where $\alpha > 0$ is the potential sweep speed, t_1 is the point in time at which the regime of the increasing potential drop is replaced by decreasing regime.

(2) for the GD regime, when the average current density, i_{av}, is set to increases from 0 to a certain value, then to decrease from this value to 0:

$$i_{av} = \begin{cases} \beta t, & t \leq t_2, \\ 2\beta t_2 - \beta t, & t > t_2, \end{cases} \qquad (2)$$

where $\beta > 0$ is the sweep speed of the current density, t_2 is the point in time at which the regime of increasing current density is replaced by decreasing.

The calculations are based on the 2D mathematical models of the overlimiting transfer enhanced by electroconvection in a flow-through ED cell for the PD [19,47] and GD [46] regimes. To simplify the numerical solution, we consider the processes in half of the ED channel at the surface of the cation-exchange membrane (CEM), Figure 4. Let x and y be the transverse and longitudinal coordinates, respectively; $x = 0$ relates to the middle of the ED channel, $x = h$ is the electrolyte solution/CEM interface; $y = 0$ corresponds to the inlet and $y = l$ to the outlet of the channel.

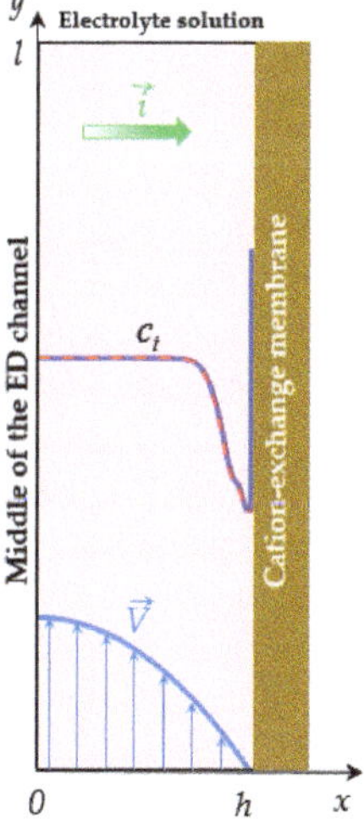

Figure 4. Scheme of the system under consideration: half of the desalination electrodialysis (ED) cell adjacent to CEM. Schematic concentration profiles of cations (c_1, solid line) and anions (c_2, dashed line), direction of the electric current $\vec{i}$, forced flow velocity $\vec{V}$ are shown.

2.1. Formulation for the PD Regime

The non-stationary process of transfer of binary electrolyte ions in membrane systems in the absence of chemical reactions, with taking into account electroconvection, is written as follows [14–19]:

$$\frac{\partial \vec{V}}{\partial t} + (\vec{V}\nabla)\vec{V} = -\nabla p + \frac{1}{Re}\Delta \vec{V} + K_{el}\Delta\varphi\nabla\varphi, \ \text{div}\,\vec{V} = 0 \tag{3}$$

$$\vec{j}_i = -z_i D_i c_i \nabla\varphi - D_i \nabla c_i + Pec_i\vec{V}, \ i = 1,\ 2 \tag{4}$$

$$\frac{\partial c_i}{\partial t} = -\frac{1}{Pe}\,\text{div}\,\vec{j}_i, \ i = 1,\ 2 \tag{5}$$

$$-\varepsilon\Delta\varphi = z_1 c_1 + z_2 c_2, \tag{6}$$

$$\vec{i} = z_1\vec{j}_1 + z_2\vec{j}_2 - \varepsilon\,Pe\,\frac{\partial}{\partial t}(\nabla\phi) \tag{7}$$

Equations (3)–(7) are given in dimensionless form. We scale time, t, by the value h/V_0; spatial coordinates, x and y, by the thickness of the considered region h (half of the ED channel thickness); velocity, $\vec{V}$, by the average velocity of the forced flow V_0; pressure, p, by the value ρV_0^2; concentration of the i-th ion, c_i, by the electrolyte concentration in the bulk solution c_0; electric potential, φ, by the value RT/F; individual ion diffusion coefficients, D_1 and D_2, by the electrolyte diffusion coefficient $D = D_1 D_2(z_1 - z_2)/(D_1 z_1 - D_2 z_2)$; current density, $\vec{i}$, by the value $Dc_0 F/h$; ion flux $\vec{j}_i$ by the value Dc_0/h. Here $Re = V_0 h/v$ is the Reynolds number, $Pe = V_0 h/D$ is the Peclet number, $\varepsilon = RT\varepsilon_0\varepsilon_r/(c_0 F^2 h^2) = 2(L_D/h)^2$ and $K_{el} = \varepsilon_0\varepsilon_r R^2 T^2/(\rho_0 V_0^2 F^2 h^2)$ are the dimensionless parameters; z_i is the charge number of the i-th ion; F is the Faraday constant; R is the gas constant; T is the absolute temperature; ε_0 is the dielectric permittivity of vacuum; ε_r is the solution relative permittivity (assumed constant); ρ_0 is the solution density (assumed constant), v is the kinematic viscosity.

$\vec{V}, p, \vec{j}_1, \vec{j}_2, c_1, c_2, \varphi, i_x, i_y$ are unknown function of t, x and y. The Navier-Stokes equations, Equations (3), describe the velocity field under the action of the forced flow and the electric body force. The equations of Nernst-Planck, Equations (4), material balance, Equations (5), and Poisson, Equation (6), describe the ion concentration and potential fields. Equation (7) is a formula for the total current density, including the conduction current, $\vec{i}_c = z_1\vec{j}_1 + z_2\vec{j}_2$, and displacement current, $\vec{i}_d = -\varepsilon\,Pe\,\frac{\partial}{\partial t}(\nabla\varphi)$. For the calculations of this article, the displacement current, i_d, is negligible (less than 10^{-7}).

The system of Equations (3)–(7) is supplemented by the boundary conditions [19,47]. At the channel inlet ($x \in [0, h]$, $y = 0$), the velocity profile is parabolic and satisfies Poiseuille's law (expressions for half of the ED channel):

$$V_x(x,0,t) = 0, \ V_y(x,0,t) = 1.5(1 - x^2). \tag{8}$$

In model from [19], the condition of uniform distribution along x for ion concentration at the channel inlet is accepted:

$$c_i(x,0,t) = 1, \ i = 1,\ 2. \tag{9}$$

In this paper, instead of condition (9), the Danckwerts' boundary condition is used, which determines that arrival rate of ions into the channel is equal to the rate with which they cross the plane $y = 0$ by the combination of flow, electromigration, and diffusion [48]:

$$\left(-z_i D_i c_i \nabla\varphi - D_i \nabla c_i + Pe\,c_i\vec{V}\right)(x,0,t) = Pe\,c\prime\,\vec{V}, i = 1,\ 2, \tag{10}$$

where $c\prime = 1$ is the input electrolyte concentration. The advantage of this condition in comparison with condition (9) is the absence of a special feature of the distribution of ion concentration near the point $(h, 0)$. At the numerical implementation with condition (10), the accuracy of fulfilling the condition that the tangential current density through the inlet vanishes, $i_y(x,0,t) = 0$, is higher than with condition (9) (Appendix A).

The condition for the electric potential is obtained from Equations (4) and (7) considering the zero tangential current density through the inlet, $i_y(x,0,t) = 0$, (the tangential component of the displacement current, $i_{d\,y}$, is negligible):

$$\frac{\partial \varphi}{\partial y}(x,0,t) = -\frac{1}{z_1^2 D_1 + z_2^2 D_2}\left(z_1 D_1 \frac{\partial c_1}{\partial y} + z_2 D_2 \frac{\partial c_2}{\partial y}\right) \tag{11}$$

At the channel outlet ($x \in [0, h]$, $y = l$) the velocity profile is again parabolic; the sum of diffusion and migration tangential components of the cation ($i = 1$) and anion ($i = 2$) fluxes is zero; the tangential derivative of the potential is set to be zero:

$$V_x(x,l,t) = 0, \ V_y(x,l,t) = 1.5(1 - x^2) \tag{12}$$

$$\left(-\frac{\partial c_i}{\partial y} - z_i c_i \frac{\partial \varphi}{\partial y}\right)(x,l,t) = 0, \ i = 1,2 \tag{13}$$

$$\frac{\partial \varphi}{\partial y}(x,l,t) = 0. \tag{14}$$

At $x = 0$, $y \in [0, l]$ (middle of the ED channel) the following conditions are applied:

$$V_x(0,y,t) = 0, \ V_y(0,y,t) = 1.5 \tag{15}$$

$$c_i(0,y,t) = 1, \ i = 1, 2 \tag{16}$$

$$\varphi(0,y,t) = 0. \tag{17}$$

At $x = 1$, $y \in [0, l]$ (the solution/membrane interface), the no-slip condition (18) is applied; the counterion concentration, c_1, is set as a constant value N_c greater than the bulk solution concentration, Equation (19), [25]; continuous flow of co-ions, Equation (20); the potential drop is set, Equation (21):

$$V_x(1,y,t) = 0, \ V_y(1,y,t) = 0 \tag{18}$$

$$c_1(1,y,t) = N_c \tag{19}$$

$$\left(-D_2 \frac{\partial c_2}{\partial x} - z_2 D_2 c_2 \frac{\partial \varphi}{\partial x}\right)(1,y,t) = \frac{(1 - T_1)}{z_2} i_x(1,y,t) \tag{20}$$

$$\varphi(1,y,t) = \Delta\varphi \tag{21}$$

The potential drop, $\Delta\varphi$, is given by Equation (1).

Thus, the formulation of the model for the PD regime includes the system of Equations (3)–(7) and boundary conditions (8), (10)–(21). The average over the channel length current density is calculated as [46]:

$$i_{av} = \frac{1}{l}\int_0^l \int_0^1 i_x dx dy \tag{22}$$

2.2. Formulation for the GD Regime

To describe the GD regime, Equations (3)–(6) and boundary conditions (8), (10)–(20) are used similarly to PD regime, but there are two differences. First, at the boundary $x = 1$, $y \in [0, l]$ (solution/membrane interface), instead of condition (21), normal to the membrane surface component of the electric field strength is specified as function of the electric current density [46]:

$$\frac{\partial \varphi}{\partial x}(1, y, t) = -\left(\frac{\left(i_x + \varepsilon\, Pe \frac{\partial^2 \varphi}{\partial x \partial t} + z_1 D_1 \frac{\partial c_1}{\partial x} + z_2 D_2 \frac{\partial c_2}{\partial x}\right)}{z_1^2 D_1 c_1 + z_2^2 D_2 c_2}\right)(1, y, t) \tag{23}$$

Condition (23) was obtained from Equations (4) and (7) [46,49].

Secondly, an additional equation is introduced to determine the distribution of current density, which is required by the boundary condition (23). For this purpose, the method of electric current flow function is used [43–46]. According to this method, the electric current stream function, η, is determined:

$$i_x = \frac{\partial \eta}{\partial y}, \; i_y = -\frac{\partial \eta}{\partial x} \tag{24}$$

Then the equation and boundary conditions for η are introduced to the mathematical formulation of the model [45,46]:

$$\Delta \eta = -\left(\left(z_1^2 D_1 \frac{\partial c_1}{\partial y} + z_2^2 D_2 \frac{\partial c_2}{\partial y}\right)\frac{\partial \varphi}{\partial x} - \left(z_1^2 D_1 \frac{\partial c_1}{\partial x} + z_2^2 D_2 \frac{\partial c_2}{\partial x}\right)\frac{\partial \varphi}{\partial y}\right) +$$
$$+ Pe\left(z_1 \frac{\partial c_1}{\partial y} + z_2 \frac{\partial c_2}{\partial y}\right)V_x - Pe\left(z_1 \frac{\partial c_1}{\partial x} + z_2 \frac{\partial c_2}{\partial x}\right)V_y + Pe(z_1 c_1 + z_2 c_2)\left(\frac{\partial V_x}{\partial y} - \frac{\partial V_y}{\partial x}\right), \tag{25}$$

$$\frac{\partial \eta}{\partial x}(0, y, t) = 0, \; \frac{\partial \eta}{\partial x}(1, y, t) = 0, \; \eta(x, 0, t) = 0, \; \eta(x, l, t) = i_{av} l \tag{26}$$

The boundary conditions (26) were derived under the simplifying assumption that the current through the channel outlet $i_y(x,l,t) \approx 0$ (due to its smallness, Figure A2). Therefore, average current density, i_{av}, can be used as a parameter determining the electrical regime in the system:

$$i_{av} = \frac{1}{l}\int_0^l i_x(0, y, t)dy = \frac{1}{l}\int_0^l i_x(1, y, t)dy \tag{27}$$

Thus, current density i_x in boundary condition (23) is determined by Formula (24).

Thus, the formulation of the model for the GD regime includes the system of Equations (3)–(6), (25) and boundary conditions (8), (10)–(20), (23) and (26).

2.3. Numerucal Implementation

Numerical solutions were found by the finite element method using Comsol Multiphysics 5.1 software package. The results presented below were obtained using a non-uniform unstructured triangular computational grid consisting of about 55,000 elements. The density of the mesh elements was increased near the solution/membrane boundary: 1000 elements were set using the "Distribution" node. The influence of the quality of the computational mesh was tested by comparing solutions for two meshes consisting of about 41,000 elements (when "Distribution" node set 700 elements on the solution/membrane boundary) and 55,000 elements (with 1000 elements on the boundary). The difference in the values of the threshold potential drop of the transition to the overlimiting current mode (both in increasing and decreasing regimes) did not exceed 2%.

The following modules are used to implement the model for the GD regime: "Laminar flow" for the Navier-Stokes Equation (3); "Transport of Diluted Species" for the anions and cations concentrations fields, Equations (4) and (5); "Poisson's equation" for the electric potential fields, Equation (6); "General form PDE" for the electric current stream function, Equation (25). For spatial discretization of the concentration, potential, and the electric current stream function fields, the quadratic Lagrange interpolation functions are used. The "Laminar flow" module has the "P2 + P1" discretization that means second order elements for the velocity components and linear elements for the pressure field [50].

For time-depended calculations a segregated node with implicit time-stepping method BDF (backward differentiation formulas) is used [50]. One segregated iteration consists of executing two segregated step: in the first step, concentration, potential and electric current stream function are calculated; on the second, speed and pressure are calculated. At each step, the multifrontal massively parallel sparse direct solver (MUMPS) method [50] is used.

The time step is automatically determined by the solver so that the requirement for the relative tolerance is met (its value was set equal to 10^{-8}). With a decrease in the relative tolerance by a factor of 10, the change in the threshold potential drop of the transition to the overlimiting mode did not exceed 1%.

The implementation of the PD regime is similar to the described for the GD regime with the difference that the equation for the electric current stream function (25) is excluded from the calculation process and the boundary condition for the potential (23) changes to (21).

3. Results

3.1. Parameters Used in Computations

The results of simulation presented here are obtained for a flow-through ED cell (Figure 4) in the case of dilute NaCl solutions. The dimensionless parameters $\varepsilon = 3.05 \times 10^{-8}$, $Pe = 589$, $Re = 1.07$, $K_{el} = 5.23 \times 10^{-4}$, which correspond to the following system parameters: the thickness of the considered region $h = 0.5H$, where $H = 0.5 \times 10^{-3}$ m is the intermembrane distance; the channel length $l = 10^{-3}$ m; the average velocity of forced flow $V_0 = 3.8 \times 10^{-3}$ m/s; the electrolyte solution density $\rho_0 = 1002$ kg/m^3; the kinematic viscosity $\nu = 0.89 \times 10^{-6}$ m^2/s; the input concentration of the electrolyte solution of NaCl $c_0 = 0.1$ mol/m^3; the temperature $T = 298$ K; the diffusion coefficients of cations $D_1 = 1.33 \times 10^{-9}$ m^2/s and anions $D_2 = 2.05 \times 10^{-9}$ m^2/s; the cation transport number in the membrane $T_1 = 0.972$ and that in the solution $t_1 = 0.395$; the ion charge numbers $z_1 = 1$, $z_2 = -1$. To simplify the numerical solution, the ratio of the counterion concentration at the solution/CEM boundary to its value in the bulk solution N_c was taken as $N_c = 1$. This value is less than in real systems, however, as Urtenov et al. [51] have shown, when $N_c \geq 1$, the value N_c does not essentially affect the distribution of concentrations and potential in the extended SCR.

The sweep speeds of the potential drop ($\alpha = 0.0064$) and average current density ($\beta = 0.0003$) are chosen sufficiently small and the solution can be considered quasi-stationary, that is, their values do not affect the CVCs trend.

3.2. Current-Voltage Curves

Figure 5 shows the CVCs calculated for the GD and PD regimes. All CVCs have a linear initial part (denoted by 1 in Figure 5a), a sloping plateau (2 in Figure 5a), and an overlimiting current (3,4 in Figure 5a), which qualitatively corresponds to the existing experimental [5,7,9,13] and theoretical [16,19,31] studies about the CVCs of membrane systems. Note that the limiting current density of the calculated CVCs, determined by the point of intersection of the tangents drawn to the initial part and to the sloping plateau of the curve is close to $i_{\lim}$, calculated using Leveque's Equation (28) (values differ by less than 2%) [47]:

$$i_{\lim} = \frac{1}{T_1 - t_1}\left(1.47\left(\frac{4h^2 V_0}{lD}\right)^{1/3} - 0.2\right) \tag{28}$$

At the underlimiting and limiting current modes (regions *1* and *2* on Figure 5a, respectively) of the CVCs calculated for PD and GD regimes coincide with high accuracy (the difference is less than $0.01i_{lim}$). In these modes (at current densities $i_{av}/i_{lim} \leq 1$ or potential drop $\Delta\varphi < V_{cr1}$), electroconvective vortices are not observed in the fluid flow (Figure 6a).

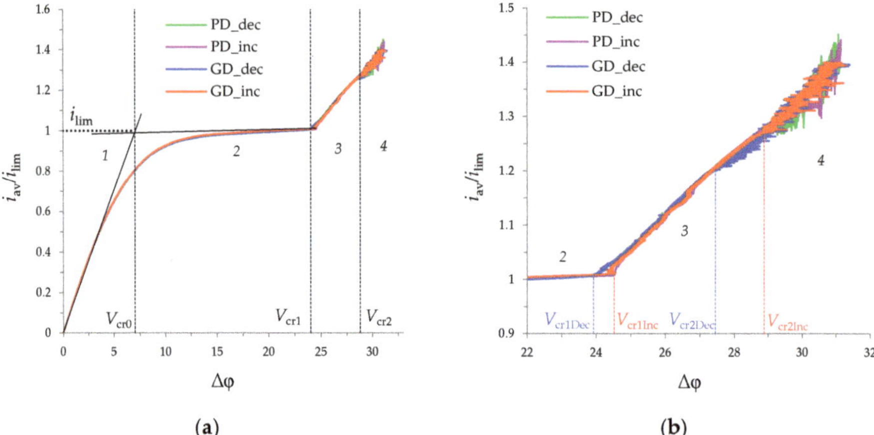

Figure 5. (**a**) CVCs calculated for the potentiodynamic (PD) (increasing $\Delta\varphi$—purple line, decreasing $\Delta\varphi$—green line) and galvanodynamic (GD) (increasing i_{av}—red line, decreasing i_{av}—blue line) regimes. The dotted line shows the limiting current density, i_{lim}, calculated using Leveque's Equation (28). The dashed lines V_{cr0}, V_{cr1}, V_{cr2} indicate changes in the CVC regions: underlimiting current **1**, plateau of the limiting current **2**, overlimiting **3**, overlimiting with chaotic oscillations **4**. (**b**) enlarged fragment of (**a**).

At the overlimiting current modes of the CVCs calculated for the both regimes single electroconvective vortices rotating in the same direction (for region 3 on Figure 5a; Figure 6b,c) and large vortex complexes consisting of several vortices rotating in opposite directions (for region 4 on Figure 5a; Figure 6d, e) are formed in the region near the membrane surface. Movement of the vortices causes current density fluctuations in the PD regime and potential drop fluctuations in the GD regime (regions 3 and 4 on Figure 5a). At the same time, the trends of the overlimiting current regions of the CVCs in both regimes approximately coincide (Figure 5b).

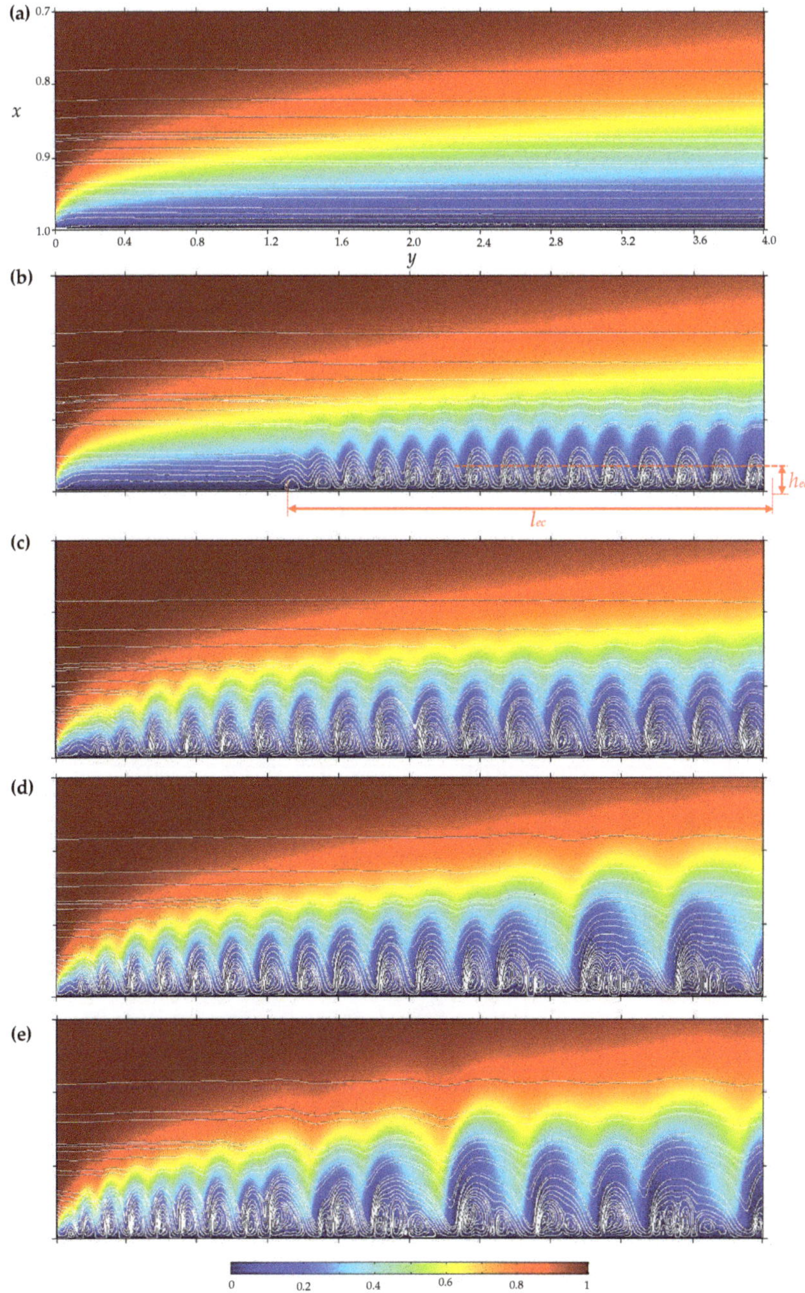

Figure 6. Distribution of cation concentration (the magnitude is shown by different colors), solution streamlines (white lines) in the area at the membrane surface. Calculation for the increasing GD regime at $i_{av}/i_{lim} = 1$ (**a**), 1.1 (**b**), 1.25 (**c**), 1.3 (**d**), 1.35 (**e**). To improve the visibility of the electroconvective vortex layer, the scale along the x axis is set larger than the y axis, thus the shape of the vortices is deformed.

3.3. Electroconvective Vortex Layer

To quantitatively describe the electroconvection vortex layer, parameters such as the thickness, h_{ec}, and length, l_{ec}, and density of vortices, d_{ec}, of this layer were determined. For systems with forced flow, the vortex sizes are not stable and depend on its position in the channel [17,19,47]; therefore, at a given point in time the thickness, h_{ec}, was determined as the distance from the membrane surface to the farthest edge of the closed streamline forming the biggest vortex [47] (Figure 6b). At each moment of time, the electroconvective vortex layer is a set of successive vortices and vortex structures. Wherein, this layer appears in the region at the channel outlet. Therefore, length, l_{ec}, was defined as the distance from the outlet to the farthest edge of the closed streamline forming the first (from the inlet) vortex (Figure 6b). Thus, h_{ec} and l_{ec} characterize the dimensions of the electroconvective vortex layer, that is the maximum transverse dimension of the biggest vortex and the length of the entire layer at a moment in time. Another important parameter characterizing the electroconvective vortex layer is the density of vortices, d_{ec}; that is, the number of vortices per unit of length.

The values of h_{ec}, l_{ec}, d_{ec}, were calculated for the PD regime for the potential drop $\Delta\varphi = 23.4$, $23.6, \ldots , 31$ (the results are indicated by crosses and trend lines in Figure 7a,c,e). Figure 7b,d,f show values of h_{ec}, l_{ec}, d_{ec}, calculated for the GD regime at the current density $i_{av}/i_{lim} = 1, 1.01, \ldots , 1.4$. The increase in the length of the electroconvective vortex layer, l_{ec}, is limited by the moment ($\Delta\varphi \approx 28$ or $i_{av}/i_{lim} \approx 1.24$), when this layer occupies almost the entire length of the channel ($l = 4$), Figure 7a,b. The thickness of the electroconvective vortex layer, h_{ec}, increases approximately linearly with increasing potential drop (or current density) everywhere in the considered range of $\Delta\varphi$ (or i_{av}) values, except for the initial region of rapid growth at $\Delta\varphi \approx V_{cr1}$ (or $i_{av}/i_{lim} \approx 1$), Figure 7c,d. Saturation of the thickness, h_{ec}, is not observed.

Figure 7e,f show a decrease in the vortices density, d_{ec}, in the range of the potential drop (or current density) corresponding to a rapid increase in the size of the electroconvective vortex layer; and the increase density, d_{ec}, in the range of the development of the large vortex complexes consisting of several vortices rotating in opposite directions (Figure 6d,e).

The described behavior of the system is characteristic of both the PD and GD regimes, both for increasing and decreasing cases. To compare the parameters of the electroconvective vortex layer in the GD and PD regimes, the average current densities, $\overline{i_{av}}$ corresponding to $\Delta\varphi = 23.4, 23.6, \ldots , 31$ for the PD regime and the average values of the potential drop, $\overline{\Delta\varphi}$, corresponding to $i_{av}/i_{lim} = 1, 1.01, \ldots$ 1.4 were calculated for GD regime. Figure 7 also show the dependences h_{ec}, l_{ec}, d_{ec}, on $\overline{\Delta\varphi}$, calculated for the GD regime and on $\overline{i_{av}}$, calculated for the PD regime. Figure 7 shows that the dependences of the parameters of the electroconvective vortex layer on the potential drop and current density in the PD and GD regimes are approximately the same.

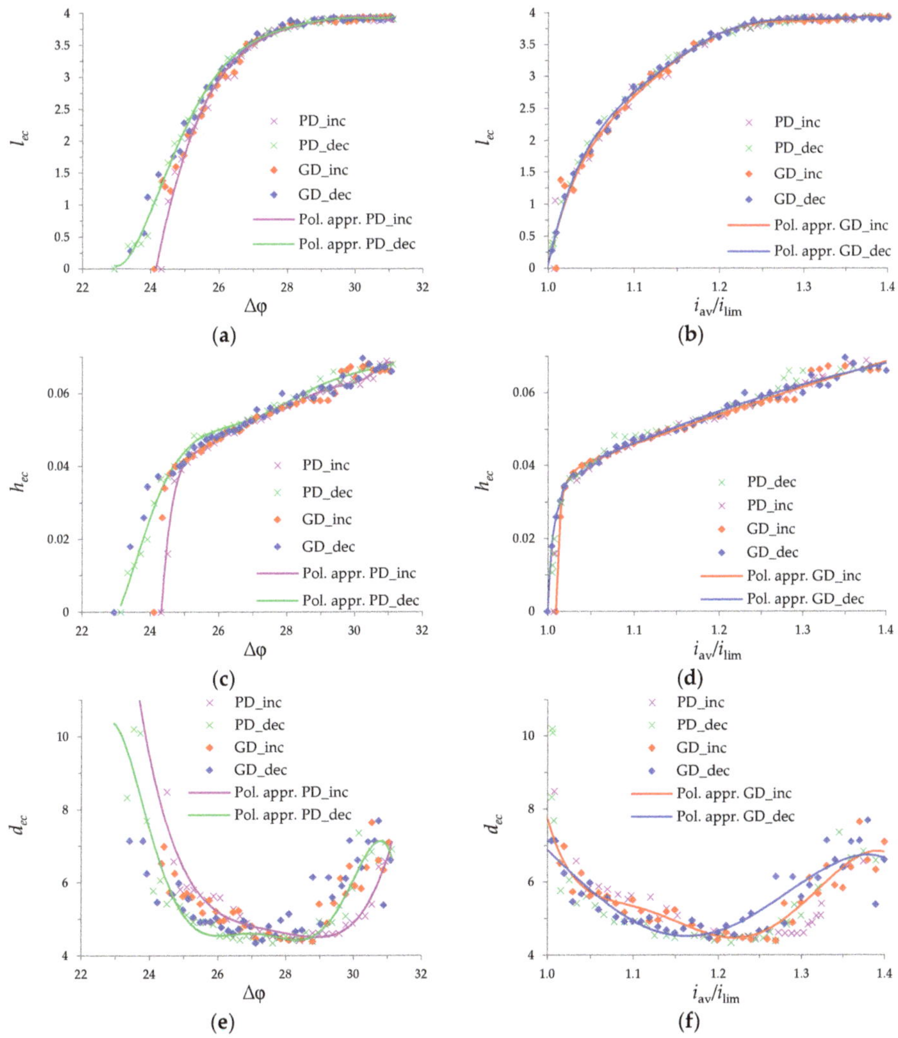

Figure 7. Dependences of the length, l_{ec}, thickness, h_{ec}, and vortex density, d_{ec}, of the electroconvective vortex layer on the potential drop, respectively (**a**), (**c**), (**e**), and current density (**b**), (**d**), (**f**). Calculations for the PD regime are indicated by crosses and for the GD regime by rhombuses. The solid lines (Pol. appr.) indicate polynomial approximation of the corresponding data.

3.4. Comparison of Increasing and Decreasing Regimes (Hysteretic Behavior)

The differences in the CVCs calculated with increasing and decreasing potential drop (or average current density) are manifested in the overlimiting current mode both in the PD and GD cases.

The critical potential drop of the transition to the overlimiting current mode in the increasing regime, V_{cr1Inc}, is larger than the corresponding value in the decreasing regime, V_{cr1Dec} (Figure 5b): $V_{\mathrm{cr1Inc}} \approx 24.52$ and $V_{\mathrm{cr1Dec}} \approx 23.92$ (these values approximately coincide for the PD and DG regimes). The hysteresis amplitude ($\Delta V_{\mathrm{cr1}} = V_{\mathrm{cr1Inc}} - V_{\mathrm{cr1Dec}} \approx 0.6$) is less than the difference in the potential drops correspond for the appearance and disappearance of vortices (determined by the values of h_{ec}, l_{ec}), which is about 1.2. This is due to the fact that the transition between the limiting and overlimiting modes on CVC appears only when the thickness h_{ec} exceeds approximately 0.02. The electroconvective vortex layer of the smaller thickness only causes fluctuations of small amplitude in the CVC.

The calculations in this article confirm the existence of the hysteretic behavior for flow-through channels in both PD and GD regimes. In the transition region between the limiting and overlimiting modes at the fixed potential drop in the decreasing regime, the length, l_{ec}, and thickness, h_{ec}, of the electroconvective vortex layer are greater than in the increasing regime (Figure 7a,c). At the fixed average current density, the length, l_{ec}, and thickness, h_{ec}, of the electroconvective vortex layer in the decreasing and increasing regimes approximately coincide (Figure 7b,d). In this region ($\Delta\varphi \approx V_{cr1}$, $i_{av}/i_{lim} \approx 1$), the density of vortices, d_{ec}, is higher in the increasing regime compared to the decreasing (Figure 7e,f).

In addition, the critical potential drop of the transition to chaotic oscillations in the increasing regime, V_{cr2Inc}, is also larger than the corresponding value in the decreasing regime, V_{cr2Dec} (Figure 5b): $V_{cr2Inc} \approx 28.88$ and $V_{cr2Dec} \approx 27.44$ for GD case; $V_{cr2Inc} \approx 30.36$ and $V_{cr2Dec} \approx 28.61$ for PD case. In the region of chaotic oscillations of the CVCs, the length and thickness of the electroconvective vortex layer oscillate in the same range for the increasing and decreasing regimes, but the density of vortices in the decreasing regime is higher. This is due to the fact that vortex complexes consisting of many vortices are maintained at the lower potential drop in the decreasing regime compared to increasing one.

4. Conclusions

On the CVCs calculated for the PD and DG regimes, four main current modes can be distinguished: underlimiting, limiting, overlimiting, and chaotic overlimiting. The influence of the electric field regime is manifested in the overlimiting current modes when a significant electroconvection vortex layer develops in the channel. The slipping of vortices along the membrane surface under the action of the forced flow leads to fluctuations in the current density at the PD regime and oscillations in the potential drop at the GD regime. The trend lines of the overlimiting sections of the CVCs for the PD and GD regimes are approximately the same, since the values of the parameters of the electroconvective vortex layer at the same values of the potential drop (or current density) in these modes are quite close.

At the fixed potential drop, the length and thickness of the electroconvective vortex layer in the decreasing regime (PD or GD) is greater than in an increasing one. This leads to the formation of a hysteresis loop in the transition region between the limiting and overlimiting regions of the CVCs. There is also a difference in the critical potential drop of the transition to the chaotic oscillations mode in the increasing and decreasing regimes.

Thus, the development of electroconvection determines the influence of the electric field regime on the processes of ion transfer in membrane systems.

Author Contributions: Formal Analysis, Funding Akquisition, Investigation, Visualization, Writing—original draft, A.U.; Methodology, A.U. and M.U.; Writing—review & editing, A.U. and M.U. All authors have read and agreed to the published version of the manuscript.

Funding: This research was funded by Russian Foundation for Basic Research grant number 18-38-00572.

Conflicts of Interest: The author declares no conflict of interest.

Appendix A

Figure A1 shows the concentration profiles for the channel inlet, $y = 0$, and at a short distance from it, $y = 0.001$, calculated with the condition of the uniform ion distribution (9) and with Danckwerts' condition (10) at $\Delta\varphi = 19.5$. In the first case, the concentration profiles vary significantly in the longitudinal direction: for the section $y = 0.001$, they are lower than for $y = 0$; in the second case, concentration profiles practically coincide (maximal difference less than 0.01). As a result, in the calculation with condition (9), a stationary vortex is formed at the inlet at $\Delta\varphi < V_{cr1}$, that is, earlier than that for the rest of the channel (Figure A2). When condition (10) is used, vortices at the inlet appear at $\Delta\varphi > V_{\mathrm{cr1}}$ during the growth of the electroconvective vortex layer, the formation of which begins at the outlet (Figure 6e).

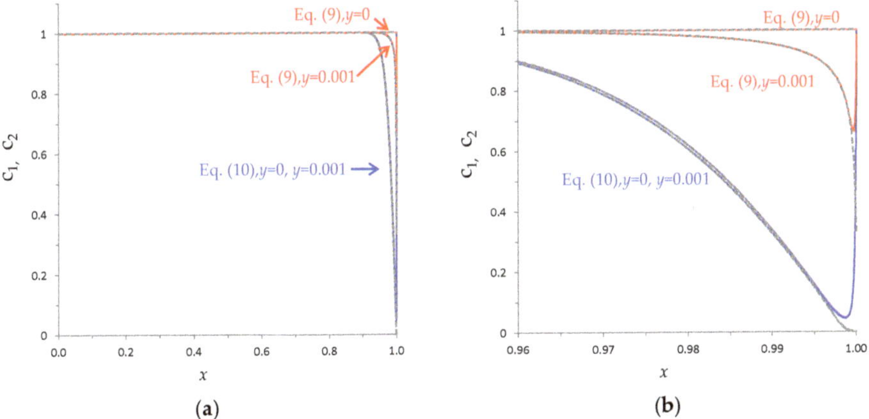

Figure A1. (**a**) Concentration profiles of cations (c_1, solid lines) and anions (c_2, dashed lines) in sections $y = 0$ and $y = 0.001$. Calculation for the PD regime at $\Delta\varphi = 19.5$ with condition (9) (red lines) and with Danckwerts' condition (10) (blue lines). (**b**) enlarged fragment of (**a**).

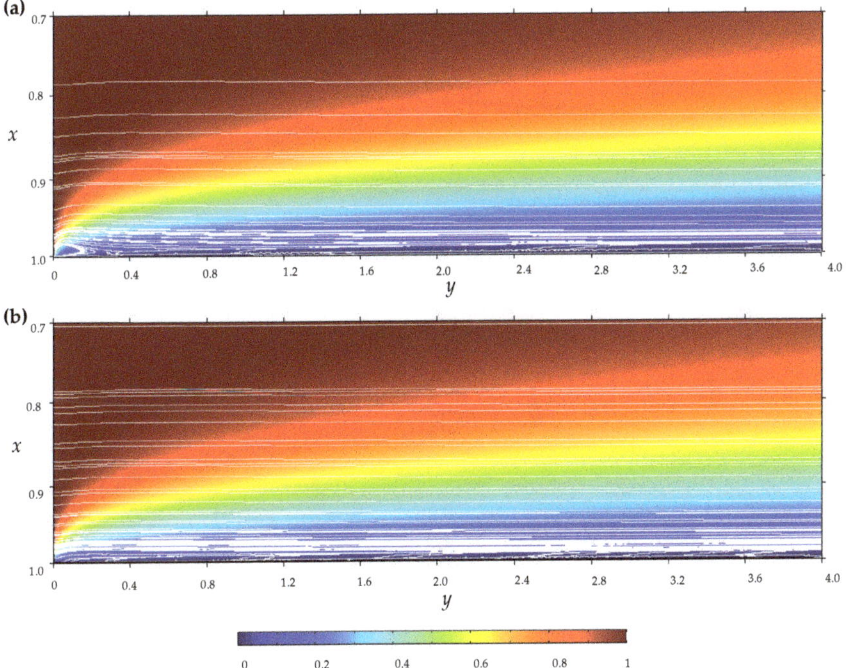

Figure A2. Distribution of cation concentration (the magnitude is shown by different colors), solution streamlines (white lines) in the area at the membrane surface. Calculation for the PD regime at $\Delta\varphi = 19.5$ with condition (9) (**a**) and with Danckwerts' condition (10) (**b**). To improve the visibility of the electroconvective vortex layer, the scale along the x axis is set larger than the y axis, thus the shape of the vortices is deformed.

In the Figure A3a,b, the dependences of the average tangential current density through the inlet, $i_{y\,av}(x,0,t)$, and outlet, $i_{y\,av}(x,l,t)$, boundaries on time are shown. In the considered range of the potential drop, the current $i_{y\,av}(x,0,t)$, calculated with condition (10), does not exceed $5 \times 10^{-5} i_{lim}$; for condition (9), this value reaches $0.09\ i_{lim}$. As a result, at the calculations with condition (10), the plateau angle of the limiting current decreases; the overlimiting region of the CVC lies lower (Figure A3a).

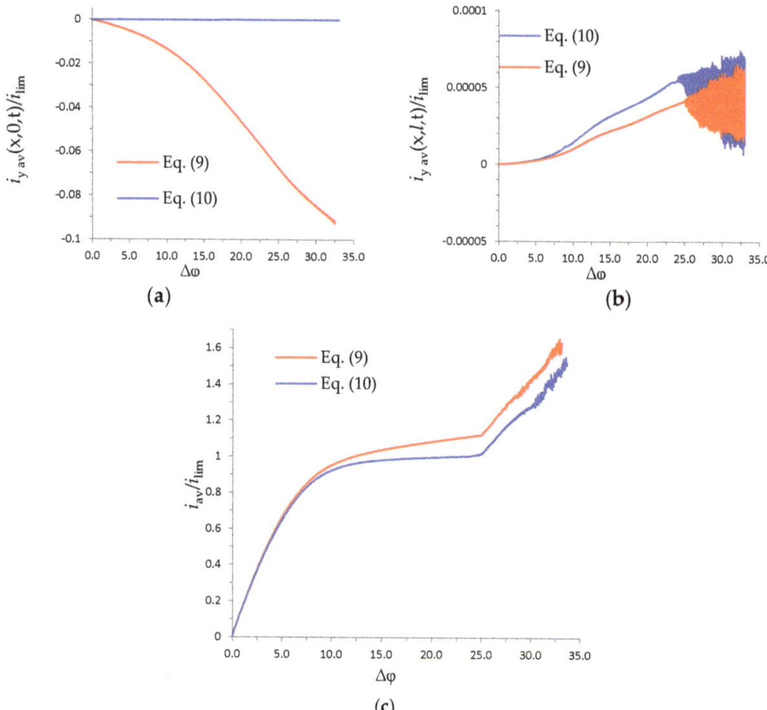

Figure A3. Average tangential current density through the inlet (**a**) and outlet (**b**); CVCs (**c**). The calculation results with condition (9) (red lines) and Danckwerts' condition (10) (blue lines) for the PD regime at $\alpha = 0.025$, other parameters as in p. 3.1.

References

1. Shannon, M.A.; Bohn, P.W.; Elimelech, M.; Georgiadis, J.G.; Mariñas, B.J.; Mayes, A.M. Science and technology for water purification in the coming decades. *Nature* **2008**, *452*, 301–310. [CrossRef] [PubMed]
2. Kim, S.-J.; Ko, S.-H.; Kang, K.H.; Han, J. Direct seawater desalination by ion concentration polarization. *Nature Nanotechnol.* **2010**, *5*, 297–301. [CrossRef] [PubMed]
3. Strathmann, H. Electrodialysis, a mature technology with a multitude of new applications. *Desalination* **2010**, *264*, 268. [CrossRef]
4. Elimelech, M.; Phillip, W.A. The Future of Seawater Desalination: Energy, Technology, and the Environment. *Science* **2011**, *333*, 712–717. [CrossRef] [PubMed]
5. Belova, E.I.; Lopatkova, G.Y.u.; Pismenskaya, N.D.; Nikonenko, V.V.; Larchet, C.; Pourcelly, G. The effect of anion-exchange membrane surface properties on mechanisms of overlimiting mass transfer. *J. Phys. Chem. B* **2006**, *110*, 13458. [CrossRef] [PubMed]
6. Nikonenko, V.V.; Kovalenko, A.V.; Urtenov, M.K.; Pismenskaya, N.D.; Han, J.; Sistat, P.; Pourcelly, G. Desalination at overlimiting currents: State-of-the-art and perspectives. *Desalination* **2014**, *342*, 85–106. [CrossRef]

7. Maletzki, F.; Rosler, H.W.; Staude, E. Ion transfer across electrodialysis membranes in the overlimiting current range: Stationary voltage current characteristics and current noise power spectra under different conditions of free convection. *J. Membr. Sci.* **1992**, *71*, 105–116. [CrossRef]

8. Zabolotsky, V.I.; Nikonenko, V.V.; Pismenskaya, N.D.; Laktionov, E.V.; Urtenov, M.K.; Strathmann, H.; Wessling, M.; Koops, G.H. Coupled transport phenomena in overlimiting current electrodialysis. *Sep. Purif. Technol.* **1988**, *14*, 255–267. [CrossRef]

9. Rubinshtein, I.; Zaltzman, B.; Pretz, J.; Linder, C. Experimental Verification of the Electroosmotic Mechanism of Overlimiting Conductance Through a Cation Exchange Electrodialysis Membrane. *Russ. J. Electrochem.* **2002**, *38*, 853–863. [CrossRef]

10. Pismenskaya, N.D.; Nikonenko, V.V.; Belova, E.I.; Lopatkova, G.Y.; Sistat, P.; Pourcelly, G.; Larchet, C. Coupled convection of solution near the surface of ion-exchange membranes in intensive current regimes. *Russ. J. Electrochem.* **2007**, *43*, 307. [CrossRef]

11. Rubinstein, S.M.; Manukyan, G.; Staicu, A.; Rubinstein, I.; Zaltzman, B.; Lammertink, R.G.H. Direct Observation of a Nonequilibrium Electro-Osmotic Instability. *Phys. Rev. Lett.* **2008**, *101*, 236101. [CrossRef] [PubMed]

12. Kwak, R.; Guan, G.; Peng, W.K.; Han, J. Microscale electrodialysis: Concentration profiling and vortex visualization. *Desalination* **2013**, *308*, 138–146. [CrossRef]

13. Bellon, T.; Polezhaev, P.; Vobecká, L.; Svoboda, M.; Slouka, S. Experimental observation of phenomena developing on ion-exchange systems during current-voltage curve measurement. *J. Membr. Sci.* **2019**, *572*, 607–618. [CrossRef]

14. Rubinstein, I.; Zaltzman, B. Electro-osmotically induced convection at a permselective membrane. *Phys. Rev. E* **2000**, *62*, 2238–2251. [CrossRef]

15. Demekhin, E.A.; Shelistov, V.S.; Polyanskikh, S.V. Linear and nonlinear evolution and diffusion layer selection in electrokinetic instability. *Phys. Rev. E* **2011**, *84*, 036318. [CrossRef]

16. Pham, S.V.; Li, Z.; Lim, K.M.; White, J.K.; Han, J. Direct numerical simulation of electroconvective instability and hysteretic current-voltage response of a permselective membrane. *Phys. Rev. E* **2012**, *86*, 046310. [CrossRef]

17. Kwak, R.; Pham, V.S.; Lim, K.M.; Han, J. Shear Flow of an Electrically Charged Fluid by Ion Concentration Polarization: Scaling Laws for Electroconvective Vortices. *Phys. Rev. Lett.* **2013**, *110*, 114501. [CrossRef]

18. Druzgalski, C.L.; Andersen, M.B.; Mani, A. Direct numerical simulation of electroconvective instability and hydrodynamic chaos near an ion-selective surface. *Phys. Fluids* **2013**, *25*, 110804. [CrossRef]

19. Urtenov, M.K.; Uzdenova, A.M.; Kovalenko, A.V.; Nikonenko, V.V.; Pismenskaya, N.D.; Vasil'eva, V.I.; Sistat, P.; Pourcelly, G. Basic mathematical model of overlimiting transfer enhanced by electroconvection in flow-through electrodialysis membrane cells. *J. Membr. Sci.* **2013**, *447*, 190–202. [CrossRef]

20. Shaposhnik, V.A.; Vasileva, V.I.; Praslov, D.B. Concentration Fields of Solutions under Electrodialysis with Ion-Exchange Membranes. *J. Membr. Sci.* **1995**, *101*, 23–30. [CrossRef]

21. Shaposhnik, V.A.; Vasil'eva, V.I.; Grigorchuk, O.V. The interferometric investigations of electromembrane processes. *Adv. Colloid Interface Sci.* **2008**, *139*, 74–82. [CrossRef] [PubMed]

22. Frilette, V.J. Electrogravitational Transport at Synthetic Ion Exchange Membrane Surfaces. *J. Phys. Chem.* **1957**, *61*, 168–174. [CrossRef]

23. Balster, J.; Yildirim, M.H.; Stamatialis, D.F.; Ibanez, R.; Lammertink, R.G.H.; Jordan, V.; Wessling, M. Morphology and microtopology of cation-exchange polymers and the origin of the overlimiting current. *J. Phys. Chem. B* **2007**, *111*, 2152–2165. [CrossRef] [PubMed]

24. Nikonenko, V.V.; Mareev, S.A.; Pis'menskaya, N.D.; Uzdenova, A.M.; Kovalenko, A.V.; Urtenov, M.K.; Pourcelly, G. Effect of Electroconvection and Its Use in Intensifying the Mass Transfer in Electrodialysis (Review). *Russ. J. Electrochem.* **2017**, *53*, 1122–1144. [CrossRef]

25. Rubinstein, I.; Shtilman, L. Voltage against current curves of cation exchange membranes. *J. Chem. Soc. Faraday Trans.* **1979**, *75*, 231–246. [CrossRef]

26. Dydek, E.V.; Zaltzman, B.; Rubinstein, I.; Deng, D.S.; Mani, A.; Bazant, M.Z. Overlimiting current in a microchannel. *Phys. Rev. Lett.* **2011**, *107*, 118301. [CrossRef]

27. Abu-Rjal, R.; Prigozhin, L.; Rubinstein, I.; Zaltzman, B. Equilibrium Electro-Convective Instability in Concentration Polarization: The Effect of Non-Equal Ionic Diffusivities and Longitudinal Flow. *Russ. J. Electrochem.* **2017**, *53*, 903–918. [CrossRef]

28. Demekhin, E.A.; Nikitin, N.V.; Shelistov, V.S. Direct numerical simulation of electrokinetic instability and transition to chaotic motion. *Phys. Fluids* **2013**, *25*, 122001. [CrossRef]

29. Demekhin, E.A.; Ganchenko, G.S.; Kalaydin, E.N. Transition to Electrokinetic Instability near Imperfect Charge-Selective Membranes. *Phys. Fluids* **2018**, *30*, 082006. [CrossRef]

30. Pham, S.V.; Kwon, H.; Kim, B.; White, J.K.; Lim, G.; Han, J. Helical vortex formation in three-dimensional electrochemical systems with ion-selective membranes. *Phys. Rev. E* **2016**, *93*, 033114. [CrossRef]

31. Karatay, E.; Druzgalski, C.L.; Mani, A. Simulation of Chaotic Electrokinetic Transport: Performance of Commercial Software versus Custom-built Direct Numerical Simulation Codes. *J. Colloid Interface Sci.* **2015**, *446*, 67–76. [CrossRef] [PubMed]

32. Magnico, P. Spatial distribution of mechanical forces and ionic flux in electro-kinetic instability near a permselective membrane. *Phys. Fluids* **2018**, *30*, 014101. [CrossRef]

33. Magnico, P. Electro-Kinetic Instability in a Laminar Boundary Layer Next to an Ion Exchange Membrane. *Int. J. Mol. Sci.* **2019**, *20*, 2393. [CrossRef] [PubMed]

34. Davidson, S.M.; Wessling, M.; Mani, A. On the Dynamical Regimes of Pattern-Accelerated Electroconvection. *Sci. Rep.* **2016**, *6*, 22505. [CrossRef] [PubMed]

35. Andersen, M.; Wang, K.; Schiffbauer, J.; Mani, A. Confinement effects on electroconvective instability. *Electrophoresis* **2017**, *38*, 702–711. [CrossRef] [PubMed]

36. Kirii, V.A.; Shelistov, V.S.; Demekhin, E.A. Hydrodynamics of spatially inhomogeneous real membranes. *J. Appl. Mech. Tech. Phy.* **2017**, *58*, 635–640. [CrossRef]

37. Dukhin, S.S.; Mishchuk, N.A. Unlimited current growth through the ion exchanger granule. *Colloid J.* **1988**, *49*, 1047.

38. Dukhin, S.S. Electrokinetic phenomena of the 2nd kind and their applications. *Adv. Colloid Interface Sci.* **1991**, *35*, 173–196. [CrossRef]

39. Mishchuk, N.A. Concentration polarization of interface and non-linear electrokinetic phenomena. *Adv. Colloid Interface Sci.* **2010**, *160*, 16–39. [CrossRef]

40. Chang, H.-C.; Demekhin, E.A.; Shelistov, V.S. Competition Between Dukhin's and Rubinstein's Electrokinetic Modes. *Phys. Rev. E* **2012**, *86*, 046319. [CrossRef]

41. Uzdenova, A.M.; Kovalenko, A.V.; Urtenov, M.K. *Mathematical Models of Electroconvection in Electromembrane Systems*; KCHGU: Karachaevsk, Russia, 2011.

42. Kwak, R.; Pham, V.S.; Han, J. Sheltering the perturbed vortical layer of electroconvection under shear flow. *J. Fluid Mech.* **2017**, *813*, 799–823. [CrossRef]

43. Mareev, S.; Kozmai, A.; Nikonenko, V.; Belashova, E.; Pourcelly, G.; Sistat, P. Chronopotentiometry and impedancemetry of homogeneous and heterogeneous ion-exchange membranes. *Desalin. Water Treat.* **2014**, *56*, 1–4. [CrossRef]

44. Mareev, S.A.; Nichka, V.S.; Butylskii, D.Y.; Urtenov, M.K.; Pismenskaya, N.D.; Apel, P.Y.; Nikonenko, V.V. Chronopotentiometric Response of Electrically Heterogeneous Permselective Surface: 3D Modeling of Transition Time and Experiment. *J. Phys. Chem. C* **2016**, *120*, 13113–13119. [CrossRef]

45. Mareev, S.A.; Nebavskiy, A.V.; Nichka, V.S.; Urtenov, M.K.; Nikonenko, V.V. The nature of two transition times on chronopotentiograms of heterogeneous ion exchange membranes: 2D modelling. *J. Membr. Sci.* **2019**, *575*, 179–190. [CrossRef]

46. Uzdenova, A. 2D mathematical modelling of overlimiting transfer enhanced by electroconvection in flow-through electrodialysis membrane cells in galvanodynamic mode. *Membranes* **2019**, *9*, 39. [CrossRef]

47. Nikonenko, V.V.; Vasil'eva, V.I.; Akberova, E.M.; Uzdenova, A.M.; Urtenov, M.K.; Kovalenko, A.V.; Pismenskaya, N.D.; Mareev, S.A.; Pourcelly, G. Competition between diffusion and electroconvection at an ion-selective surface in intensive current regimes. *Adv. Colloid Interface Sci.* **2016**, *235*, 233–246. [CrossRef]

48. Danckwerts, P.V. Continuous flow systems. Distribution of residence times. *Chem. Eng. Sci.* **1953**, *2*, 1–13. [CrossRef]

49. Uzdenova, A.; Kovalenko, A.; Urtenov, M.; Nikonenko, V. 1D Mathematical Modelling of Non-Stationary Ion Transfer in the Diffusion Layer Adjacent to an Ion-Exchange Membrane in Galvanostatic Mode. *Membranes* **2018**, *8*, 84. [CrossRef] .

50. Comsol Multiphysics Reference Manual. Available online: https://doc.comsol.com/5.4/doc/com.comsol.help.comsol/COMSOL_ReferenceManual.pdf (accessed on 10 March 2020).
51. Urtenov, M.A.K.; Kirillova, E.V.; Seidova, N.M.; Nikonenko, V.V. Decoupling of the Nernst-Planck and Poisson equations, Application to a membrane system at overlimiting currents. *J. Phys. Chem. B* **2007**, *11151*, 14208–14222. [CrossRef]

 *membranes*

Article

Reasons for the Formation and Properties of Soliton-Like Charge Waves in Membrane Systems When Using Overlimiting Current Modes

Makhamet Urtenov [1,*], Natalia Chubyr [2] and Vitaly Gudza [1]

1   Department of Applied Mathematics, Kuban State University, Krasnodar 350040, Russia; flash.wetal@mail.ru
2   Department of Applied Mathematics, Kuban State Technological University, Krasnodar 350042, Russia; chubyr-natalja@mail.ru
*   Correspondence: urtenovmax@mail.ru

Received: 22 July 2020; Accepted: 13 August 2020; Published: 16 August 2020

Abstract: The study of ion transport in membrane systems in overlimiting current modes is an important problem of physical chemistry and has an important application value. The influence of the space charge on the transport of salt ions under overlimiting current modes was first studied in the work of Rubinstein and Shtilman and later in the works of many authors. The purpose of this research is to study, using the method of mathematical modeling, the reasons of formation and properties of the local maximum (minimum) space charge in membrane systems under overlimiting current conditions. It is shown that, in the diffusion layer of the cation-exchange membrane (CEM), the local maximum of the space charge appears due to the limited capacity (exchange capacity) of the membrane at a given potential jump, i.e., the local maximum of space charge appears due to the presence of a local minimum of space charge at the surface of the CEM. The local maximum of the space charge moves as a single soliton-like wave into the depth of the solution. Unlike real solitons, this charged wave changes its size and shape, albeit quite slowly. In the section of the desalination channel, the situation is completely different. First, the space charge of the anion-exchange membrane (AEM) has a negative value, so we should be talking about the local minimum (or the maximum of the absolute value of the charge). However, this is an insignificant clarification. Secondly, the space charge waves of different signs begin to interact, which leads to a new effect, namely the effect of the breakdown of the space charge. The dependence of the local maximum on the input parameters—the cation diffusion coefficient, the growth rate of the potential jump, and the initial and boundary concentrations—is studied.

Keywords: ion-exchange membrane; mathematical modelling; using overlimiting current modes; membrane systems; cation-exchange membrane; effect of the breakdown of the space charge

1. Introduction

A thorough development of membrane devices has been done over the decades [1]. Of course, progress in separation through phase boundaries/interfaces has been accompanied by issues. The problem of ion transport across phase boundaries is one of the fundamental problems of physical chemistry and electrochemistry and is also important for membrane technologies. Studies [2–7] have shown the prospects of using intensive current modes.

The influence of space charge on the structure of the diffusion layer was first studied by Rubinstein and Shtilman [8]. Instead of the traditional equations of electroneutrality to the system of equations of the Nernst–Planck diffusion layer, the authors introduced the Poisson equation, and the ion-exchange membrane was taken as selective (the effective transport number of counterions was taken independent of current density). The problem was solved numerically. In the future, various mathematical methods and approaches to solving the Nernst–Planck–Poisson equations for electromembrane systems under extreme current conditions were developed [9–14].

The study of the non-stationary transfer of binary electrolyte in the diffusion layer is interesting because it allows us to determine the structure of the diffusion layer and its change over time, which is necessary, for example, for the asymptotic analysis of problems and the establishment of simple engineering calculation formulas for the dependence of the concentration distribution and electric field strength on the parameters of the problem. Research on non-stationary problems is limited to studies [15–18]. In these articles, the main attention is paid to the overlimiting potential dynamic mode and the analysis of the setting time depending on the parameters of the problem.

Studies [19,20] are devoted to the study of «shock electrodialysis», in which a deionization wave propagates through a microchannel or a porous medium with a sharp boundary between the concentrated and depleted zones. The deionization waves can be compared with the charge wave, since the deionization region actually coincides with the space charge region and a «powerful gradient» occurs in the desalination channel. Shock electrodialysis is a newly developed method for the desalination of water and deionization in micro-scale pores near an ion-selective element. In contrast to «shock electrodialysis», we study electro-membrane systems (the diffusion layer and the section of the desalination channel) with macroscale dimensions of the order of millimeters. In addition, it is shown that, in these systems, the interaction of charge waves is possible, up to their destruction (breakdown).

2. Materials and Methods

Mathematical Model of One-Dimensional Non-Stationary Ion Transport in Membrane Systems

Two electromembrane systems will be considered below: The depleted diffusion layer at the cation-exchange membrane (CEM), then $x = 0$, the beginning of the diffusion layer, and $x = H$ is the conditional boundary of the solution/CEM. Additionally, the section of the desalination channel, in this case $x = 0$, is the conditional boundary of the anion-exchange membrane (AEM)/solution, and $x = H$ has the same meaning. The transfer of 1:1 salt ions in both cases is described by the same equations—the difference in the boundary conditions at $x = 0$. System of equations.

The non-stationary transport of salt ions for a 1:1 electrolyte is described by the following system of equations:

$$\frac{\partial C_i}{\partial t} = -\frac{\partial j_i}{\partial x} \qquad i = 1,2 \tag{1}$$

$$j_i = -\frac{F}{RT} z_i D_i C_i \frac{\partial \phi}{\partial x} - D_i \frac{\partial C_i}{\partial x} \qquad i = 1,2 \tag{2}$$

$$\frac{\partial^2 \phi}{\partial x^2} = -\frac{F}{\varepsilon_a}(C_1 - C_2) \tag{3}$$

$$I_c = F(j_1 - j_2) \tag{4}$$

Here, (1) is the equation of material balance, (2) is the equation the Nernst–Planck for fluxes of sodium $i = 1 \leftrightarrow Na^+$ and chloride $i = 2 \leftrightarrow Cl^-$ ions, the charge number of cations $z_1 = 1$ and anions $z_2 = -1$, (3) is a Poisson equation for the electric field potential, (4) is the equation of the current flow, which means that the current flowing through the diffusion layer is determined by the flow of ions, ε_a is the dielectric permeability of the solution, F is the Faraday number, R is the universal gas constant, ϕ is the potential, $E = -\frac{\partial \phi}{\partial x}$ is the electric field intensity, and C_i, j_i, D_i I_c are concentration, flux, the diffusion coefficient of the i-th ion, and current density determined by the flux of ions, respectively.

Boundary conditions for the diffusion layer.

The boundary conditions for a diffusion layer consist of the following boundary and initial conditions:

$$C_1(t,0) = C_0, \ C_2(t,0) = C_0, \ \phi(t,0) = 0 \tag{5}$$

$$C_1(t,H) = C_{1m}(t), \ \left(\frac{\partial C_2}{\partial x} + \frac{F}{RT}C_2 E\right)(t,H) = 0, \phi(t,H) = \Delta_r(t),$$

$$C_{10}(x) = C_0, \ C_{20}(x) = C_0, \ \phi_0(x) = 0,$$

where $\Delta_r(t)$ is the potential jump. As a rule, $\Delta_r(t)$ is either constant, or $\Delta_r(t) = -d{\cdot}t$, where d is the growth rate of the potential jump and has the dimension V/s.

Boundary conditions for the desalination channel.

To model the ion transport, replace the boundary condition (5) with the following:

$$C_2(t,0) = C_{2m}, \ \left(\frac{\partial C_1}{\partial x} - \frac{F}{RT}C_1 E\right)(t,H) = 0, \ \phi(t,0) = 0 \tag{6}$$

3. Results

3.1. Reasons for the Formation of a Local Maximum (Minimum) Space Charge in the Extended Space Charge Region (Extended SCR)

When the current passes through the membrane, the concentration of cations decreases and reaches its minimum at the left border of the border layer. When using overlimiting current densities, this minimum is preserved, but the minimum value is greatly reduced, and there are practically no anions in the area of the minimum point. The size of the space charge $\rho = F(C_1 - C_2)$ is almost completely determined by the concentration of cations $\rho \approx FC_1$(Figure 1b). Therefore, the space charge also has a local minimum other than zero at this point. The minimum value depends on the applied potential jump (or on the current that is passed through the system). On the other hand, in the depth of the solution, where the condition of electroneutrality is met, the value of the space charge is almost zero. Therefore, the value of the space charge must have a local maximum between them (Figure 1a). In [9,12], it is shown that $x^* = \frac{I_{np}}{I}H$; therefore, $[0, \frac{I_{np}}{I}H)$ is the region of electroneutrality, and $(\frac{I_{np}}{I}H, H)$ is the extended space charge region (SCR). In addition, in the field of electroneutrality, a balance is observed between the processes of diffusion and electromigration, and the currents of diffusion and electromigration (ohmic) are equal. At the same time, in the SCR, the electromigration is an order of magnitude greater than the diffusion one. Thus, the current pattern changes at the local maximum point.

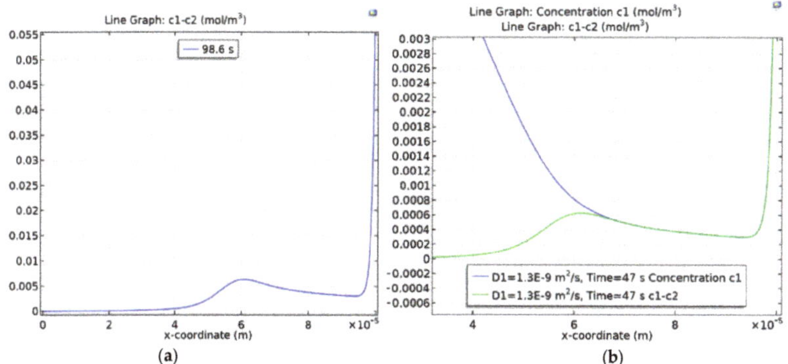

Figure 1. (a) Graph of the space charge in the diffusion layer; (b) Comparison with the graph of the concentration of C_1 cations near the cation-exchange membrane (CEM) when using overlimiting current modes in the dimensional form.

Using the above assumptions, you can analytically determine the solution in the extended SCR. Indeed, we put in the system of equations $C_2(t,x) = 0$, $j_2(t,x) = 0$, $\left|\frac{F}{RT_0}z_iC_i\frac{\partial\phi}{\partial x}\right| >> \left|\frac{\partial C_i}{\partial x}\right|$, then we get:

$$\frac{\partial j_1}{\partial x} = 0 \tag{7}$$

$$j_1 = -\frac{F}{RT}D_1C_1\frac{\partial\phi}{\partial x} \tag{8}$$

$$\frac{\partial^2\phi}{\partial x^2} = -\frac{F}{\varepsilon_a}C_1 \tag{9}$$

$$I_c = Fj_1$$

From Equation (7), we get:

$$j_1 = j_1(t) = \frac{I_c(t)}{F}$$

Multiply Equation (9) by $\frac{\partial\phi}{\partial x}$, then, taking into account equation (8), we have:

$$\frac{\partial\phi}{\partial x}\frac{\partial^2\phi}{\partial x^2} = \frac{RT}{\varepsilon_aD_1}j_1$$

$$\frac{1}{2}\left[\frac{\partial\phi}{\partial x}\right]^2 = \frac{RT}{\varepsilon_aD_1}j_1x + \beta$$

or

$$\frac{\partial\phi}{\partial x} = -\sqrt{2\frac{RT}{\varepsilon_aD_1}j_1x + \beta} \tag{10}$$

where $\beta > 0$ is the integration constant. From (8) and (10), we have:

$$C_1(t,x) = \frac{j_1RT}{FD_1\sqrt{2\frac{RT}{\varepsilon_aD_1}j_1x + \beta}}$$

or

$$C_1(t,x) = \frac{RT}{F^2D_1\sqrt{\frac{2RT}{\varepsilon_aFD_1}I_c(t)x + \beta}}I_c(t)$$

From this formula, it can be seen that $C_1(t,x)$ monotonously decreases in x in the extended SCR and reaches a minimum on its right border. In addition, as the current density increases, the minimum value decreases, which coincides with the numerical solution.

In order to understand the role and significance of the occurrence and growth of a local maximum, it is necessary to study its dependence on the input parameters of the problem, such as the initial and boundary concentrations, the rate of growth of the potential jump, the composition of the electrolyte solution, etc.

3.2. Dependence of the Local Maximum in the Diffusion Layer on the Boundary Concentration of C_{1m}

Consider the dependence of the local maximum on the parameter C_{1m}, which characterizes the exchange capacity of the membrane.

The value in Figure 2 increases by the formula $C_{1m} = 0.5 \cdot t$, with the local maximum appearing and disappearing. The point of the local maximum is shifted to the right, but the value does not change much (Figure 2a,b). At the same time, the local minimum is gradually filled with increasing C_{1m} (the value of the local minimum increases) and it is first compared and then surpassed (Figure 2c,d).

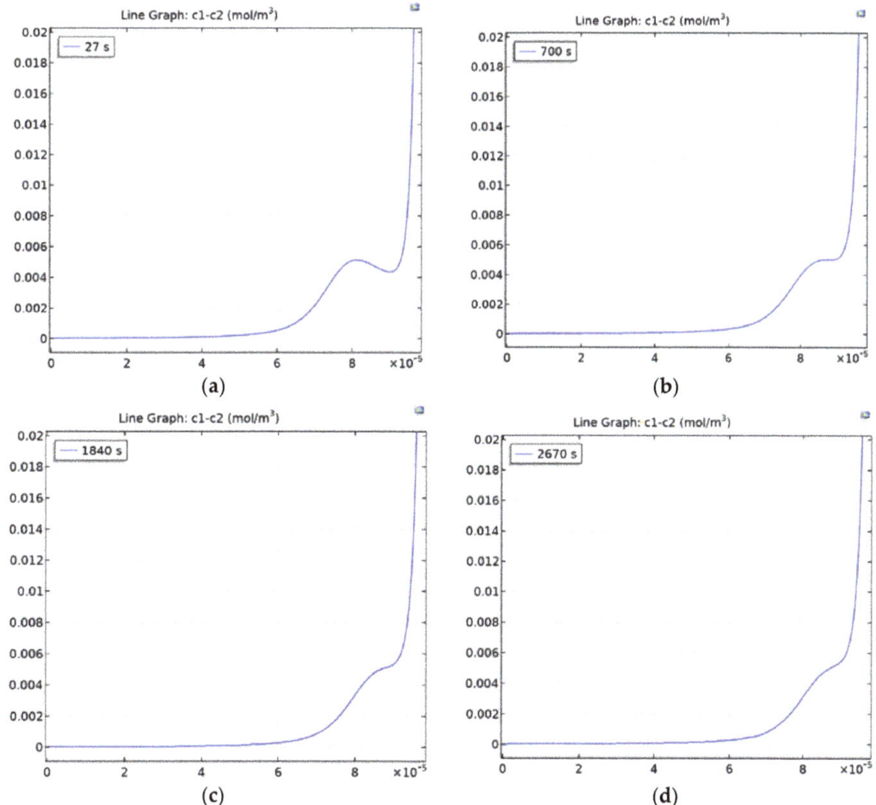

Figure 2. Dependence of the local maximum on the exchange capacity of the CEM (NaCl, $\Delta_r\phi = 0.5$): (a) $C_{1m} = 14$, (b) $C_{1m} = 350$, (c) $C_{1m} = 920$, (d) $C_{1m} = 1335$.

Thus, it can be concluded that the local maximum of the space charge appears due to the limited exchange capacity of the membrane at a given potential jump, i.e., the local maximum of the space charge appears due to the presence of a local minimum of the space charge at the surface of the CEM.

3.3. Soliton-like Charge Wave. Dependence of the Local Maximum (Minimum) Charge on the Growth Rate of the Potential Jump d

Consider a non-stationary model of transport in the diffusion layer of the CEM (1–5) with a linear growth of the potential jump $\Delta_r\phi(t, H) = -d{\cdot}t$.

The calculations are at times greater than a certain critical t_k, at which point the potential jump $\Delta_r\phi(t_k, H) = -d{\cdot}t_k$ corresponds to the limiting current density and a local maximum of the space charge appears as a single soliton-like wave begins to move into the depth of the solution. Unlike true solitons, this wave has a charge. In addition, it changes its size and shape.

With a decrease in the growth rate of the potential jump d, the space charge wave becomes less pronounced, but remains.

3.4. Dependence of Charge Waves on the Cation Diffusion Coefficient

3.4.1. Diffusion Layer

In order to study the dependence of the local maximum on the diffusion coefficient of the cation, solutions of NaCl and KCl are considered. Calculations show that, despite the difference in the diffusion coefficients of sodium and potassium by 1.5 times, the positions of the local minimum and maximum and their value in the diffusion layer of CEM differ slightly (Figure 3).

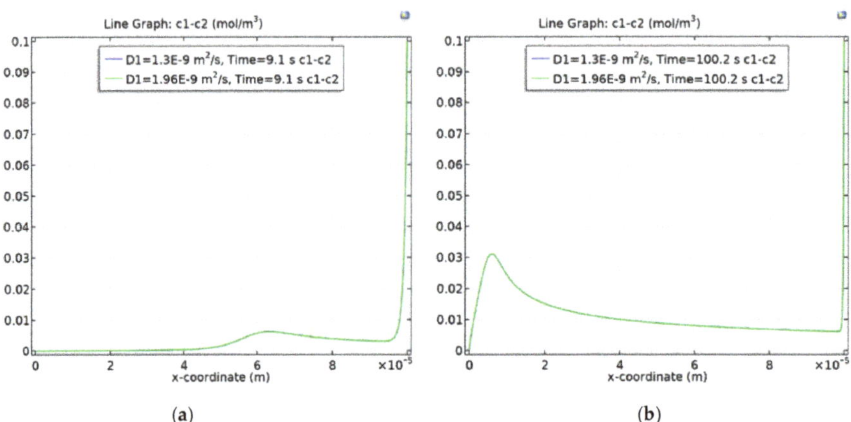

Figure 3. Wave charge when $d = 0.1$ when $t \approx 9c$ (**a**) and $t \approx 100c$ (**b**). The blue and green lines (they merge with great accuracy) correspond to solutions of NaCl and KCl.

3.4.2. Section of the Desalination Channel

Let us analyze the issue (1–4), (6). In the section of the desalination channel, the situation is completely different. First, the space charge of AEM has a negative value, so we must talk about the local minimum (or the local maximum of the absolute value of the charge). However, this is an insignificant clarification. Secondly, the space charge waves of different signs begin to interact, which leads to a new effect, namely the effect of the breakdown of the space charge.

Let us first consider the behavior of the local maximum in the KCl solution at d = 0.1. As can be seen from Figure 4a–f, two symmetrical soliton-like lone waves are formed, which move towards each other. Unlike real solitons, these waves have, as noted above, charges, the left wave negative charges and the right positive charges. At first, they practically do not interact, but as they approach each other they begin to attract and their speed of convergence increases and at the moment before contact there is a practical instantaneous breakdown, and they are discharged. A further increase in the potential jump does not lead to the formation of a new wave of local maximum, since the concentration of the solution practically becomes zero (Figure 4f–h), with the exception of narrow border layers in AEM and CEM, due to the fact that the concentration of anions and cations is maintained constant at the boundaries of AEM/solution and solution/CEM. As the potential jump increases, the width of the border layers decreases very slowly. A similar scenario is realized for the NaCl solution, except that the local minimum of negative space charge is generated much later (Figure 4f) than for the KCl solution. This is due to the fact that the current density in the first case is greater than in the second.

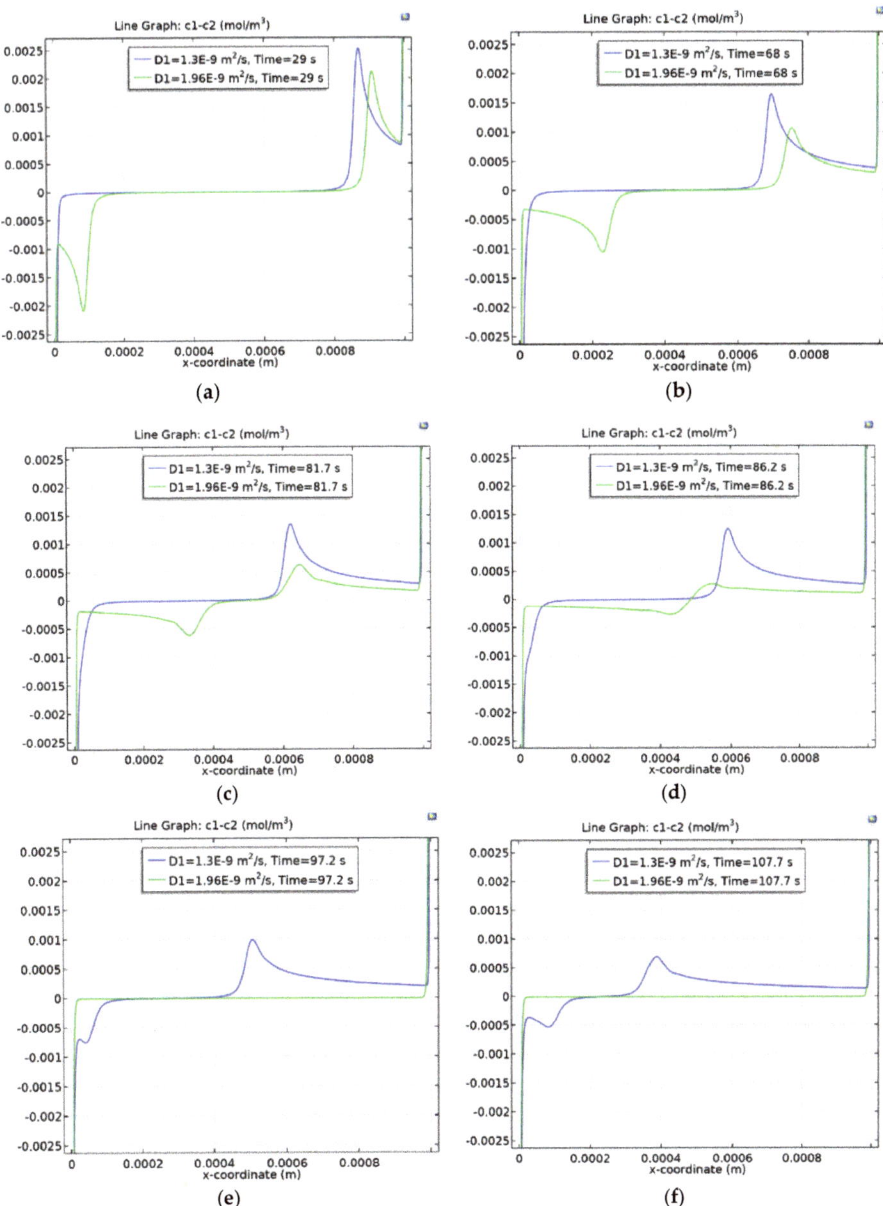

Figure 4. *Cont.*

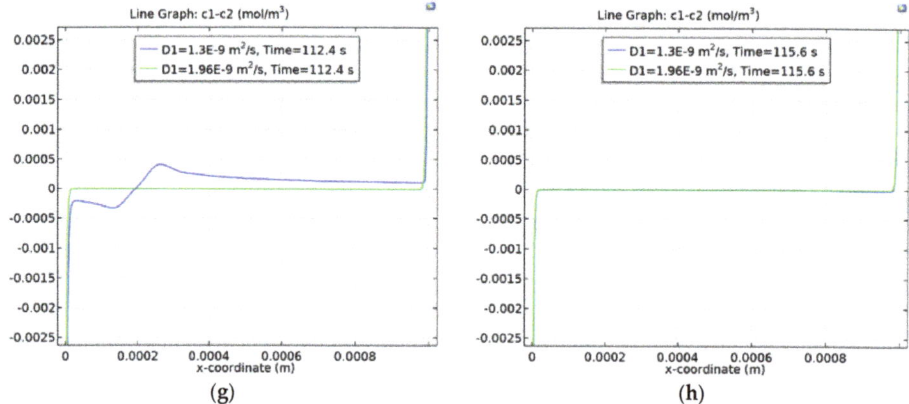

(g)   (h)

Figure 4. Graphs of the space charge at $d = 0.1$, $C_0 = 0.1\,mol/m^3$. $t \approx 29c$ (**a**); $t \approx 68c$ (**b**); $t \approx 82c$ (**c**); $t \approx 86c$ (**d**); $t \approx 97c$ (**e**); $t \approx 108c$ (**f**); $t \approx 112c$ (**g**); $t \approx 116c$ (**h**).

At $d = 0.005$ in NaCl, a single wave occurs in the CEM, which reaches the region of negative space charge in the AEM, where there is no local minimum and maximum, and is discharged at 436 s. In KCl, two almost symmetrical waves are formed, one in the CEM and the other in the AEM, which meet almost in the middle and discharge at 282 c. Thus, the behavior of charge waves at $d = 0.005$ in NaCl and KCl solutions is completely different.

3.5. Dependence of Charge Waves on the Initial Concentration of C_0

The comparison of space charge graphs for $C_0 = 0.01\,mol/m^3$ and for $C_0 = 0.1\,mol/m^3$ (see Figures 4a and 5a) at $t = 29\,s$ shows that the less concentrated solution is desalinated faster and the charge waves are already very different, and, at $t = 54.3\,s$ (Figure 5b), the charge waves for KCl began to discharge.

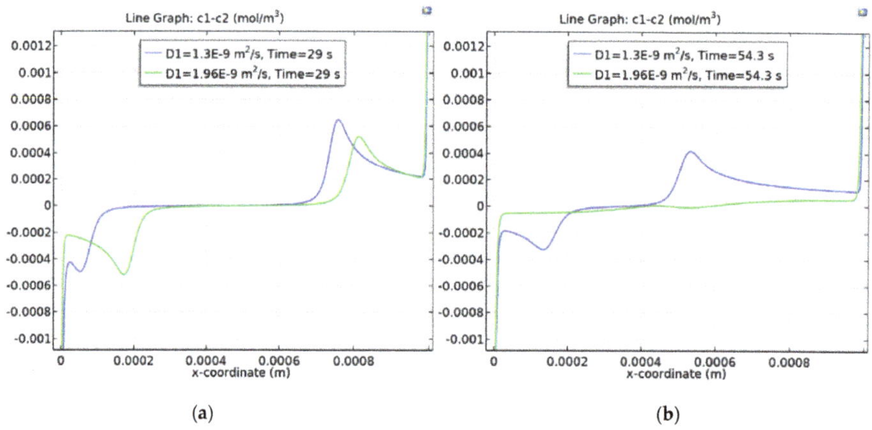

(a)   (b)

Figure 5. Space charge graph at $C_0 = 0.01\,mol/m^3$. $t \approx 29c$ (**a**); $t \approx 54c$ (**b**).

4. Conclusions

In this paper, the reasons for the formation and properties of the local maximum (minimum) space charge in membrane systems when using overlimiting current conditions are investigated. The depleted diffusion layer at the cation-exchange membrane and the section of the desalination channel are considered as membrane systems. It is shown that the local maximum and minimum of space charge appear due to the limited capacity of ion-exchange membranes at a given potential jump. It is shown that the local maximum of the space charge in the diffusion layer moves as a single soliton-like wave into the depth of the solution, slowly changing its size and shape. In the section of the desalination channel, the space charge waves of different signs begin to interact, which leads to a new effect, namely the effect of discharge (breakdown) of the space charge. The fundamental laws of this phenomenon are studied.

Author Contributions: Conceptualization, M.U., N.C., V.G.; Investigation, M.U., N.C.; Formal Analysis, M.U., V.G.; Visualization, M.U., N.C., V.G. All authors have read and agreed to the published version of the manuscript.

Funding: This research was funded by Russian Foundation for Basic Research project number 20-58-12018 NNIO_a: The influence of electroconvection, water dissociation, and geometry of spacers on electrodialysis desalination in intensive current regimes.

Acknowledgments: The study was carried out with the financial support of the Russian Foundation for Basic Research grant number 20-58-12018 NNIO_a: The influence of electroconvection, water dissociation, and geometry of spacers on electrodialysis desalination in intensive current regimes.

Conflicts of Interest: The authors declare no conflict of interest.

References

1. Tabani, H.; Khodaei, K.; Varanusupakul, P.; Alexovič, M. Gel electromembrane extraction: Study of various gel types and compositions toward diminishing the electroendosmosis flow. *Microchem. J.* **2020**, *153*, 104520. [CrossRef]

2. Rubinshtein, I.; Zaltzman, B.; Prez, J.; Linder, C. Experimental verification of the electroosmotic mechanism for the formation of overlimiting current in a system with a cation-exchange electrodialysis membrane. *Electrochemistry* **2002**, *38*, 956–967.

3. Nikonenko, V.; Kovalenko, A.V.; Urtenov, M.K.; Pismenskaya, N.D.; Han, J.; Sistat, P.; Pourcelly, G. Desalination at overlimiting currents: State-of-the-art and perspectives. *Desalination* **2014**, *342*, 85–106. [CrossRef]

4. Urtenov, M.K.; Uzdenova, A.M.; Kovalenko, A.V.; Nikonenko, V.V.; Pismenskaya, N.D.; Vasil'eva, V.I.; Sistat, P.; Pourcelly, G. Basic mathematical model of overlimiting transfer enhanced by electroconvection in flow-through electrodialysis membrane cells. *J. Membr. Sci.* **2013**, *447*, 190. [CrossRef]

5. Vasilieva, V.I.; Zhiltsova, A.V.; Malykhin, M.D.; Zabolotsky, V.I.; Lebedev, K.A.; Chermit, R.H.; Sharafan, M.V. Influence of the chemical nature of ionogenic groups of ion-exchange membranes on the size of the region of electroconvective instability in high-intensity current modes. *Electrochemistry* **2014**, *50*, 134.

6. Vasilieva, V.I.; Shaposhnik, V.A.; Zabolotsky, V.I.; Lebedev, K.A.; Petrunya, I.P. Diffusive boundary layers at the ion exchange membrane/solution boundary under high-intensity electrodialysis regimes. *Sorpt. Chromatogr. Process.* **2005**, *5*, 111–127.

7. Pismenskaya, N.D.; Nikonenko, V.V.; Belova, E.I.; Lopatkova, G.Y.; Sista, F.; Purseli, Z.; Larshe, K. Conjugate convection of a solution at the surface of ion-exchange membranes under intense current conditions. *Electrochemistry* **2007**, *43*, 325–343.

8. Rubinstein, I.; Shtilman, L. Voltage against current curves of cation exchange membranes. *J. Chem. Soc. Faraday Trans.* **1979**, *75*, 231–246. [CrossRef]

9. Urtenov, M.K.; Nikonenko, V.V. Analysis of electrodiffusion equations in decomposition form. *Electrochemistry* **1996**, *32*, 207–214.

10. Nikonenko, V.V.; Zabolotsky, V.I.; Gnusin, N.P. Electric transfer of ions through a diffusive layer with disturbed electroneutrality. *Electrochemistry* **1989**, *25*, 301–306.

11. Listovnichy, A.V. Concentration polarization of the ionite membrane-electrolyte solution system in the overlimiting mode. *Electrochemistry* **1991**, *27*, 316–323.

12. Urtenov, M.A.K.; Nikonenko, V.V. Analysis of solutions for the boundary value problem for the Nernst-Planck-Poisson equations: The 1:1 electrolyte Case. *Electrochemistry* **1993**, *29*, 239–245.

13. Zabolotsky, V.I.; Manzanares, H.A.; Mafe, S.; Nikonenko, V.V.; Lebedev, K.A. Accounting for electroneutrality violations in mathematical modeling of stationary ion transport through a three-layer membrane system. *Electrochemistry* **2002**, *38*, 921–929.

14. Zabolotsky, V.I.; Manzanares, J.A.; Mafe, S.; Nikonenko, V.V.; Lebedev, K.A. Mathematical simulation of a stationary electrodiffusion kinetics in multilayer ion-exchange membrane systems with the help of the numerical shooting parallel method continued by parameters. *Desalination* **2002**, *147*, 387–392. [CrossRef]

15. Manzanares, J.A.; Murphy, W.D.; Mafe, S.; Reiss, H. Numerical simulation of the non-equilibrium diffuse double layer in ion-exchange membranes. *J. Phys. Chem.* **1993**, *97*, 8524–8530. [CrossRef]

16. Zabolotsky, V.I.; Nikonenko, V.V. Electrodialysis of dilute solutions of electrolytes. Some theoretical and applied aspects. *Electrochemistry* **1996**, *32*, 246–254.

17. Urtenov, M.A.K.; Kirillova, E.V.; Seidova, N.M.; Nikonenko, V.V. Decoupling of the Nernst–Planck and Poisson Equations. Application to a Membrane System at Overlimiting Currents. *J. Phys. Chem.* **2007**, *111*, 14208–14222. [CrossRef] [PubMed]

18. Uzdenova, A.; Kovalenko, A.V.; Urtenov, M.; Nikonenko, V. 1D Mathematical Modelling of Non-Stationary Ion Transfer in the Diffusion Layer Adjacent to an Ion-Exchange Membrane in Galvanostatic Mode. *Membranes* **2018**, *8*, 84. [CrossRef] [PubMed]

19. Deng, D.; Aouad, W.; Braff, W.A.; Schlumpberger, S.; Suss, M.E.; Bazant, M.Z. Water purification by shock electrodialysis: Deionization, filtration, separation, and disinfection. *Desalination* **2015**, *357*, 77–83. [CrossRef]

20. Mani, A.; Bazant, M.Z. Deionization shocks in microstructures. *Phys. Rev. E* **2011**, *84*, 061504. [CrossRef] [PubMed]

 *membranes*

Article

Characterization of Poly(Acrylic) Acid-Modified Heterogenous Anion Exchange Membranes with Improved Monovalent Permselectivity for RED

Ivan Merino-Garcia, Francis Kotoka, Carla A.M. Portugal, João G. Crespo and Svetlozar Velizarov *

Associated Laboratory for Green Chemistry—Clean Technologies and Processes (LAQV), REQUIMTE, Chemistry Department, FCT, Universidade Nova de Lisboa, 2829-516 Caparica, Portugal; ime.garcia@fct.unl.pt (I.M.-G.); franciskotoka90@gmail.com (F.K.); cmp@fct.unl.pt (C.A.M.P.); jgc@fct.unl.pt (J.G.C.)
* Correspondence: s.velizarov@fct.unl.pt

Received: 28 May 2020; Accepted: 24 June 2020; Published: 26 June 2020

Abstract: The performance of anion-exchange membranes (AEMs) in Reverse Electrodialysis is hampered by both presence of multivalent ions and fouling phenomena, thus leading to reduced net power density. Therefore, we propose a monolayer surface modification procedure to functionalize Ralex-AEMs with poly(acrylic) acid (PAA) in order to (i) render a monovalent permselectivity, and (ii) minimize organic fouling. Membrane surface modification was carried out by putting heterogeneous AEMs in contact with a PAA-based aqueous solution for 24 h. The resulting modified membranes were firstly characterized by contact angle, water uptake, ion exchange capacity, fixed charge density, and swelling degree measurements, whereas their electrochemical responses were evaluated through cyclic voltammetry. Besides, their membrane electro-resistance was also studied via electrochemical impedance spectroscopy analyses. Finally, membrane permselectivity and fouling behavior in the presence of humic acid were evaluated through mass transport experiments using model NaCl containing solutions. The use of modified PAA-AEMs resulted in a significantly enhanced monovalent permselectivity (sulfate rejection improved by >35%) and membrane hydrophilicity (contact angle decreased by >15%) in comparison with the behavior of unmodified Ralex-AEMs, without compromising the membrane electro-resistance after modification, thus demonstrating the technical feasibility of the proposed membrane modification procedure. This study may therefore provide a feasible way for achieving an improved Reverse Electrodialysis process efficiency.

Keywords: anion exchange membranes; poly(acrylic) acid modification; monovalent permselective membranes; antifouling strategies; reverse electrodialysis

---

1. Introduction

The continuous rise of worldwide electricity demand has led to an increasing global interest in the study and development of green technologies capable of generating sustainable and renewable power [1,2]. In this respect, Reverse Electrodialysis (RED) represents an attractive technology due to the possibility of harvesting renewable energy from salinity gradients (e.g., between seawater and river water) through the use of alternating anion exchange membranes (AEMs) and cation exchange membranes (CEMs) forming cell pairs, where the different compartments between these membranes are fed with streams of different salinity (feedwaters with high and low salt concentration) [3–5]. Salinity gradient is, therefore, the driving force for the transport of ions from one compartment to the adjacent ones, thus creating an ionic current, which can be converted into electrical current by using electrodes, at which reversible redox reactions occur [6]. However, the practical application of this technology is currently limited by the presence of divalent ions in natural streams [2,4,7,8], which decreases the obtainable net power density according to Nernst equation as follows:

$$OCV = \frac{NRT}{F}\left[\frac{\alpha_{CEM}}{z_{CEM}}\ln\frac{\gamma_{c,CEM}\,c_c}{\gamma_{d,CEM}\,c_d} + \frac{\alpha_{AEM}}{z_{AEM}}\ln\frac{\gamma_{c,AEM}\,c_c}{\gamma_{d,AEM}\,c_d}\right] \tag{1}$$

where OCV represents the open circuit voltage, N is the number of cell pairs, R the universal gas constant, T the absolute temperature, F the Faraday constant, α is equal to the permselectivity of the corresponding ion exchange membrane, γ represents the activity coefficient (where subscripts c and d stand for concentrate and dilute saline solutions, respectively), c is the molar concentration, and z is equal to the valence of the anion/cation that crosses the corresponding membrane. Accordingly, the higher the z, the lower the OCV, thus leading to reduced obtainable power output. As a result, the development of mono-selective AEMs and CEMs is crucial for an improved RED process efficiency. Besides, new efforts focusing on optimizing the operation variables of RED systems and their effects on the overall internal resistance, gross power and OCV, among others, have been recently considered to move forward into the large-scale implementation of this technology [9,10].

Moreover, different fouling-based phenomena such as organic fouling and scaling that negatively affects ion exchange membranes performance, promotes a significant loss (with time) of the generated power density [11]. Therefore, fouling control represents one of the main challenges to be addressed for a successful industrial implementation of the RED technology [6,12–14]. Although fouling issues can be investigated at stack level [15], most research is mainly focused on membrane level [13,16,17], especially regarding AEMs due to the negative charge of natural organic matter (NOM) such as humic acids, creating undesirable interactions between the fixed positively charged groups of AEMs and such foulant materials that hamper the performance of the process. In this respect, the presence of NOM has been demonstrated to present a larger impact on the obtainable net power density than the ionic composition [18].

In this context, surface modification of AEMs represents one of the most promising strategies to render a monovalent permselectivity as well as increasing fouling resistance [19–21]. Although different modification techniques such as polymerization by UV-irradiation [22] or chemical oxidation [23], among others, have been addressed in literature in an attempt to overcome the so-mentioned limitations, the possibility of incorporating a negative hydrophilic layer (or multilayers) on an AEM surface represents an attractive option for the electro-membrane processes field [24], because such a layer would act not only as a multivalent ions rejection wall, but also favoring monovalent ions passage owing to the Donnan effect [17,25], and preventing fouling because of its hydrophilic properties and negative charge at the same time [26]. As recently reviewed [1], direct casting, dip coating, immersion, and layer-by-layer (LbL) deposition, among others, are the most common available approaches considered to incorporate a beneficial hydrophilic layer on a membrane surface. In addition, due to the importance of selecting an appropriate surface modifying agent, a wide variety of interesting and feasible alternatives, ranging from polymers and biopolymers to nanoparticles and ionic liquids, have been already proposed. For example, focusing on the development of polymers and biopolymers, on the one hand, both poly(sodium 4-styrene sulfonate) (PSS) and poly(ethylenimine) (PEI)-based aqueous solutions were used as AEM modifying agents via the LbL method, demonstrating an improvement of mono-selectivity and antifouling properties, highlighting the importance of controlling both modifying agent concentration and deposition time, with the purpose of producing thinner membranes with lower resistance for an improved RED power performance [2]. On the other hand, the application of biopolymers such as chitosan-based materials are emerging for functionalizing AEMs, owing to their easiness to create thin layers on membrane surface as well as their multiple beneficial associated properties such as biodegradability, stability, and low toxicity, among others, leading to an enhanced selectivity for monovalent anions [27,28].

Moreover, several studies have reported an improved rejection of multivalent ions such as sulfate (usually expressed in RED studies as the permselectivity between Cl^- and SO_4^{2-}) by different modified AEMs [29]. For instance, a Fujifilm AEM Type I was modified via LbL deposition, reaching an improved permselectivity between Cl^- and SO_4^{2-} from 0.81 to 47.04 after modification [20]. On the other hand, a Neosepta AMX AEM modified with polydopamine (PDA) by dip coating was reported to be capable of decreasing the permselectivity between SO_4^{2-} and Cl^- from 1.2 to 0.22 [30], thus denoting that the transport of SO_4^{2-} across AEMs can be controlled by their surface modification.

The scientific community is, however, continuously seeking to investigate innovative approaches in an attempt to design tailor-made modified AEMs respecting greener and more sustainable preparation/utilization ways. Thus, the development of cheaper, environmentally friendly, non-toxic, stable, hydrophilic, and durable materials for AEMs functionalization still represents a real challenge to be solved. In this context, as the LbL method represents a subsequent addition of negatively and positively charged layers on a membrane surface, more toxic polymers/substances are involved in the formation of positively charged layers, which leads to a certain loss of sustainability related aspects.

Therefore, we propose here the monolayer modification of AEMs with poly(acrylic) acid (PAA), which is a cheap, eco-friendly, and non-hazardous substance. In addition, due to the negative charge of most of PAA chains in aqueous solutions at neutral pH, this polyelectrolyte can be easily used to create a negative hydrophilic layer on AEMs, which may exhibit antifouling features and improved monovalent anion permselectivity. PAA has been previously used as a model foulant [31] as well as for improving membrane hydrophilicity [32,33]. However, to the best of our knowledge, our study is the first attempt reported in literature to functionalize the surface of AEMs with PAA for RED related purposes. In this respect, a comprehensive characterization of membrane behavior is essential to move forward on the development of novel modified membranes for blue energy harvesting [34].

Consequently, this work focuses on the comprehensive characterization of PAA-modified heterogeneous Ralex-AEMs, which are strongly basic AEMs with quaternized ammonium functional groups [35] with a significantly lower cost compared to that of homogenous AEMs. As known, one of the main current limitations for commercialization of the RED process is the relatively high costs of the IEMs, thus cheaper materials have to be considered. The monovalent anion permselectivity and fouling of the prepared membranes in the presence of humic acid (HA) through mass transport experiments is also assessed. The behavior of the modified membranes as a function of PAA concentration is compared with the performance of unmodified commercial AEMs, thus providing new insights and knowledge for the continuous research, design, and development of functionalized AEMs for an improved RED process operation.

2. Materials and Methods

2.1. Membrane Preparation

Commercial polyester-based heterogeneous Ralex-AEMs (MEGA, Stráž pod Ralskem, Czech Republic) were modified by putting them in contact with PAA-based solutions for 24 h, where both conductivity and pH of the PAA-modifier solution were monitored with time during modification. Trizma® (Sigma-Aldrich, St. Louis, MO, USA) buffer aqueous solutions (0.1 M) including different PAA concentrations (from 1 to 5 g/L) were used to create a negatively charged monolayer onto the membrane surface. After functionalization, the solution was replaced by fresh 0.1 M Trizma® solution to carry out membrane cleaning for 24 h. The resulting modified membranes were kept in water (total immersion in deionized water) before characterization and/or use. Table 1 shows the classification, modification conditions, and nomenclature of the membranes under investigation.

Table 1. Anion-exchange membranes (AEMs) under study: classification and characteristics.

List of Membranes	Modification Type/Modified Sides	PAA Concentration (g/L)	Nomenclature
(1)	Unmodified	-	Unmodified
(2)	Monolayer/one	1	One side 1 g/L PAA
(3)	Monolayer/one	3	One side 3 g/L PAA
(4)	Monolayer/both	3	Both sides 3 g/L PAA
(5)	Monolayer/one	5	One side 5 g/L PAA

2.2. Membrane Surface Characterization

Surface hydrophilicity was evaluated by using a goniometer (CAM 100, KSV Instruments Ltd., Helsinki, Finland) together with a software for drop shape analysis. In the experiment, a droplet of deionized water was provided by means of a syringe onto membrane surface (membranes dried at 35 °C for 24 h were used), where different (at least four) surface points were taken into consideration to reach an averaged contact angle value in each case, including the standard deviation.

The water uptake (*WU*) of the prepared AEMs was measured by weighing membrane mass at dry (m_{dry}) and wet (m_{wet}) conditions, respectively. Firstly, AEMs were dried at 35 °C for 24 h, followed by using a desiccator for 1 day to remove traces of water. Finally, the membranes were totally immersed in deionized water for 24 h to obtain the wet mass of the corresponding AEMs, after removing the excess of water from membrane surface with a tissue paper. The *WU* percentage is then calculated as follows:

$$WU\ (\%) = \frac{m_{wet} - m_{dry}}{m_{dry}} \times 100 \tag{2}$$

The amount of fixed charged groups per unit weight (g) of dry polymer in the prepared AEMs, that is, their ion exchange capacity (*IEC*), was firstly measured by adapting the Mohr titration method [36]. Wet AEM samples with known masses were immersed in a 0.4 M NaCl aqueous solution for 24 h. The anion exchange is carried out by replacing the former solution by 0.2 M Na_2SO_4 aqueous solution, which was kept in contact with the AEMs for 3 h, thus replacing Cl^- by SO_4^{2-}. The resulting solution containing the released Cl^- was finally titrated using a volumetric 0.1 M $AgNO_3$ aqueous solution (potassium chromate was utilized as indicator) to calculate *IEC* values in mmol per mass (g) of dry membrane as a function of the type of AEM under study as follows:

$$IEC\left(mmol/g_{dry}\right) = \frac{V_{AgNO_3} \times N_{AgNO_3}}{m_{dry}} \tag{3}$$

A spectrophotometric IEC determination method, proposed in [37], was also performed for the sake of comparison. The membrane samples were immersed in 1 M KNO_3 aqueous solution for 24 h, followed by immersion in 0.1 M NaCl for 12 h in order to exchange Cl^- for NO_3^-, which is released to the solution. The concentration of NO_3^- was measured using a UV-visible spectrophotometer (Thermo Scientific Evolution 201, ThermoFisher Scientific, Waltham, MA, USA) that operates at a wavelength of 300 nm, thus determining the IEC taking into consideration the measured NO_3^- (mmol) and the averaged mass of the dry membranes.

The membrane fixed charge density (*CD_{fix}*), which represents the concentration of fixed charge groups per unit volume of water in the membrane under study, was estimated by the relation between the *IEC* and the *WU* as follows: $CD_{fix} = IEC/WU$.

The swelling effect was also evaluated by measuring the thickness (Elcometer 124 Thickness gauge, Elcometer Instruments, Manchester, UK) and diameter of the prepared membranes in both wet (AEMs in contact with deionized water for 24 h) and dry (AEMs at 35 °C 24 h, followed by using a desiccator for 1 day) conditions.

The analysis of functional groups at the membrane surface was carried out by Fourier attenuated atomic force microscopy (*ATR-FTIR*) technique using a Spectrum Two FT-IR Spectrometer (PerkinElmer®, Waltham, MA, USA). Due to operational requirements, the analyzed AEMs were dried (in an oven at 35 °C for 24 h) and kept in a desiccator for 1 day before use. The response of unmodified heterogeneous Ralex-AEM was also measured as a reference. At least three points (different positions) of a membrane surface were analyzed to obtain reproducible spectra in each case.

2.3. Electrochemical Characterization

The prepared AEMs were firstly electrochemically characterized by cyclic voltammetry (*CV*) using a divided two-compartment diffusion cell, where the effect of the type of electrodes and their relative position was studied in order to reach ideal resistor performance. For that purpose, copper (Cu), graphite, and silver (Ag) rods were taken into consideration as electrodes in this study. The feed compartment (containing the counter electrode) was filled with an aqueous solution of 1 g/L NaCl + 0.1 g/L Na_2SO_4 (including the effect of the presence of 25 ppm of HA), whereas 30 g/L NaCl (mimicking a seawater salinity) was utilized at the receiver compartment (containing the working electrode) to perform the electrochemical measurements from −0.6 to 0.6 V (a scan rate of 200 mV/s was used). The potential was controlled by using a potentiostat/galvanostat (Ivium Technologies, Eindhoven, Netherlands). Three repeated scans were carried out in all tests to study membrane and diffusion stability. In this context, the obtained electrochemical responses are related to the overall transport of ions crossing the corresponding AEM under study. On the other hand, the current obtained at the maximum applied voltage (0.6 V) was evaluated in four different feed aqueous solutions with the same molar concentration (i.e., 0.017 M + 0.0007 M) such as KCl + Na_2SO_4, KCl + K_2SO_4, NaCl + Na_2SO_4, and LiCl + Na_2SO_4, respectively, in order to study the effect of the nature of the co-ion.

A dedicated, compact, and robust electrochemical flow cell designed by Østedgaard-Munck et al. [38,39] was used to characterize the AEMs through electrochemical impedance spectroscopy (*EIS*) measurements, aiming at studying membrane electro-resistance. The redox flow cell consists of two symmetrical halves separated by the membrane under study. The geometric active area of the cell is 6.25 cm^2. One stainless steel end plate is placed as a final element at the end of each cell side. The end plates are electrically insulated from gold plate-based current collectors by using a Viton sheet. Graphite blocks act as electrodes, which are intentionally designed with an interdigitated flow pattern based on 12 channels with a dimension of 25 × 25 × 2 mm. Additionally, two Teflon gaskets were located between the electrodes and the membrane to ensure the correct adjustment of the key element of the system. The two feed streams (0.5 M NaCl aqueous solutions) were supplied in co-flow mode to the cell at 20 mL/min by using a peristaltic pump (Masterflex, Cole-Palmer, Chicago, IL, USA). Besides 0.5 M NaCl solutions, 0.5 M KCl and 0.5 M LiCl solutions were also considered to evaluate the effect of changing the co-ion on the membrane electro-resistance. Finally, the cell is specifically tightened at a torque force of 3.0 Nm using a torque wrench, which allows an optimal contact between the electrodes, the membrane, and the different elements in the redox flow cell. Impedance analyses were carried out at room temperature and constant voltage (50 mV) with an amplitude of 0.1 V using the same Ivium potentiostat described for *CV* measurements, with frequencies ranging from 0.5 MHz to 100 Hz. Different EIS measurements at the same conditions were run to reach an averaged membrane electro-resistance, including blank experiments (EIS experiments without membrane). The data was fitted by means of equivalent circuit analysis to determine the effect of the electric double-layer (EDL) in each case, which relates to the structure of charge accumulation and charge separation that occurs at the interface between the membrane and the aqueous solution-based electrolyte, as shown in Figure 1. Due to the high frequency range used in this study, the effect of the diffusion boundary layer (DBL), also represented in Figure 1, should be lower than the one associated with the membrane resistance, even though it has also been assessed for the sake of clarity. Therefore, we focus the *EIS* investigation on (i) membrane electro-resistance (R_M), and (ii) electric double-layer resistance (R_{EDL}).

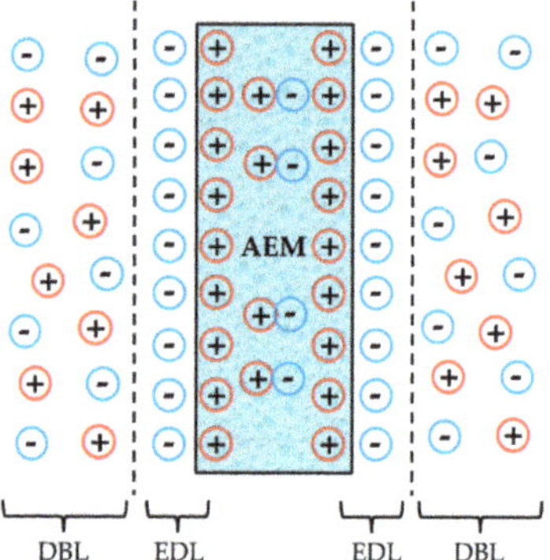

Figure 1. Anion exchange membrane diagram including both electric double-layer (EDL) and diffusion boundary layer (DBL) effects, adapted with permission from [40]. Copyright 2016 Elsevier.

2.4. Mass Transport Experiments

With the purpose of evaluating the counter-ion permselectivity and fouling behavior of the prepared AEMs, the same two-compartment diffusion cell used for membrane modification and CV analyses was utilized to carry out mass transport studies. Thus, the feed and the receiver compartments were filled with model streams of low (i.e., river water) and high (i.e., seawater) salt concentrations, while 25 ppm of HA (Fluka, Ign. residue: ≈20%) was introduced in the feed solution in several experiments as model organic foulant to evaluate fouling behavior. Figure 2 shows the diffusion cell layout, where it is worth noting that no potential difference is applied to the system to conduct the diffusion experiments. The time evolutions of concentrations of the ions present (Na^+, SO_4^{2-} and Cl^-) in both compartments were followed by taking and analyzing samples for 24 h. Na^+ and SO_4^{2-} concentrations were determined using an Inductively Coupled Plasma (*ICP*) Spectroscopy technique (Na and S determination) and further calculations for SO_4^{2-}, whereas Cl^- was calculated using charge balance difference. In the presence of HA, the absorbance of feed and receiver solutions was studied by using the UV-visible spectrophotometer mentioned above to demonstrate that HA (absorbance at 280 nm) is not permeating through the membrane from feed to receiver compartment.

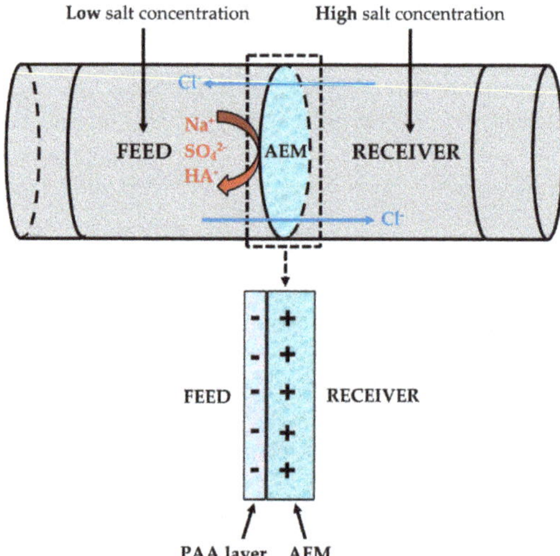

Figure 2. Two-compartment diffusion cell layout.

3. Results and Discussion

3.1. Membrane Surface Characterization: Contact Angle, Water Uptake, Ion Exchange Capacity, Fixed Charge Density, Swelling, and Fourier Attenuated Atomic Force Microscopy

Surface hydrophilicity was evaluated by using a goniometer integrated with a software for drop shape analysis, where the lower the contact angle of a membrane, the higher its hydrophilicity. Thus, Figure 3 shows contact angle values for different one side PAA modified AEMs and their comparison with the unmodified (commercial) membrane, demonstrating an improved membrane hydrophilicity after modification with 1, 3, and 5 g/L of PAA-based solution respectively. In particular, the deposition of a 1 g/L of PAA-based layer involves an improvement of 15% in hydrophilicity in comparison with the contact angle value achieved for the unmodified heterogeneous AEM, while increasing the PAA concentration up to 3 g/L during modification step results in modified AEMs with higher hydrophilic properties (31% enhanced). However, a further increase in the PAA concentration to 5 g/L did not enhance the hydrophilicity results reached at 3 g/L, which might be possibly due to surface saturation at higher PAA concentrations. It is worth highlighting that different positions on the surface of the dried membranes under investigation were considered to reach averaged contact angle values.

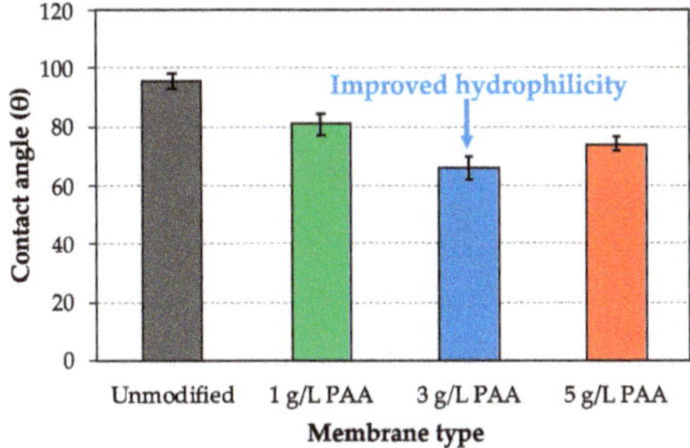

Figure 3. Hydrophilicity analysis: contact angle data.

The membrane hydrophilicity improvement for the PAA-modified AEMs was confirmed by water uptake (*WU*) analyses. Table 2 summarizes *WU*, ion exchange capacity (*IEC*) via two different methods, and fixed charge density (CD_{fix}) results of the prepared membranes as a function of PAA concentration, including the data associated with the unmodified membrane. In this regard, the modified AEMs present higher WU values compared to the unmodified membrane, denoting the higher hydrophilic properties of the membranes, for which an additional negative PAA layer was incorporated onto their surfaces, even though alterations in PAA concentration did not involve significant (taken into account the standard deviation) changes in WU values, thus denoting comparable water absorption properties of the prepared PAA-modified AEMs.

Table 2. Water uptake (*WU*), ion exchange capacity (*IEC*), and fixed charge density (CD_{fix}) results of the prepared AEMs.

Membrane Type	WU (%)	IEC [1] (mmol/g)	IEC [2] (mmol/g)	CD_{fix} [1] (mmol/g)	CD_{fix} [2] (mmol/g)
(1) Unmodified	59.0 ± 1.8	0.949 ± 0.18	1.369 ± 0.05	1.605 ± 0.25	2.321 ± 0.08
(2) One side 1 g/L PAA	67.4 ± 6.0	0.681 ± 0.03	1.625 ± 0.05	1.017 ± 0.14	2.413 ± 0.08
(3) One side 3 g/L PAA	62.7 ± 3.8	0.433 ± 0.01	1.751 ± 0.05	0.692 ± 0.05	2.793 ± 0.08
(4) Both sides 3 g/L PAA	61.5 ± 0.9	0.378 ± 0.01	1.751 ± 0.05	0.614 ± 0.02	2.847 ± 0.08
(5) One side 5 g/L PAA	63.3 ± 5.0	0.352 ± 0.02	1.760 ± 0.05	0.560 ± 0.08	2.779 ± 0.08

[1] Titration method; [2] Spectrophotometric method.

Membrane composition affects *IEC* due to the presence of different fixed functional groups, which can be divided into weak and strong ion-exchangers according to their dissociation constants [37]. Two distinct methods were considered to evaluate the *IEC* of the prepared membranes, as shown in Table 2. The results obtained demonstrate that the determined *IEC* values are dependent on the method applied, which suggests that care should be taken when selecting the most appropriate analytical technique in case of surface modified membranes. With the Mohr titration technique (method 1), *IECs* from 0.3 to 0.95 mmol/g were obtained. As expected, the PAA modified AEMs presented lower *IEC* values, in comparison with the unmodified membrane. As the concentration of PAA during modification step increases, a stronger repulsion effect of the negative PAA layer on SO_4^{2-} occurs, leading to a reduced anion exchange/replacing between Cl^- and SO_4^{2-} and, respectively, lower *IEC* values determined in this way.

On the other hand, when spectrophotometric measurements (method 2) are conducted, *IEC* values from 1.3 to 1.8 mmol/g were achieved as PAA concentration increases. In this regard, *IEC* values are slightly higher for the modified AEMs compared to those for the unmodified membrane, even though the influence of PAA concentration seems to be negligible at higher concentrations, which involves a similar replacing between Cl^- and NO_3^- in the different modified membranes. Therefore, this study highlights the importance of the method adopted to evaluate *IEC* in ion exchange membranes, thus demonstrating that *IEC* depends on the selected ion for replacement/exchange. In this context, Mohr titration and visualization methods usually involve higher errors due to the difficulty to determine the final equivalent point by a naked eye. In this relation, spectrophotometric methods could determine more accurate *IEC* values (similar to those obtained via elemental analysis) according to a study, in which several methods for determining *IEC* of AEMs have been discussed and compared [37]. As a result, we recommend spectrophotometric approaches to determine the *IEC* of surface modified AEMs. Regarding the CD_{fix} results, this characterization parameter is affected by both *IEC* and *WU*. Since different *IEC* values were observed as a function of the method used, CD_{fix} results follow the same tendency as the one observed for the *IEC*.

Membrane swelling is another essential parameter that may negatively affect the performance of RED because, for instance, the thickness of the membrane may increase its electrical resistance, leading to reduced power output from RED. Despite the fact that swelling degree is often measured as water uptake in literature [36,41], it is worth mentioning that membrane swelling must also be quantified in terms of membrane dimensional changes. Therefore, in this work the effect of swelling on both membrane thickness and diameter is evaluated. Thus, Table 3 shows not only the study of the mentioned dimensions for the prepared PAA-modified AEMs after and before swelling, that is, under wet and dry conditions, respectively, but also including the behavior of the unmodified commercial Ralex membrane.

Table 3. Swelling study results.

Membrane Type	Thickness Wet (μm)	Thickness Dry (μm)	Diameter Wet (mm)	Diameter Dry (mm)
(1) Unmodified	643.3 ± 5.8	472.7 ± 2.3	45.7 ± 0.6	43.3 ± 1.2
(2) One side 1 g/L PAA	664.0 ± 0.0	487.7 ± 4.0	45.0 ± 1.7	44.8 ± 2.0
(3) One side 3 g/L PAA	654.7 ± 4.6	473.3 ± 2.3	47.3 ± 1.2	44.8 ± 0.8
(4) Both sides 3 g/L PAA	650.0 ± 20.4	473.7 ± 6.0	47.7 ± 1.5	45.5 ± 0.5
(5) One side 5 g/L PAA	654.8 ± 21.3	476.0 ± 5.3	47.3 ± 0.6	45.3 ± 0.6

The thickness of the modified PAA-AEM is slightly higher at both dry and wet conditions in comparison with the values obtained for unmodified membranes, denoting the effect of the additional negative PAA layer incorporated onto membrane surface. Significant changes can be observed in the thickness of the membranes under investigation after swelling, denoting an increase of the thickness of swelled membranes of around 36–38% compared to their thicknesses at dry conditions, which demonstrates the essential role of the membrane operating conditions for improved RED performance, since higher thicknesses might result in increased membrane electro-resistance, thus reducing the obtainable net power density. On the other hand, focusing on the study of diameter differences, a similar trend is observed, even though the values are increased after swelling by 5% in most of the cases. These dimension changes suggest that either the thickness and diameter of the membranes might be affected by swelling conditions, which may involve alterations in membrane electro-resistance, permselectivity and, therefore, RED process efficiency. Overall, this analysis demonstrates the importance of controlling the dimensions of AEMs with the purpose of optimizing membrane design and operating conditions for RED applications.

The analysis of functional groups via *ATR-FTIR* spectra is shown in Figure 4, where both the responses of unmodified and one side/both sides PAA-modified AEMs are presented in an attempt to demonstrate successful membrane modification and stability.

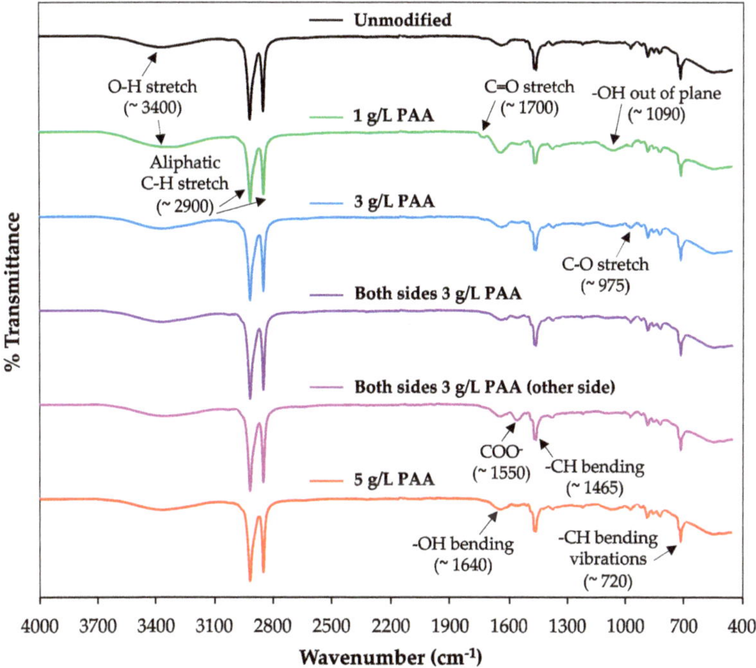

Figure 4. Fourier attenuated atomic force microscopy (*ATR-FTIR*) spectra of unmodified and poly(acrylic) acid (PAA)-modified AEMs.

Similar *ATR-FTIR* profiles can be observed for the modified and unmodified membranes, which might be explained partially by taking into account that the amount of the modifying agent at the membrane is not enough to produce significant changes in *FTIR* spectra. This fact suggests the absence of chemical reactions between the PAA and the membrane, thus demonstrating that the attachment is electrostatic. Consequently, since PAA is not covalently bound to the membrane, the amount of PAA before membrane washing with Trizma® solution must be higher and, therefore, *FTIR* signals for the PAA-modified AEMs should be stronger, helping to easily identify the expectable contribution of PAA to the FTIR spectra. However, the washing step is essential because the goal is to remove the loosely attached PAA, avoiding the release of this modifying agent during the process while ensuring membrane stability. Additionally, further difficulties in accessing the presence of PAA are present because both, Ralex membrane and PAA, have similar groups.

Thus, the different bands/peaks observed in the spectra correspond to the polyester fabric (Ralex AEM) and the PAA used for creating the negative monolayer on membrane surface. For instance, the first band that can be observed at $\approx$3400 cm^{-1} is related to the O–H stretch of the carboxylic group that is present in both polyester and PAA, whereas the next two peaks at around 2900 cm^{-1} are associated with the C–H stretch of the two substances, corresponding to –CH$_2$ and –CH$_3$ groups. At lower wavenumbers, the C=O stretch can also be identified ($\approx$1700 cm^{-1}), which relates to polyester and to the carboxylic group of the PAA. Besides, the sharp bands at around 1640 and 1465 cm^{-1} correspond to PAA and polyester as –OH and –CH bendings, respectively [42,43]. At 1090 cm^{-1}, the "shaped like U" peak might be either associated to -OH out of plane (carboxylic group) or O=C–O–C stretching of the main polymer of the membrane [43,44]. In this regard, the band is more visible in the modified membranes compared to the response achieved for unmodified AEMs (black spectrum), which the authors relate to the inclusion of an additional –OH group during surface modification with PAA. On the other hand, the C–O stretch referred to the glycol was also observed in the six spectra at 975 cm^{-1}, approximately [44], including the signal of C=C stretching related to the benzene ring of polyester at a similarly wavenumber. Moreover, –CH bending vibrations signal was also identified at 720 cm^{-1} [45]. Since the pH of the solution during modification was higher than 6.4, the dissociation of COOH into COO$^-$ and H$^+$ is clearly presented in the PAA-modified membranes at $\approx$1550 cm^{-1} [46], which evidences the presence of PAA on the modified membrane surfaces. This *ATR-FTIR* analysis, therefore, demonstrates the high chemical stability of the polyester-based AEMs after adding a negative PAA monolayer onto their surfaces.

3.2. Electrochemical Characterization: Cyclic Voltammetry and Impedance Spectroscopy Measurements

The electrochemical characterization of the prepared AEMs was firstly evaluated through *CV* measurements. Firstly, the one side modified membrane with a PAA concentration of 1 g/L was selected (owing to its higher water uptake properties) in an attempt to evaluate its electrochemical response as a function of (i) type of electrodes, (ii) relative electrodes position in the diffusion cell (feed or receiver compartment), and (iii) HA presence, with the purpose of determining the optimum conditions for RED process operation. In this regard, Cu electrodes were first employed in this study, but due to the problem of Cu oxidation at this potential window (i.e., −0.6–0.6 V) as well as the low current responses achieved, the application of graphite and Ag electrodes was further evaluated. However, the application of graphite electrodes in both compartments resulted in poor electrochemical responses as well, which can be associated with the low conductivity of this material at room temperature, even though the combination of a graphite electrode with a silver rod one resulted in an improved membrane behavior in terms of current-voltage response. Nevertheless, none of the possible combinations between Ag and graphite rods allowed us to reach the ideal resistor performance. This behavior, however, was reached at Ag-Ag electrodes due to the high stability and conductivity of this material in aqueous solutions. Besides, it is important to highlight that this combination is able to reach zero current at zero voltage, approximately, which is essential from an electro-membrane process point of view. Moreover, the presence/absence of HA did not affect the current voltage profile, denoting the negligible impact on the overall transport of ions across the membrane under investigation, although the monovalent permselectivity of the membrane can be reduced (negatively affected) in the presence of HA. Therefore, this Ag-Ag combination was selected to be used in further *CV* characterization analyses.

Thus, the effect of PAA concentration during modification step as well as the comparison between the different one side modified AEMs and the unmodified one in terms of current-voltage (I-E) behavior was also investigated in the presence/absence of HA, as presented in Figure 5. No significant changes were observed in terms of current-voltage profiles for all membranes, even though the PAA-modified AEMs presented higher current responses due to their improved hydrophilic properties, as demonstrated by the contact angle and water uptake data obtained, which results in a higher overall transport of ions through the corresponding modified membranes. It is also worth noting that very similar currents were achieved at −0.6 V for the AEMs modified with 3 and 5 g/L of PAA, respectively. The best membrane performance, that is, the one which is able to reach higher currents at −0.6 V, was observed to be the modified AEM with 3 g/L of PAA-based solution (in the absence of HA), denoting that there is an optimum PAA concentration level in terms of overall ions transport and behavior for RED. Moreover, the low overall current-voltage responses achieved in the presence of HA (25 ppm at the feed compartment), especially at high PAA concentrations (i.e., 3 and 5 g/L) may be explained by considering the fact that part of the HA is deposited onto membrane surface, this reducing the overall transport of ionic species through the membrane under investigation. This issue is further evaluated at the end of this section via mass transport experiments. It is also worth noting that linear sweep voltammetry (*LSV*) tests were carried out at the same conditions to confirm that the I-E curves behave the same as the stable cycles reached via *CV* analyses.

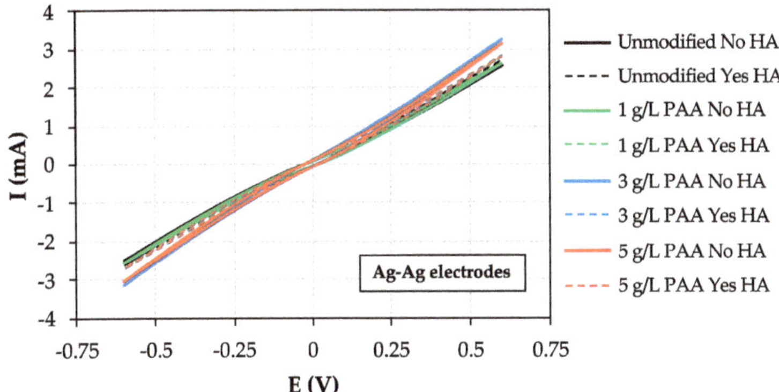

Figure 5. Cyclic voltammetry analyses: effect of PAA concentration in AEM modification at Ag electrodes.

Besides, with the purpose of evaluating the effect of the nature of the co-ion present in the feed compartment (simulating river water streams), Table 4 reports the current obtained at the maximum applied voltage (i.e., 0.6 V) for different membranes (i.e., unmodified, one side 3 g/L modified and one side 5 g/L modified AEMs) as a function of the monovalent salt involved in the feed electrolyte composition.

Table 4. Currents obtained at the maximum applied voltage of 0.6 V as a function of the feed composition and the type of membrane. Receiver solution: 0.5 M NaCl.

Membrane Type	Feed Solution	Current (mA) at 0.6 V
(1) Unmodified	$KCl + K_2SO_4$	3.5
	$KCl + Na_2SO_4$	3.2
	$NaCl + Na_2SO_4$	2.7
	$LiCl + Na_2SO_4$	2.6
(3) One side 3 g/L PAA	$KCl + K_2SO_4$	3.4
	$KCl + Na_2SO_4$	3.3
	$NaCl + Na_2SO_4$	3.2
	$LiCl + Na_2SO_4$	2.8
(3) One side 5 g/L PAA	$KCl + K_2SO_4$	3.8
	$KCl + Na_2SO_4$	3.4
	$NaCl + Na_2SO_4$	3.2
	$LiCl + Na_2SO_4$	3.0

As it can be observed by comparing the currents reached for the different feed compositions due to the increasing hydrated ionic radii in the order $K^+ < Na^+ < Li^+$, the registered currents decreased in the same order (Table 4). This behavior is clearly shown regardless whether the membranes were modified or not, which could be attributed to the more efficient Donnan exclusion of more hydrated (i.e., with bigger sizes) co-ions from the AEMs.

Furthermore, *EIS* experiments were carried out at a constant voltage of 50 mV and high frequency levels (0.5 MHz to 100 Hz) to focus on the following two key parameters: (i) membrane electro-resistance (R_M) and (ii) electric double-layer resistance (R_{EDL}), both obtained at high and moderate frequencies considering the frequency range of the scope of this study, respectively. Although the diffusion boundary layer resistance (R_{DBL}) is often neglected under these operating conditions, its effect has also been evaluated in this comprehensive study for the sake of clarity.

The equivalent circuit analysis tool was used to fit the results obtained from EIS measurements in order to obtain the three target parameters. In this respect, the equivalent circuit which better fits the data obtained with high accuracy for the unmodified and the modified membranes is associated with the series and parallel combination of a resistor (R) and two capacitors (C), respectively, as shown in Figure 6. At higher frequencies (i.e., ≥ 10 kHz), the real impedance measured at zero phase shift represents R_M.

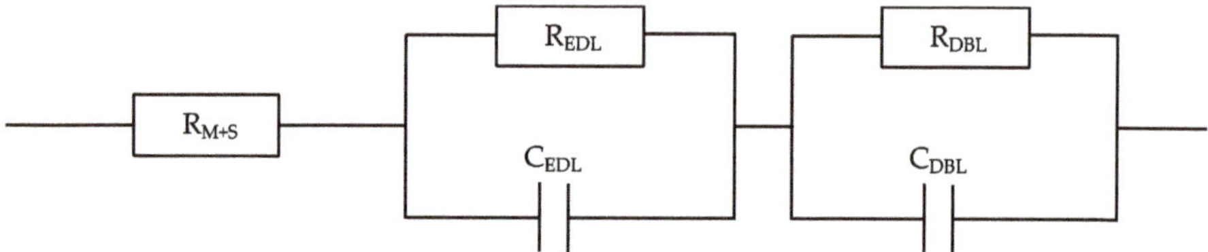

Figure 6. Equivalent circuit diagram showing the combination of membrane, electric double-layer, and diffusion boundary layer resistances.

In this context, R_M was firstly experimentally calculated by subtracting the electrical resistance of the blank experiment (solution flowing without membrane in the redox flow cell) to the combined resistance between the membrane and the solution (R_{M+S}) at zero phase shift. Each experimentally measured R_M value was compared and validated with the value obtained by using the equivalent circuit model provided by the Ivium apparatus incorporated software. The evolution of the experimentally measured impedance and phase shift (φ) of the unmodified AEM as a function of frequency range is represented by the Bode plot (Figure 7a) as an example, where three replicates (same membrane) at the same conditions were run to obtain the key parameters with accuracy (i.e., averaged R_M, R_{EDL}, and R_{BDL}), including the fitting results for the sake of comparison. Taking into account the experimental data and the equivalent circuit analysis at zero phase shift, the electrical resistance of the unmodified membrane (R_M), considering the geometric active area of the redox flow cell was 5.01 $\pm$ 0.52 $\Omega \cdot cm^2$ (at 1.7×10^4 Hz). Subsequently, both R_{EDL} and R_{DBL} were obtained using the equivalent circuit tool provided by the Ivium software of the potentiostat. In this regard, the R_{EDL} was found to be significantly lower than membrane resistance, reaching an average value of 1.83 $\pm$ 0.23 $\Omega \cdot cm^2$ for the frequency range from 500 to 1.7×10^4 Hz, which denotes the fact that the restriction of ions transfer is higher in the membrane. Finally, the effect of the DBL (observed in the frequency range of 100–500 Hz) was also quantified in terms of resistance (R_{DBL}), achieving an averaged value of 0.74 $\pm$ 0.15 $\Omega \cdot cm^2$, which represents a considerably lower resistance value compared either with R_M and R_{EDL}, as previously expected due to the frequency range considered.

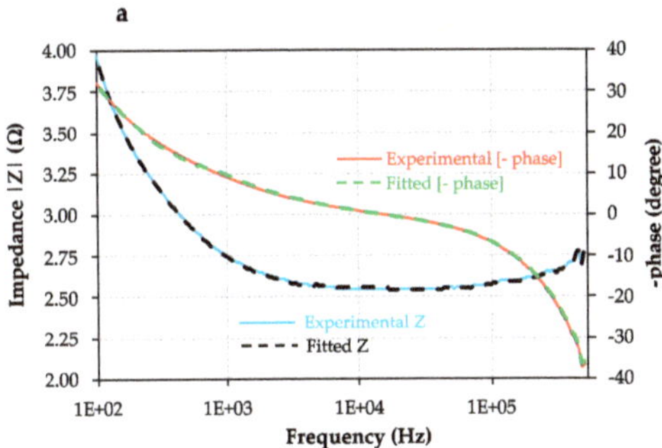
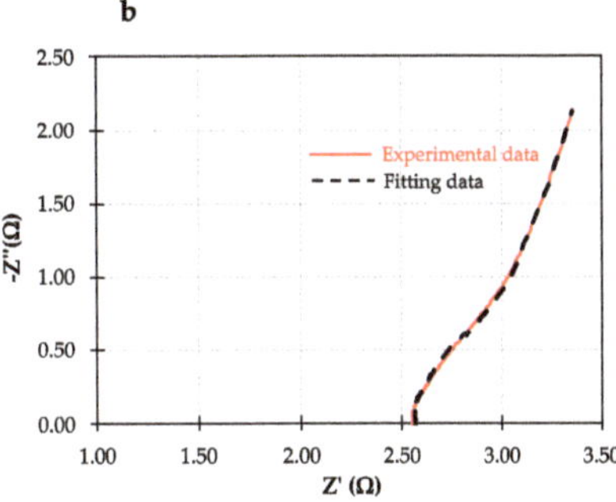

Figure 7. Electrochemical impedance spectroscopy (*EIS*) study of the unmodified membrane including experimental (continuous line) and fitting data (dotted lined): (**a**) Bode plot; (**b**) Nyquist plot.

On the other hand, Figure 7b shows through the Nyquist plot both real (Z′) and imaginary (Z″) impedances, which were measured for the unmodified membrane.

In this work, a part of the typical well-defined semicircle in Nyquist plots is appearing due to the frequency range from 0.5 MHz to 100 Hz considered in this study in order to avoid possible salt accumulation in the graphite blocks of the electrochemical cell at low frequencies (personal communication to the authors). However, the mentioned accuracy of the equivalent circuit considered for the determination of the target parameters was high enough as demonstrated by the X^2 error function (0.002 approximately).

The averaged membrane electro-resistances of the different tested membranes are shown in Table 5, including both averaged EDL and DBL effects. The small increase in R_M when introducing PAA onto the membrane surface might be associated with the higher thicknesses of the PAA-modified AEMs. Besides, R_{EDL} values for the PAA-modified AEMs are closer to the parameter achieved for the unmodified AEM (around 1.8 $\Omega \cdot cm^2$). Therefore, these small differences in both R_M and R_{EDL} are insignificant, thus indicating that the addition of PAA monolayers with different modifying concentrations onto heterogeneous AEM surfaces is not compromising neither the electrical conductivity nor the ohmic/non-ohmic resistances of the different prepared PAA-AEMs under investigation, which is crucial for improving the obtainable net power density from RED. On the other hand, although both R_M and R_{EDL} are the dominant resistances of the system, increased R_{DBL} values of up to 0.9 $\Omega \cdot cm^2$ were expectedly reached at lower frequencies (100–500 Hz) for all membranes. As well known, in RED applications this effect can be reduced by either increasing the flow rate or inducing turbulence [47].

Table 5. AEM electro-resistances, electric double-layer, and diffusion boundary layer effects in 0.5 M NaCl aqueous solutions, measured through equivalent circuit model tool.

Membrane Type	R_M ($\Omega \cdot cm^2$)	R_{EDL} ($\Omega \cdot cm^2$)	R_{DBL} ($\Omega \cdot cm^2$)	C_{EDL} (μF)	C_{DBL} (μF)	X^2 Error Function
(1) Unmodified	5.01 ± 0.52	1.83 ± 0.23	0.74 ± 0.15	115 ± 16	46 ± 10	0.0023
(2) One side 1 g/L PAA	5.14 ± 0.50	1.98 ± 0.08	0.89 ± 0.14	200 ± 19	93 ± 16	0.0028
(3) One side 3 g/L PAA	5.21 ± 0.04	1.56 ± 0.18	0.67 ± 0.08	188 ± 12	67 ± 8	0.0023
(4) Both sides 3 g/L PAA	5.33 ± 0.16	1.92 ± 0.17	0.53 ± 0.01	72 ± 12	19 ± 4	0.0020
(5) One side 5 g/L PAA	5.36 ± 0.18	1.58 ± 0.02	0.68 ± 0.02	140 ± 15	42 ± 5	0.0023

It is also worth noting that one of the prepared membranes (i.e., one side 3 g/L PAA) was selected to evaluate the reproducibility of the modification procedure proposed, which represents an important aspect to be considered for the practical application of the developed modified AEMs. In this context, three different membrane samples were modified (independently) under the same conditions with 3 g/L

PAA-based solutions. The three modified membranes were fully characterized (physicochemically and electrochemically), and the standard deviations of the different results are considered in the results referred to the membrane (one side 3 g/L PAA) in the whole manuscript, highlighting the high reproducibility (small standard deviation) of the results obtained for this membrane.

Thus, the results obtained suggest that the PAA is distributed uniformly on the membrane surface, which may lead to a reduced disorderliness and surface heterogeneity, as well as decreased charge transfers. As a result, the capacitance of the electric double-layer and the diffusion boundary layer (C_{EDL} and C_{DBL}, respectively) can be obtained from the selected equivalent circuit model with high accuracy according to the low values achieved for the X^2 error function.

The unmodified membrane and the one side 3 g/L PAA-modified AEM were selected in this study to evaluate the effect of the co-ion through EIS analyses. Table 6 summarizes the *EIS* data obtained as a function of the used aqueous salt solution.

Table 6. EIS results via equivalent circuit model tool in 0.5 M NaCl, 0.5 M KCl, and 0.5 M LiCl aqueous solutions.

Membrane Type		R_M ($\Omega \cdot cm^2$)	R_{EDL} ($\Omega \cdot cm^2$)	R_{DBL} ($\Omega \cdot cm^2$)	C_{EDL} (μF)	C_{DBL} (μF)	X^2 Error Function
	LiCl	5.35 ± 0.03	2.32 ± 0.44	0.35 ± 0.00	147 ± 4	59 ± 3	0.0013
(1) Unmodified	NaCl	5.01 ± 0.52	1.83 ± 0.23	0.74 ± 0.15	115 ± 16	46 ± 10	0.0023
	KCl	4.88 ± 0.02	2.17 ± 0.57	0.30 ± 0.00	166 ± 6	66 ± 5	0.0011
	LiCl	5.49 ± 0.02	1.79 ± 0.30	0.49 ± 0.00	145 ± 3	55 ± 2	0.0013
(3) One side 3 g/L PAA	NaCl	5.21 ± 0.04	1.56 ± 0.18	0.67 ± 0.08	188 ± 12	67 ± 8	0.0023
	KCl	4.87 ± 0.01	1.27 ± 0.11	0.39 ± 0.00	162 ± 2	60 ± 1	0.0011

As expected, the membrane electro-resistances decrease following the order LiCl > NaCl > KCl, corresponding to the increasing ionic mobility (i.e., decreasing ionic hydrated radii) in the order Li^+ < Na^+ < K^+. The most relevant data in terms of RED are those for NaCl because of its abundance in seawater.

Finally, in order to compare the membrane resistance results of the present study with the performance of different modified commercial AEMs in terms of this key parameter, Table 7 shows the membrane resistance values (before and after modification) reported in several studies, focusing not only on immersion modification methods (similar strategies to the one proposed in this work), but also taking into account alternative modification approaches.

The application of heterogeneous membranes generally involves a higher membrane electro-resistance in comparison with homogeneous membranes. In this context, the change in their resistance after modification is quite low (see Table 5). By contrast, although the initial (unmodified) membrane resistance of homogeneous AEMs is lower due to their lower thickness and uniform distribution of fixed functional groups, the change in this parameter after modification is considerably higher; in some cases, even higher than that typical for heterogenous membranes. In any case, it is also worth noting, that AEMs with monovalent selective properties are advantageous for RED due to the negative impact of multivalent ions presence on the obtainable net power density. Therefore, there is a trade-off between membrane monovalent permselectivity and membrane electro-resistance which must be optimized in each particular case.

Table 7. Comparison of the membrane resistance values obtained in this study with values reported in literature for other commercial AEMs before and after modification.

AEM Type	Modification Approach	Modifying Agent	R_M Before Modification ($\Omega \cdot cm^2$)	R_M After Modification ($\Omega \cdot cm^2$)	Reference
Heterogeneous Ralex AM-PES (Mega a.s.)	Direct contact/immersion	Poly(acrylic) acid	5.01	5.1–5.4	This study
Heterogeneous AEM (Zhe-jiang Qianqiu Environmental Protection & Water Treatment Co. Ltd.)	LbL deposition	Glutaraldehyde and poly(ethyleneimine)	4.5	4.8	[48]
Neosepta AMX (Astom Corp.)	Dip coating	Polydopamine (PDA)	1.2	2.9	[30]
Neosepta AMX (Astom Corp.)	Immersion	PDA	2.5	5.0	[49,50]
Homogeneous Neosepta ASE (Astom Corp.)	Immersion (co-deposition)	PDA and poly (sodium 4-styrene sulfonate)	3.6	4.5	[51]
Homogeneous JAM-II-07 (Yanrun)	Coating by deposition	Sulfonated reduced graphene oxide nanosheets	3.1	3.7	[52]
Homogeneous Type I (Fujifilm)	Self-adhesion deposition	Sulfonated polydopamine	1.0	6.8	[53]
AEM * (Ionics)	Coating by adsorption	Olygourethane surfactants and disodium salt α, ω-oligooxipropylene-bis(o-urethane-2.4, 2.6 tolueneurylbenzene sulphonic acid)	2.5	5.7	[54]

* No specific membrane name is reported.

3.3. Mass Transport Experiments: Sulfate Rejection Study

Mass transport experiments were carried out to evaluate the behavior of the different negatively charged monolayers incorporated onto unmodified AEM surfaces in terms of sulfate rejection, leading to an improved monovalent permselectivity, which would affect favorably the RED process performance. Figure 8 reports the evolution of sulfate concentration with time in both compartments (feed, *F*, and receiver, *R*) as a function of the investigated AEM, including unmodified and modified PAA-AEMs in the absence of HA.

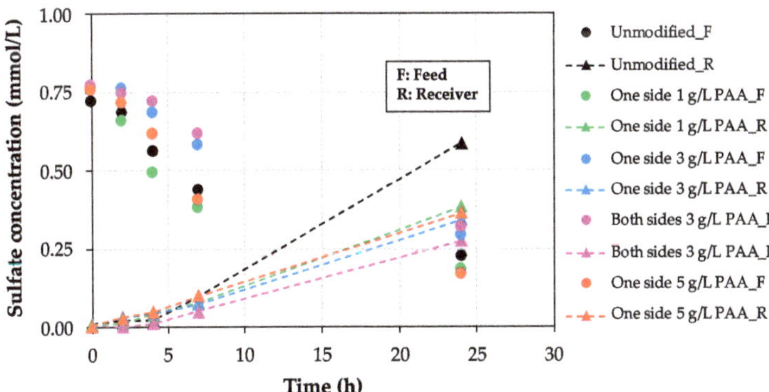

Figure 8. Sulfate evolution with time during mass transport experiments as a function of the AEM used in the absence of humic acid (HA) in the feed compartment.

Since at the beginning of the experiment sulfate is only present in the feed (low salt concentration) compartment, the analysis is focused on the evolution of sulfate concentration in the receiver compartment (high salt concentration) for 24 h (dotted lines) in an attempt to follow the sulfate transport through the membrane, thus focusing on sulfate rejection. As can be seen, sulfate rejection is improved as the concentration of PAA increases, although the modified AEM with the maximum PAA concentration (i.e., 5 g/L) did not show the best behavior, which can be associated with hydrophilicity losses as shown in contact angle analysis (see Figure 3). The highest sulfate rejection value is achieved when both sides of the membrane are modified with 3 g/L PAA-based solution (pink dotted line). Thus, the rejection of sulfate is 36%, 42%, 39%, and 54% enhanced for the one side 1 g/L, 3 g/L, 5 g/L, and both sides 3 g/L PAA-modified AEMs, respectively, in comparison with the reached value for the commercial unmodified Ralex-AEM, clearly demonstrating the positive effect of modifying heterogeneous AEMs with PAA solutions to improve the monovalent permselectivity of these membranes for RED applications.

The presence of HA in the feed compartment during mass transport tests was also investigated, as shown in Figure 9. The presence of 25 ppm of HA in the feed solution involves a negative effect on sulfate rejection owing to fouling phenomena. Surprisingly, the behavior of commercial Ralex AEMs was improved in the presence of HA comparing the concentration of sulfate after 24 h of experiment with the value observed in its absence for the same membrane, denoting behavior changes when a model organic foulant is brought into play. Nonetheless, the performance of the PAA-modified AEMs is worse than that achieved for the unmodified membrane, which demonstrates the high negative effect of fouling phenomena on sulfate rejection and, therefore, on membrane permselectivity. As a result, the concentration of HA in both compartments was monitored with time by measuring the absorbance of each sample at a wavelength of 280 nm. In this respect, it was demonstrated that HA is not crossing the corresponding AEM from the feed compartment to the adjacent one (absorbance very close to zero) in any of the tests examined, which evidences the fact that HA was not present in that compartment. The observed small decrease of HA concentration in the feed solution with time is therefore associated with HA attachment on membrane surface. This phenomenon was confirmed by observing a slight change in the color of the membrane surface (darker orange-like color) that was in contact with the feed solution containing 25 ppm of HA. In short, this study denotes that the behavior of the PAA-modified AEMs is clearly negatively affected by the presence of organic foulants, which would reduce the efficiency of the RED process performance. In this context, the application and study of real natural streams of different salinity, which contains different multivalent ions (i.e., Ca^{2+} and Mg^{2+}) as well as different foulants, is essential for further understanding of membrane behavior in order to develop and implement the RED technology at industrial scale.

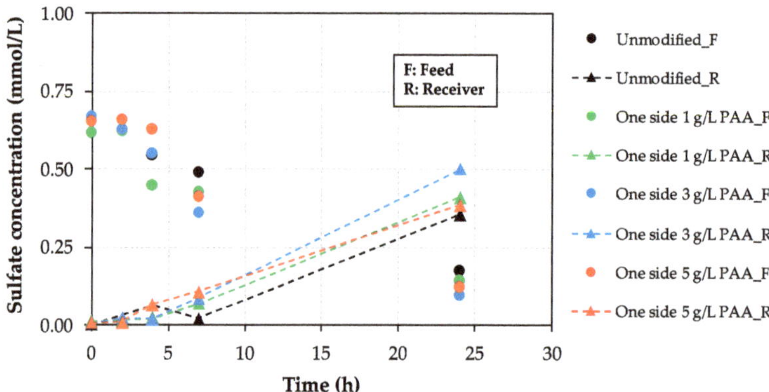

Figure 9. HA studies: sulfate evolution with time during mass transport experiments as a function of the one side modified AEM used.

Moreover, in order to further support the feasibility of the PAA-based modification procedure, the achieved sulfate fluxes (mmol/(m²·h)) were estimated through the sulfate concentration time profiles in the receiver compartment for the one side modified AEMs. The concentration differences between 7 and 24 h (when the evolution of the sulfate concentration in time is linear (Figure 8)) were considered for the respective sulfate fluxes calculations and the data obtained are presented in Table 8.

Table 8. Summary of the sulfate flux results for the one side PAA-modified AEMs as a function of the absence/presence of humic acid in the feed chamber.

Membrane Type	Sulfate Flux (mmol/(m^2·h))
(1) Unmodified	3.6
(2) One side 1 g/L PAA	2.2
(3) One side 3 g/L PAA	1.9 ± 0.1
(5) One side 5 g/L PAA	1.9

The flux of sulfate was nearly halved for the 3 and 5 g/L PAA modified membranes compared to that for the unmodified membrane, thus confirming the improved Cl^-/SO_4^{2-} permselectivity under these operating conditions. The optimal modifying agent (PAA) concentration was equal to 3 g/L.

The future outlook of this research will cover the design, set-up, and long-term operation of a RED stack in order to evaluate the obtainable net power density by using the proposed modified AEMs under real conditions, i.e., using natural feedwaters, which is essential to move forward towards the large-scale implementation of the RED technology.

4. Conclusions

In this work, we investigated the complete characterization of poly(acrylic) acid-modified monovalent-anion-permselective membranes for Reverse Electrodialysis applications, where the effect of poly(acrylic) acid concentration during the membrane modification step (from 1 to 5 g/L) was evaluated through several characterization techniques, including mass transport experiments. The following insights can be derived from the results obtained in this study:

- Improved membrane hydrophilicity properties are shown via contact angle analyses for the poly(acrylic) acid (optimal concentration of 3 g/L) modified membranes in comparison with the behavior of the unmodified one.
- The importance of the method used to evaluate the ion exchange capacity of anion exchange membranes is demonstrated, depending on the nature of the replacing anions.
- The swelling effect was investigated in terms of dimension changes (i.e., thickness and diameter). The thickness of swelled membranes is increased by 27%, whereas the diameter is widened by 5% in most of the cases at the same conditions. This analysis highlights the essential role of both thickness (higher thicknesses might result in increased membrane electro-resistance) and diameter (modifications in membrane area may affect process efficiency) to optimize membrane design for RED applications.
- The analysis of functional groups present on membrane surfaces demonstrates the high chemical stability of the polyester-based anion exchange membranes after adding a negative poly(acrylic) acid monolayer onto their surfaces, suggesting the absence of chemical reactions between the modifying agent and the membrane, thus demonstrating that the attachment is electrostatic.
- The use of silver electrodes in cyclic voltammetry measurements allowed to reach ideal resistor behavior. The modified membranes present higher current-voltage responses due to improved hydrophilic properties, which involves a higher overall transport of anions through the corresponding modified membrane, even though the presence of humic acid as model foulant involved a certain decrease in this overall transport owing to its attachment onto the membrane surface.
- The membrane electro-resistances, double-layer resistances, and diffusion boundary layer resistances of the different modified membranes were in the same order of magnitude compared to

the unmodified anion exchange membrane (i.e., 5.0–5.4 $\Omega \cdot cm^2$, 1.6–2.0 $\Omega \cdot cm^2$, and 0.5–0.9 $\Omega \cdot cm^2$, respectively) in 0.5 M NaCl aqueous solutions. The small difference observed in the modified ones might be associated with their higher membrane thicknesses. Therefore, the electrical conductivity of the different prepared modified membranes is not compromised by the addition of a negative monolayer onto their surfaces with uniform characteristics, which might involve a reduction of both surface heterogeneity and disorderliness. The membrane electro-resistances decrease in external electrolytes following the order LiCl > NaCl > KCl, owing to the increasing hydrated radii (decreased ionic mobility) in the order $K^+ < Na^+ < Li^+$. The membrane electro-resistance results obtained in the present study after membrane modification are comparable to those reported in literature for both modified heterogeneous and homogeneous anion exchange membranes.

- Mass transport tests finally prove that the rejection of sulfate (monovalent permselectivity) is improved in the absence of humic acid as the concentration of poly(acrylic) acid increases up to 3 g/L. In this respect, when both sides of the membrane are modified (3 g/L), sulfate rejection is enhanced by 54% compared to the performance of the unmodified membrane, thus suggesting an improved reverse electrodialysis process performance. Nevertheless, the behavior of the modified samples is clearly negatively affected by the presence of organic foulants such as humic acid. The sulfate flux results show that the optimal modifying agent concentration is equal to 3 g/L of poly(acrylic) acid.

Although this study provides new insights and fundamental knowledge for the continuous development of hydrophilic, environmentally friendly, stable, and durable functionalized anion exchange membranes for an enhanced reverse electrodialysis performance, the following challenges, among others, still need to be addressed to make this electro-membrane process feasible/preferred at industrial scale: (i) anion exchange membrane fouling understanding including fouling mechanisms and collective behavior of foulants under natural saline streams conditions, (ii) development of appropriate pre-treatment and cleaning strategies to increase membrane durability and its re-use, (iii) design of greener and cheaper tailor-made anion exchange membranes and modification procedures.

Author Contributions: Conceptualization, S.V.; methodology: S.V., C.A.M.P. and I.M.-G.; software: I.M.-G. and F.K.; validation: I.M.-G. and F.K.; formal analysis: I.M.-G., F.K., C.A.M.P. and S.V.; investigation: I.M.-G. and F.K.; resources: S.V.; data curation: I.M.-G. and F.K.; writing—original draft preparation: I.M.-G.; writing—review and editing: F.K., C.A.M.P, J.G.C. and S.V.; visualization: I.M.-G. and F.K.; supervision: S.V.; project administration: S.V.; funding acquisition: S.V and J.G.C. All authors have read and agreed to the published version of the manuscript.

Funding: This research was funded by "Programa Operacional Regional de Lisboa, na componente FEDER" and "Fundação para a Ciência e Tecnologia, I.P.". (FCT) through research project PTDC/EQU-EPQ/29579/2017. This work was also supported by the Associate Laboratory for Green Chemistry-LAQV, which is financed by national funds from FCT/MCTES (UID/QUI/50006/2019).

Acknowledgments: Ivan Merino-Garcia would like to acknowledge his post-doctoral contract in the frame of Research Project PTDC/EQU-EPQ/29579/2017. Francis Kotoka acknowledges his scholarship awarded by the European Commission-Education, Audiovisual and Culture Executive Agency (EACEA), under the program: Erasmus Mundus Master in Membrane Engineering for a Sustainable World—EM3E-4SW.

References

1. Pawlowski, S.; Huertas, R.M.; Galinha, C.F.; Crespo, J.G.; Velizarov, S. On operation of reverse electrodialysis (RED) and membrane capacitive deionisation (MCDI) with natural saline streams: A critical review. *Desalination* **2020**, *476*, 114183. [CrossRef]

2. Gao, H.; Zhang, B.; Tong, X.; Chen, Y. Monovalent-anion selective and antifouling polyelectrolytes multilayer anion exchange membrane for reverse electrodialysis. *J. Membr. Sci.* **2018**, *567*, 68–75. [CrossRef]

3. Vermaas, D.A.; Kunteng, D.; Veerman, J.; Saakes, M.; Nijmeijer, K. Periodic feedwater reversal and air sparging as antifouling strategies in reverse electrodialysis. *Environ. Sci. Technol.* **2014**, *48*, 3065–3073. [CrossRef]

4. Gómez-coma, L.; Ortiz-martínez, V.M.; Carmona, J.; Palacio, L.; Prádanos, P.; Fallanza, M.; Ortiz, A.; Ibañez, R.; Ortiz, I. Modeling the influence of divalent ions on membrane resistance and electric power in reverse electrodialysis. *J. Membr. Sci.* **2019**, *592*, 117385. [CrossRef]

5. Veerman, J.; Vermaas, D.A. Reverse electrodialysis: Fundamentals. In *Sustainable Energy from Salinity Gradients*; Elsevier Ltd: Amsterdam, The Netherlands, 2016.

6. Pintossi, D.; Saakes, M.; Borneman, Z.; Nijmeijer, K. Electrochemical impedance spectroscopy of a reverse electrodialysis stack: A new approach to monitoring fouling and cleaning. *J. Power Source* **2019**, *444*, 227302. [CrossRef]

7. Güler, E.; van Baak, W.; Saakes, M.; Nijmeijer, K. Monovalent-ion-selective membranes for reverse electrodialysis. *J. Membr. Sci.* **2014**, *455*, 254–270. [CrossRef]

8. Pintossi, D.; Chen, C.; Saakes, M.; Nijmeijer, K.; Borneman, Z. Influence of sulfate on anion exchange membranes in reverse electrodialysis. *NPJ Clean Water* **2020**, *3*, 29. [CrossRef]

9. Ortiz-Martínez, V.M.; Gómez-Coma, L.; Tristán, C.; Pérez, G.; Fallanza, M.; Ortiz, A.; Ibañez, R.; Ortiz, I. A comprehensive study on the effects of operation variables on reverse electrodialysis performance. *Desalination* **2020**, *482*, 114389. [CrossRef]

10. Ortiz-Imedio, R.; Gomez-Coma, L.; Fallanza, M.; Ortiz, A.; Ibañez, R.; Ortiz, I. Comparative performance of Salinity Gradient Power-Reverse Electrodialysis under different operating conditions. *Desalination* **2019**, *457*, 8–21. [CrossRef]

11. Luque Di Salvo, J.; Cosenza, A.; Tamburini, A.; Micale, G.; Cipollina, A. Long-run operation of a reverse electrodialysis system fed with wastewaters. *J. Environ. Manag.* **2018**, *217*, 871–887. [CrossRef]

12. Pawlowski, S.; Crespo, J.G.; Velizarov, S. Profiled ion exchange membranes: A comprehensible review. *Int. J. Mol. Sci.* **2019**, *20*, 165. [CrossRef] [PubMed]

13. Rijnaarts, T.; Moreno, J.; Saakes, M.; de Vos, W.M.; Nijmeijer, K. Role of anion exchange membrane fouling in reverse electrodialysis using natural feed waters. *Colloids Surf. A Physicochem. Eng. Asp.* **2019**, *560*, 198–204. [CrossRef]

14. Mikhaylin, S.; Bazinet, L. Fouling on ion-exchange membranes: Classification, characterization and strategies of prevention and control. *Adv. Colloids Interface Sci.* **2016**, *229*, 34–56. [CrossRef] [PubMed]

15. Moreno, J.; de Hart, N.; Saakes, M.; Nijmeijer, K. CO_2 saturated water as two-phase flow for fouling control in reverse electrodialysis. *Water Res.* **2017**, *125*, 23–31. [CrossRef] [PubMed]

16. Park, J.S.; Lee, H.J.; Choi, S.J.; Geckeler, K.E.; Cho, J.; Moon, S.H. Fouling mitigation of anion exchange membrane by zeta potential control. *J. Colloids Interface Sci.* **2003**, *259*, 293–300. [CrossRef]

17. Hao, L.; Liao, J.; Jiang, Y.; Zhu, J.; Li, J.; Zhao, Y.; Van der Bruggen, B.; Sotto, A.; Shen, J. "Sandwich"-like structure modified anion exchange membrane with enhanced monovalent selectivity and fouling resistant. *J. Membr. Sci.* **2018**, *556*, 98–106. [CrossRef]

18. Kingsbury, R.S.; Liu, F.; Zhu, S.; Boggs, C.; Armstrong, M.D.; Call, D.F.; Coronell, O. Impact of natural organic matter and inorganic solutes on energy recovery from five real salinity gradients using reverse electrodialysis. *J. Membr. Sci.* **2017**, *541*, 621–632. [CrossRef]

19. Fernandez-Gonzalez, C.; Kavanagh, J.; Dominguez-Ramos, A.; Ibañez, R.; Irabien, A.; Chen, Y.; Coster, H. Electrochemical impedance spectroscopy of enhanced layered nanocomposite ion exchange membranes. *J. Membr. Sci.* **2017**, *541*, 611–620. [CrossRef]

20. Zhao, Y.; Zhu, J.; Ding, J.; Van der Brugge, B.; Shen, J.; Gao, C. Electric-pulse layer-by-layer assembled of anion exchange membrane with enhanced monovalent selectivity. *J. Membr. Sci.* **2018**, *548*, 81–90. [CrossRef]

21. Mulyati, S.; Takagi, R.; Fujii, A.; Ohmukai, Y.; Matsuyama, H. Simultaneous improvement of the monovalent anion selectivity and antifouling properties of an anion exchange membrane in an electrodialysis process, using polyelectrolyte multilayer deposition. *J. Membr. Sci.* **2013**, *431*, 113–120. [CrossRef]

22. Deng, J.; Wang, L.; Liu, L.; Yang, W. Developments and new applications of UV-induced surface graft polymerizations. *Prog. Polym. Sci.* **2009**, *34*, 156–193. [CrossRef]

23. Saracco, G.; Zanetti, M.C. Ion Transport Through Monovalent-Anion-Permselective Membranes. *Ind. Eng. Chem. Res.* **1994**, *33*, 96–101. [CrossRef]

24. Nebavskaya, X.; Sarapulova, V.; Butylskii, D.; Larchet, C.; Pismenskaya, N. Electrochemical properties of homogeneous and heterogeneous anion exchange membranes coated with cation exchange polyelectrolyte. *Membranes* **2019**, *9*, 13. [CrossRef] [PubMed]

25. Zhang, H.; Ding, R.; Zhang, Y.; Shi, B.; Wang, J.; Liu, J. Stably coating loose and electronegative thin layer on anion exchange membrane for efficient and selective monovalent anion transfer. *Desalination* **2017**, *410*, 55–65. [CrossRef]

26. Li, Y.; Shi, S.; Cao, H.; Zhao, Z.; Su, C.; Wen, H. Improvement of the antifouling performance and stability of an anion exchange membrane by surface modification with graphene oxide (GO) and polydopamine (PDA). *J. Membr. Sci.* **2018**, *566*, 44–53. [CrossRef]

27. An, Y.; Jiang, G.; Ren, Y.; Zhang, L.; Qi, Y.; Ge, Q. An environmental friendly and biodegradable shale inhibitor based on chitosan quaternary ammonium salt. *J. Pet. Sci. Eng.* **2015**, *135*, 253–260. [CrossRef]

28. Zhao, Y.; Tang, K.; Liu, H.; Van der Bruggen, B.; Sotto Díaz, A.; Shen, J.; Gao, C. An anion exchange membrane modified by alternate electro-deposition layers with enhanced monovalent selectivity. *J. Membr. Sci.* **2016**, *520*, 262–271. [CrossRef]

29. Besha, A.T.; Tsehaye, M.T.; Aili, D.; Zhang, W.; Tufa, R.A. Design of monovalent ion selective membranes for reducing the impacts of multivalent ions in reverse electrodialysis. *Membranes* **2020**, *10*, 7. [CrossRef]

30. Vaselbehagh, M.; Karkhanechi, H.; Takagi, R.; Matsuyama, H. Surface modification of an anion exchange membrane to improve the selectivity for monovalent anions in electrodialysis - experimental verification of theoretical predictions. *J. Membr. Sci.* **2015**, *490*, 301–310. [CrossRef]

31. Kedem, O.; Schechtmann, L.; Mirsky, Y.; Saveliev, G.; Daltrophe, N. Low-polarisation electrodialysis membranes. *Desalination* **1998**, *118*, 305–314. [CrossRef]

32. Daraei, P.; Madaeni, S.S.; Ghaemi, N.; Ahmadi Monfared, H.; Khadivi, M.A. Fabrication of PES nanofiltration membrane by simultaneous use of multi-walled carbon nanotube and surface graft polymerization method: Comparison of MWCNT and PAA modified MWCNT. *Sep. Purif. Technol.* **2013**, *104*, 32–44. [CrossRef]

33. Madaeni, S.S.; Zinadini, S.; Vatanpour, V. A new approach to improve antifouling property of PVDF membrane using in situ polymerization of PAA functionalized TiO2 nanoparticles. *J. Membr. Sci.* **2011**, *380*, 155–162. [CrossRef]

34. Sarapulova, V.; Shkorkina, I.; Mareev, S.; Pismenskaya, N.; Kononenko, N.; Larchet, C.; Dammak, L.; Nikonenko, V. Transport characteristics of fujifilm ion-exchange membranes as compared to homogeneous membranes AMX and CMX and to heterogeneous membranes MK-40 and MA-41. *Membranes* **2019**, *9*, 84. [CrossRef] [PubMed]

35. Melnikov, S.; Shkirskaya, S. Transport properties of bilayer and multilayer surface-modified ion-exchange membranes. *J. Membr. Sci.* **2019**, *590*, 117272. [CrossRef]

36. Guler, E.; Zhang, Y.; Saakes, M.; Nijmeijer, K. Tailor-made anion-exchange membranes for salinity gradient power generation using reverse electrodialysis. *ChemSusChem.* **2012**, *5*, 2262–2270. [CrossRef]

37. Karas, F.; Hnát, J.; Paidar, M.; Schauer, J.; Bouzek, K. Determination of the ion-exchange capacity of anion-selective membranes. *Int. J. Hydrogen Energy* **2014**, *39*, 5054–5062. [CrossRef]

38. Østedgaard-Munck, D.N.; Catalano, J.; Birch Kristensen, M.; Bentien, A. Data on flow cell optimization for membrane-based electrokinetic energy conversion. *Data BR* **2017**, *15*, 1–11. [CrossRef]

39. Østedgaard-Munck, D.N.; Catalano, J.; Kristensen, M.B.; Bentien, A. Membrane-based electrokinetic energy conversion. *Mater. Today Energy* **2017**, *5*, 118–125. [CrossRef]

40. Zhang, W.; Ma, J.; Wang, P.; Wang, Z.; Shi, F.; Liu, H. Investigations on the interfacial capacitance and the diffusion boundary layer thickness of ion exchange membrane using electrochemical impedance spectroscopy. *J. Membr. Sci.* **2016**, *502*, 37–47. [CrossRef]

41. Villafaña-lópez, L.; Reyes-valadez, D.M.; González-vargas, O.A. Custom-Made Ion Exchange Membranes at Laboratory Scale for Reverse Electrodialysis. *Membranes* **2019**, *9*, 145. [CrossRef]

42. Manatunga, D.C.; De Silva, R.M.; De Silva, K.M.N. Double layer approach to create durable superhydrophobicity on cotton fabric using nano silica and auxiliary non fluorinated materials. *Appl. Surf. Sci.* **2016**, *360*, 777–788. [CrossRef]

43. Ahmadi, E.; Zarghami, N.; Jafarabadi, M.A.; Alizadeh, L.; Khojastehfard, M.; Yamchi, M.R.; Salehi, R. Enhanced anticancer potency by combination chemotherapy of HT-29 cells with biodegradable, pH-sensitive nanoparticles for co-delivery of hydroxytyrosol and doxorubicin. *J. Drug Deliv. Sci. Technol.* **2019**, *51*, 721–735. [CrossRef]

44. Memon, A.A.; Arbab, A.A.; Sahito, I.A.; Sun, K.C.; Mengal, N.; Jeong, S.H. Synthesis of highly photo-catalytic and electro-catalytic active textile structured carbon electrode and its application in DSSCs. *Sol. Energy* **2017**, *150*, 521–531. [CrossRef]

45. Parvinzadeh, M.; Ebrahimi, I. Influence of atmospheric-air plasma on the coating of a nonionic lubricating agent on polyester fiber. *Radiat. Eff. Defects Solids* **2011**, *166*, 408–416. [CrossRef]
46. Alvarez-Gayosso, C.; Canseco, M.A.; Estrada, R.; Palacios-Alquisira, J.; Hinojosa, J.; Castano, V. Preparation and microstructure of cobalt(III) poly (acrylate) hybrid materials. *Int. J. Basic Appl. Sci.* **2015**, *4*, 255–263. [CrossRef]
47. Długołecki, P.; Ogonowski, P.; Metz, S.J.; Saakes, M.; Nijmeijer, K.; Wessling, M. On the resistances of membrane, diffusion boundary layer and double layer in ion exchange membrane transport. *J. Membr. Sci.* **2010**, *349*, 369–379. [CrossRef]
48. Wang, M.; Wang, X.; Liang, J.; Xiang, Y.; Liu, X. An attempt for improving electrodialytic transport properties of a heterogeneous anion exchange membrane. *Desalination* **2014**, *351*, 163–170. [CrossRef]
49. Vaselbehagh, M.; Karkhanechi, H.; Takagi, R.; Matsuyama, H. Biofouling phenomena on anion exchange membranes under the reverse electrodialysis process. *J. Membr. Sci.* **2017**, *530*, 232–239. [CrossRef]
50. Vaselbehagh, M.; Karkhanechi, H.; Mulyati, S.; Takagi, R.; Matsuyama, H. Improved antifouling of anion-exchange membrane by polydopamine coating in electrodialysis process. *Desalination* **2014**, *332*, 126–133. [CrossRef]
51. Cao, R.; Shi, S.; Li, Y.; Xu, B.; Zhao, Z.; Duan, F.; Cao, H.; Wang, Y. The properties and antifouling performance of anion exchange membranes modified by polydopamine and poly (sodium 4-styrenesulfonate). *Colloids Surf. A Physicochem. Eng. Asp.* **2020**, *589*, 124429. [CrossRef]
52. Zhao, Y.; Tang, K.; Ruan, H.; Xue, L.; Van der Bruggen, B.; Gao, C.; Shen, J. Sulfonated reduced graphene oxide modification layers to improve monovalent anions selectivity and controllable resistance of anion exchange membrane. *J. Membr. Sci.* **2017**, *536*, 167–175. [CrossRef]
53. Ruan, H.; Zheng, Z.; Pan, J.; Gao, C.; Van der Bruggen, B.; Shen, J. Mussel-inspired sulfonated polydopamine coating on anion exchange membrane for improving permselectivity and anti-fouling property. *J. Membr. Sci.* **2018**, *550*, 427–435. [CrossRef]
54. Grebenyuk, V.D.; Chebotareva, R.D.; Peters, S.; Linkov, V. Surface modification of anion-exchange electrodialysis membranes to enhance anti-fouling characteristics. *Desalination* **1998**, *115*, 313–329. [CrossRef]

Article

Computational Fluid Dynamics Modeling of the Resistivity and Power Density in Reverse Electrodialysis: A Parametric Study

Zohreh Jalili [1,2], Odne Stokke Burheim [1,*] and Kristian Etienne Einarsrud [2]

1   Department of Energy and Process Engineering, Norwegian University of Science and Technology (NTNU), 7491 Trondheim, Norway; zohreh.jalili@ntnu.no
2   Department of Materials Science and Engineering, Norwegian University of Science and Technology (NTNU), 7491 Trondheim, Norway; kristian.e.einarsrud@ntnu.no
*   Correspondence: burheim@ntnu.no

Received: 30 June 2020; Accepted: 25 August 2020; Published: 29 August 2020

Abstract: Electrodialysis (ED) and reverse electrodialysis (RED) are enabling technologies which can facilitate renewable energy generation, dynamic energy storage, and hydrogen production from low-grade waste heat. This paper presents a computational fluid dynamics (CFD) study for maximizing the net produced power density of RED by coupling the Navier–Stokes and Nernst–Planck equations, using the OpenFOAM software. The relative influences of several parameters, such as flow velocities, membrane topology (i.e., flat or spacer-filled channels with different surface corrugation geometries), and temperature, on the resistivity, electrical potential, and power density are addressed by applying a factorial design and a parametric study. The results demonstrate that temperature is the most influential parameter on the net produced power density, resulting in a 43% increase in the net peak power density compared to the base case, for cylindrical corrugated channels.

Keywords: reverse electrodialysis; computational fluid dynamics; power density; factorial design

---

1. Introduction

The energy economy is facing its most challenging decade, as it must transcend into a more climate-friendly one, as half of the emitted CO_2 due to energy generation and consumption has been targeted for reduction. To achieve this, the technologies used must be changed from those depending on the burning of fossil fuels into electricity and heat, towards technologies which provide electricity and store it in the form of chemical energy. Striving for renewable energy generation, energy storage systems, and renewable hydrogen production, reverse electrodialysis is one of the few technologies that could address all three of these needs [1–3].

Salinity gradient energy (SGE)—particularly RED, which harvests energy produced by mixing two aqueous solutions with different salinities,—has received great interest in the literature [2–11] since its first use, which was reported by Pattle in 1954 [12]. Concentration batteries have also been recently proposed and discussed, which couple salinity gradient energy (SGE) technologies for energy generation to their corresponding desalination technologies [2,4,13]. Jalili et al. developed mathematical models to compare three types of energy storage systems: electrodialytic, osmotic, and capacitive batteries [2]. Influential parameters, such as temperature and energy consumption of the pump, on the performance of different concentration batteries were also discussed in their work [2] applying a mathematical model. They reported that the peak power densities of the energy storage systems increase at elevated temperature [2].

A schematic of a simple RED stack is shown in Figure 1. In general, a unit cell consists of a dilute solution compartment, a concentrated solution compartment, a cation exchange membrane, and an anion exchange membrane. By repeating unit cells and connecting the end points of the stack to an anode and a cathode compartment (where the electrode rinse solutions are present), a RED stack can be completed for converting an ionic flux into an electrical one [2].

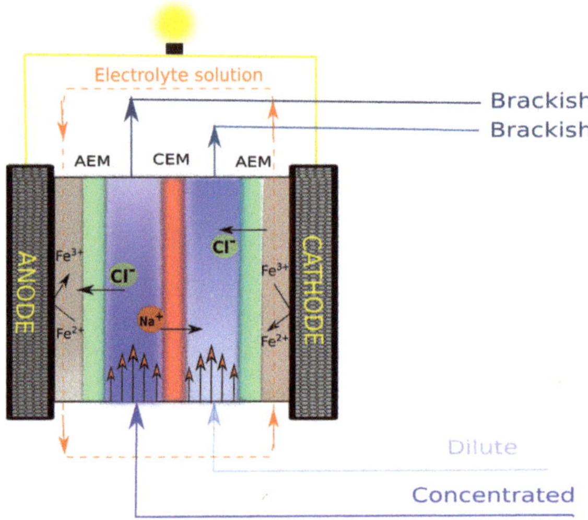

Figure 1. Schematic of a simple RED stack, containing (from the left) an anode, an anode electrolyte compartment, a unit cell, an additional membrane, a cathode electrolyte compartment, and a cathode.

The electrical potential of a RED unit cell is always lower than the open-circuit potential, due to the ohmic resistance, concentration changes in the boundary layer, and concentration changes in the bulk solutions. The last two sources can be interpreted as non-ohmic resistances [5,14]. Non-ohmic resistance is mainly controlled by concentration polarization [15], which has been investigated and discussed by several researchers in the literature [15–21].

Although it has been agreed, by some researchers that increasing the flow velocity and the introduction of flow promoters (i.e., spacers) can mitigate the concentration polarization and enhance the mass transfer by disturbing the diffusive boundary layer [16,20,22,23], Vermaas et al. [24] through an experimental work showed that at low Re numbers (less than 100), which are typically used for RED, introducing non-conductive sub-corrugation is not that beneficial to reduce the ohmic losses and increase the power density [24]. They also showed that although the non-ohmic resistance (concentration boundary layer effects) decreases significantly when increasing the Reynolds number; the ohmic resistances are almost independent of the Re number at high Re numbers and dominates the power loss [24]. Pawlowski et al. performed an extensive literature review of the development and application of corrugated membranes in electro-membrane-based processes [25]. They reported the effect of corrugated membranes in the performance of reverse electrodialysis (RED), showing that electrodialysis (ED) is significantly influenced by the shape of the corrugation, Reynolds number, and ion concentrations. For high Reynolds numbers, corrugation creates eddies which lead to enhanced mass transfer, reduced deposition of foulants, and increased diffuse boundary layer thickness. In particular, they highlighted the role of conductive spacers in lowering the resistance of the RED stack, by eliminating spacer shadow effects [25]. They foresaw the rapid progress of the design and manufacturing of corrugated membranes due to advances in CFD simulations and 3D printing technology [25]. Gurreri et al. [26] used CFD modeling to study fluid flow behavior in a reverse electrodialysis stack, aiming to address the effect of the spacer material on the pressure losses along the channel, evaluating the choice of a fiber-structure porous medium, instead of the commonly adopted net spacers, and investigated the influences of the distributor and channel configurations on fluid dynamics in a RED system [26]. They documented that the total pressure loss in a RED stack is the sum of the pressure drop relevant to the feed distributor, the pressure drop inside the channel, and the pressure drop in the discharging collector [26]. Simulations revealed that the spacer geometry may not necessarily be the main factor controlling the overall pressure drop. In addition, the pressure drop induced by a porous medium made of small fibers is larger than that for a typical net spacer; therefore, they might not be suitable for RED [26]. Pawlowski et al. [27] showed, by CFD modeling, that chevron-corrugated membranes have the highest net produced power density among several investigated profiled membranes, due to increased membrane area, reduction of the concentration polarization, and the proper trade-off between momentum and mass transfer [27]. These results were validated also through experimental comparison [28]. Cerva et al. [29] presented a coupled study of one-dimensional CFD modeling with three-dimensional finite volume modeling for a flat channel, profiled membranes, and different spacer-filled corrugations in a RED stack. Then, they validated the overall model by comparison with experimental data measured in a laboratory [29]. Their results showed that the boundary layer potential drop is significantly lower than the ohmic losses. In addition, woven spacers had the smallest boundary layer potential loss, followed by Overlapped Crossed Filaments (OCF) profiled membranes and then the flat channel, thus indicating that woven spacers provide the most efficient and effective mixing among the considered systems [29]. The highest gross power density and the highest short-circuit current density were reported for OCF profiles, followed by the woven spacers and then the flat channel. However, the highest net power density per cell pair was provided by the flat channel, followed by OCF profiled membranes and then by the woven spacers [29]. Mehdizadeh et al. [30] experimentally studied several non-conductive spacers with different geometries and properties (e.g., different diameters, angles, distances, area fractions, and volume fractions) to understand the spacer shadow effect on the membrane and solution compartment resistances in RED. They reported a correlation between the spacer shadow effect on the membrane resistance and a combined parameter of spacer area fraction and spacer diameter [30]. The spacer shadow effect on the solution compartment resistance was also correlated with the spacer area and volume fraction. They observed that the spacer area fraction had a dominant effect only for less porous spacers [30]. Jalili et al. [31,32] used CFD modeling to examine the influence of flow velocities and spacer topology with respect to the transport of mass and momentum, as well as the flow channel resistivity of a RED unit cell. They reported that the resistivity of the dilute solution channel dominates over the resistivity of the concentrated solution channel and membranes in a RED unit cell [32]. Similar observations have also been reported by Ortiz-Martinez et al. [33]. The electrical potential of a RED unit cell was enhanced by reducing the flow velocity and introducing flow promoters in a dilute solution channel, due to reduced solution resistance [32]. Introducing spacers in a concentrated solution channel or increasing the flow velocity in a dilute solution channel increases the resistivity and has adverse effects on the electrical potential [32]. They also demonstrated that the mass transfer is higher for active membrane-integrated spacers, compared to inactive spacers, under similar flow velocity and spacer topology, due to increased active membrane area [31]. They also concluded that cylindrical membrane-integrated corrugation is an optimum spacer geometry at low flow velocities, while triangular membrane-integrated corrugation is a better geometry at high flow velocities [31]. Recently Dong et al. [34] performed a CFD study of mass and momentum transfer for several types of profiled membrane channels in RED. Their work showed that conductive wavy sub-corrugations improved the mass transfer and reduced the concentration polarization (i.e., non-ohmic losses) [34]. Furthermore, they showed that single-sided wave-profiled membranes had better performance, compared to single-sided pillar-profiled membranes; while single-sided profiled membranes had a smaller impact on the performance, compared to double-sided chevron-profiled membrane and woven spacer-filled channels [34].

Long et al. reported a numerical study matched with experimental data for optimizing channel geometry and flow rate of the concentrated and diluted solutions with non-conductive spacers, to obtain maximum net power output by RED. They reported that the optimal channel thickness and flow rate in the concentrated solution compartment in a RED stack are, respectively, much less than those of the dilute solution compartment [35]. In another work, they revealed that the optimal flow rates in the dilute and concentrated solution channels in an RED stack with varying flow rates along the flow direction to achieve maximum energy efficiency were lower than the optimal flow rates to obtain the maximum net power density. Therefore, an optimization study based on the Non-dominated Sorting Genetic Algorithm II (NSGA-II) was performed, in order to analyze the compromise between the net peak power density and the energy efficiency [36]. Their work showed that the net power density at maximum energy efficiency was less than the peak power density [36].

Several researchers have highlighted the potential use of waste heat in RED systems. Luo et al. [37] reported that by using ammonium bicarbonate as a working fluid in a thermally driven electrochemical generator, waste heat could be converted to electricity [37]. A maximum power density was obtained at an overall energy efficiency of 0.33 W m^{-2}, by operating a RED system with a dilute concentration of 0.02 M [37]. Micari et al. [38] reported the conversion of waste heat into electricity by coupling RED with membrane distillation (MD), resulting in considerable system energy efficiency improvement. The construction and operation of the first lab-scale prototype unit of a thermolytic reverse electrodialysis heat engine (t-RED HE) for converting low-temperature waste heat into electricity have been reported by Giacalone et al. [39]. Ortiz-Imedio et al. [33] documented the strong dependence of the performance of RED on temperature. They reported that the membrane resistance increased when reducing the temperature, and that the perm-selectivity reduced when increasing the temperature [33]. Jalili et al. [31] showed that increasing the temperature enhanced the mass transfer of dilute and concentrated solutions, due to higher diffusivity and lower viscosity at increased temperature. In another work, they reported that the open-circuit potential increased with increasing temperature [2]. Contrary to the most of the literature, which has investigated salinity gradient energy at isothermal conditions, Long et al. [40] addressed the asymmetric temperature influence in dilute and concentrated solution channels on the performance of nanofluidic power systems, using numerical simulation by coupling the Poisson–Nernst–Planck equation and the Navier–Stokes equation, as well as the energy-conservation equation. They observed that when the temperature of the concentrated solution channel is lower than the temperature of the dilute solution channel, the ion-concentration polarization is suppressed, ion diffusion along the osmotic direction enhances, and perm-selectivity increases; thus, the membrane potential improves [40]. However when the temperature in the concentrated solution channel is higher than that of the dilute solution channel, the membrane potential reduces; although the diffusion current increases, due to the lower resistance [40]. In another work [41], they reported the influences of heat transfer and the membrane thermal conductivity in the performance of nanofluidic energy conversion systems. They reported that when the temperature of the concentrated solution channel is lower than the temperature of the dilute solution channel, a larger membrane thermal conductivity results, with reduced electrical power improvement; on the other hand, when the temperature of the concentrated solution channel is higher than the temperature of the dilute solution channel, the increased membrane thermal conductivity leads to enhanced power density [41].

Although several studies have reported the application of CFD modeling for investigating momentum and mass transfer in order to determine the trade-off between the pressure loss and mass transfer in an RED channel [16,18,19,22,23,27], there have been limited CFD studies of electrical potential in an RED channel [42,43]. To the best of our knowledge, there have been no parametric studies which assessed the relative effect of relevant parameters on the net power density for a RED cell. In particular, addressing the influence of temperature, as proposed by Jalili et al. [31], was not compared to the other parameters. The current work is an extension of the previously published works [31,32] by the current authors. We demonstrate that the electrical potential changes linearly with the height of the channel for a constant concentration profile, and that it follows a logarithmic trend with length of the channel height when the concentration profile varies linearly with the channel height [32]. Other interesting observations of this work [32] can be summarized as follows: First, the concentration gradient near the walls of the channel increase, due to reduced boundary layer thickness, with higher Re number. In fact, the concentration at the center of the channel is at its maximum for the concentrated solution channel and is at its minimum for the diluted solution channel [32]. Second, the pressure drop for the dilute solution channel is lower than that in the concentrated solution channel, given similar Re number and channel geometry [32]. This observation was also reported by Zhu et al. [21], when conducting several experiments. Third, the resistance of the dilute solution is more dominant, compared to the resistance of the concentrated solution channel, which can be seen as a limiting factor for the power density of a RED stack. Reducing the Re number (i.e., reducing the velocity at a constant temperature) or introducing corrugation in a dilute solution channel reduces the resistivity of the dilute solution channel by increasing the thickness of the boundary layer, which provides a thicker and more conductive region in the flow channel and results in improved mixing by the developing wakes downstream from the spacers [32]. An opposite trend was observed for the resistivity of the concentrated solution channel [32]. This observation was also supported by Long et al. [35].

This present work describes a numerical framework for simulation of the Navier–Stokes (NS) and Nernst–Planck (NP) system, based on the open source CFD platform OpenFOAM [44], with the aim of predicting the influence of flow velocity, temperature, and geometry on concentration, pressure drop, electrical potential drop, and net power density. Factorial design [45] is applied to address the relative effects of the parameters on the peak power density.

2. Theory and Governing Equations

The flow in the channel is considered to be two-dimensional, incompressible, steady-state, isothermal, and laminar. Physical properties such as density and viscosity are assumed to be constant. There is charge neutrality in the whole system, where only monovalent ions exist. The Navier–Stokes and Nernst–Plank equations [42,46,47] are presented by Equation (1) and Equation (2), respectively.

$$\rho \vec{u} \cdot \nabla \vec{u} = -\nabla p + \mu \nabla^2 \vec{u}. \tag{1}$$

$$\nabla \cdot [\mathcal{D}_i \nabla C_i - \vec{u} C_i + C_i \mu_{EP} \nabla \phi] = 0, \tag{2}$$

for species i, where C_i is the concentration ($[mol/m^3]$), $\mathcal{D}_i$ is the diffusivity ($[m^2/s]$), $\vec{u}$ is the fluid velocity ($[m/s]$), and

$$\mu_{EPi} = \frac{\mathcal{D}_i z_i F}{RT} \tag{3}$$

is the electrophoretic mobility ($[m^2/Vs]$), where z_i is the valency, $F = 96485.3$ C/mol is the Faraday constant, $R = 8.314$ J/K·mol is the universal gas constant, and T is the temperature (in Kelvin), while ϕ is the electrostatic potential ($[V]$).

Assuming two monovalent ionic species, denoted + and - , and using charge neutrality (i.e., $C_+ = C_- = C$), Equation (2) can be written as [31,32]:

$$(\vec{u} \cdot \nabla) C = \frac{2 \cdot \mathcal{D}_+ \cdot \mathcal{D}_-}{\mathcal{D}_+ + \mathcal{D}_-} \nabla^2 C \equiv \mathcal{D} \nabla^2 C, \tag{4}$$

where $\mathcal{D}$ is the effective diffusivity for the salt and C is the concentration. The effective diffusivity is assumed to be a function of temperature, using the published data by Bastug and Kuyucak [48].

The electrical potential can be calculated from the conservation of electrical current density $\vec{j}$ [32],

$$\nabla \cdot \vec{j} = 0. \tag{5}$$

The electrical current density is obtained by a weighted sum of the charged species, resulting in

$$\vec{j} \;=\; F\,(\,\mathcal{D}_- - \mathcal{D}_+\,)\,\nabla C - \frac{F^2 C}{RT}\,(\,\mathcal{D}_+ + \mathcal{D}_-\,)\,\nabla\phi, \tag{6}$$

where the advective flux cancels out, due to monovalent ions and charge neutrality. Combining Equations (5) and (6), we obtain the following relation [31,32]:

$$\left(\frac{\mathcal{D}_+ - \mathcal{D}_-}{\mathcal{D}_+ + \mathcal{D}_-}\right)\nabla^2 C = \frac{F}{RT}\,\nabla\cdot(\,C\nabla\phi\,), \tag{7}$$

from which the electrostatic potential can be calculated, given a known concentration field in Equation (4). The proposed framework essentially consists of four one-way coupled equations—namely the incompressible Navier–Stokes Equations (1) which, together with continuity, determine the pressure and velocity fields; the concentration Equation (4), which essentially is an advection–diffusion equation with a known velocity; and, finally, the equation for the electrostatic potential (7), which is essentially reduced to a Poisson equation with a known source term. Given the domain and boundary conditions described in the following sections, the incompressible Navier–Stokes equations are solved by means of the simpleFoam solver in OpenFOAM, modified to account for concentration and potential following the steps described, for instance, in the openfoamwiki [49].

The trade-off between maximum produced electrical potential and the current density provides the peak power density. The peak power density, P_{RED}^{peak} (W/m^2), of a RED unit cell, the principal parameter of interest in the current work, can be expressed as follows: [5,11,14]:

$$P_{RED}^{peak} = \frac{1}{r_{unit\ cell}}\,\frac{E_{OCP}^2}{4}, \tag{8}$$

where $r_{unit\ cell}$ and E_{OCP} represent the area resistance of the unit cell and the open-circuit potential of the unit cell, correspondingly. The area resistance of the unit cell can be calculated by Equation (9) [5,14]:

$$r_{unitcell} = (r_{AEM} + r_{CEM} + r_d + r_c), \tag{9}$$

where r_{AEM} and r_{CEM} are the area resistances of the AEM and CEM, respectively, and r_c and r_d are the total area resistances for concentrated and dilute solution channels, respectively. The open-circuit potential depends upon the concentrations of dilute and concentrated channels as well as temperature, each of which are assumed fixed for a given setup in the current work. Assuming constant membrane properties, the only remaining variables are the area resistances of the channels. The total area resistance of the channels is calculated by dividing area-weighted average of electrical potential difference across the channel by the current density at the peak power density of RED unit cell, as shown by Equation (10) [14,32]:

$$r_j = \left|\frac{\Delta\tilde{\Phi}}{j}\right|, \tag{10}$$

where r_j is the total area resistance (ohmic and non-ohmic) of the concentrated or dilute channels, j is the current density, and

$$\Delta\tilde{\Phi} = \frac{1}{A_{AEM}}\int_{AEM}\phi\,dA - \frac{1}{A_{CEM}}\int_{CEM}\phi\,dA, \tag{11}$$

is the difference in area-weighted average of electrical potential Φ, calculated on the active membrane. The electrostatic potential across each channel, and thereby also the resistance, can be calculated based on the coupled Nernst–Planck and Navier–Stokes framework, presented in the theory and governing equations section. The formulation used in the current work accounts for both local values and gradients in concentration, and thus accounts for both ohmic and non-ohmic contributions.

It should be noted that when dividing the potential drop by the imposed current, as in the above equation, non-ohmic contributions appear as an ohmic potential drop, although they are not of an ohmic nature [14].

When operating a RED system, the diluted and concentrated solutions are pumped through the compartments between the membranes, which inevitably leads to an energy loss. The required pump power density for each channel can be estimated by Equation (12) [14]:

$$P_{pump} = \Delta p \frac{Q}{A} = \Delta p \frac{H}{L} u,\qquad(12)$$

where A is the membrane area, Q is the volumetric flow rate through the channel, H is the height of the channel, L is the length of the channel, u is the average velocity in the channel, and Δp is the pressure drop across the channel length which will be estimated through CFD modeling. To reduce ohmic energy losses in RED systems, the channel height should be as thin as possible; however, as this leads to increased pumping losses, there is a need to find an optimum value though. There are several factors affecting the optimal thickness of the inter-membrane distance, dictated by flow velocity, salinity and hydrodynamic pressure drops, but generally 50–300 µm is considered an optimum. This is for sterile particle free systems, but also fouling and other effects in nature can affects this further [2,50].

Given the energy consumption in the pump, the net peak power density can be calculated as:

$$P_{net} = P_{RED}^{peak} - P_{pump}^{total}.\qquad(13)$$

In summary, the net peak power density can be calculated as follows:

1. Coupled flow, concentration and potential fields are calculated through Equations (1)–(7).
2. The potential difference across each channel is computed, allowing for determination the corresponding area resistances, as of Equations (11) and (10).
3. Unit cell resistances and the peak power densities are calculated based on Equations (8) and (9).
4. Pumping power is estimated using Equation (12), considering the flow velocities, and pressure drop from Equation (1).
5. The net peak power density is finally computed as of Equation (13).

3. Simulation Setup

Flat and non-conductive spacer-filled channels with cylindrical or triangular corrugation are shown in Figure 2. Jalili et al. [32] reported that introducing flow promoters in a dilute solution channel improves the performance of a RED unit cell, while it has an adverse effect in the concentrated channel. Hence, the corrugated geometries were assumed for the dilute solution compartments, while the flat geometry was considered for concentrated solution compartments in this work.

The inlet concentrations for the channels were considered to be uniform and equal to 0.016 M (close to the salinity of brackish water) for the dilute solution channel and 0.484 M (close to the average salinity of seawater) for the concentrated solution channel.

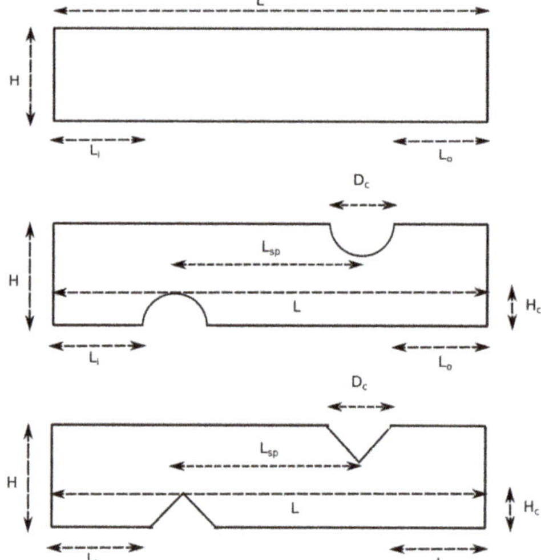

Figure 2. Schematic presentation for sections of the geometry of flat, cylindrical, and triangular corrugated channels with characteristic length scales.

3.1. Boundary Conditions

A constant molar flux, according to the following equation, was assumed in the current model [16,51]. This molar flux corresponds to a constant current density $\vec{j}$, from which the peak power density of the RED system can be obtained (see Equation (8)):

$$\vec{i_i}^m = \frac{t_i^0}{z_i F}\vec{j},\tag{14}$$

where $\vec{i_i}^m$ is the ionic flux of species i and t_i^0 is the transport number of species i. Assuming an ideal membrane from the perm-selectivity perspective and the transport properties for both cations and anions (of a monovalent binary electrolyte such as NaCl) in the solution for simplicity, we obtain [16,51]:

$$i_{IEM} = \pm\frac{0.5j}{F},\tag{15}$$

where the sign shows the incoming flux in the dilute channel or outgoing flux in the concentrated channel. Applying Fick's first law of diffusion, as given in Equation (16), and substituting it into Equation (15), we obatin a constant concentration gradient, as shown in Equation (17).

$$i_{IEM} = \mathcal{D}\frac{\partial C}{\partial n},\tag{16}$$

where n is the normal direction to the wall, $\mathcal{D}$ is the effective diffusivity, and i_{IEM} is representative of the ionic flux through the membrane. Equating Equations (15) and (16) provides us with the boundary condition for the concentration at the membranes:

$$\frac{\partial C}{\partial n} = \pm\frac{0.5j}{F\mathcal{D}}.\tag{17}$$

The boundary condition for the electrical potential on the top membrane is

$$\nabla\phi = \frac{RT}{F^2 C}\left[\frac{F\left(\mathcal{D}_- - \mathcal{D}_+\right)\nabla C - \vec{j}}{\left(\mathcal{D}_+ + \mathcal{D}_-\right)}\right].\tag{18}$$

Evidently, Equation (7) can be solved using the boundary conditions for the concentration and electrical potential, considering Equations (17) and (18).

The constant flux assumption is an approximation representing the features corresponding to an average concentration difference between the channels. Figure 3 shows the specified boundary conditions for different parts of the channel.

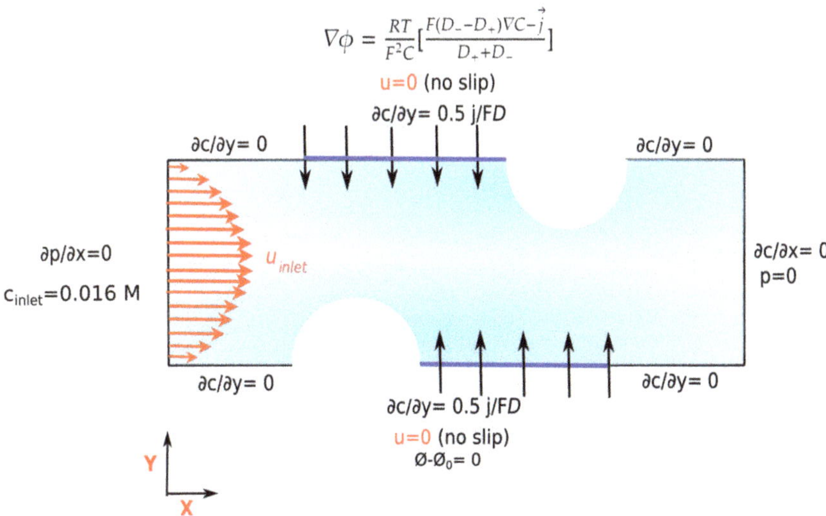

Figure 3. The boundary conditions for a section of dilute, non-conductive cylindrical spacer-filled channel. The blue line shows the active membrane section and the arrows show the diffusion direction from the top and bottom wall toward the dilute bulk. The geometry is repeated to build the full length of the compartment.

The value of the velocity at the inlet depends on the sought Reynolds number, and is given as a parabolic profile. The outlet is specified to atmospheric pressure. The membranes and spacers are set to no-slip conditions at the walls, and with zero gradient in pressure. In the case of the spacer-filled channel, the spacers are assumed to be non-ion conductive, with a corresponding zero flux boundary condition. The electric potential at the bottom wall of the channel is set to zero and the electrical potential on the top wall (active membrane) is calculated based on Equation (18).

3.2. Grid Dependence, Verification, and Validation

A grid dependence study was performed in our previous publication [32]. Local mesh refinement was used for different channel typologies, with extensive refinement near the wall of the channel and spacers, as shown in Figure 4.

Each of the simulations in the current work are based on the finest resolution identified in [32], with an average resolution of 1.13 and 0.25 μm in x- and y- directions, respectively, resulting in approximately 1 M (hexahedral) cells for the full domain. As shown in [32], this resolution introduces an error of less than 0.5%.

The flow behavior of the proposed framework was validated by comparison with the experimental measurements reported by Da Costa et al. [52] and Haaksman et al. [53]; as presented in [32]. The simulated pressure gradient with cylindrical corrugation was found to be somewhat lower than the pressure gradient for woven spacers, as reported by Gurreri et al. [26] at a given Re number; however, some discrepancies are expected, as Gurreri et al. considered the pressure drop in the collector and distributor of the RED stack, in addition to the main channel. The numerical results for the potential and concentration have been verified for a flat channel by comparison with the semi-analytical solution proposed by Lacey [43], for both dilute and concentrated channels (see, e.g., [32]), showing good agreement between the concentration profile and the corresponding electrical potential across the height of the dilute compartment and our numerical solution [32].

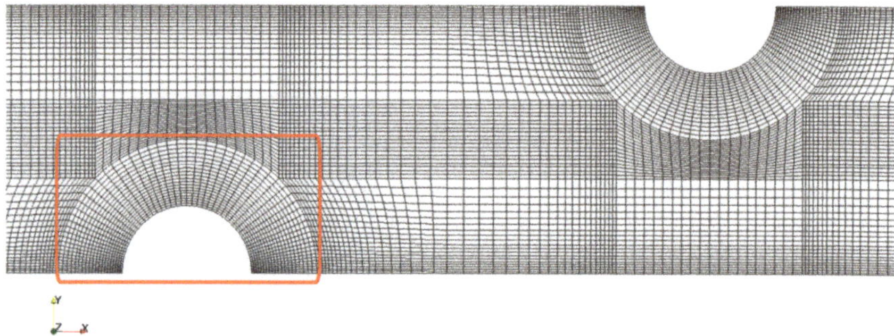

Figure 4. Local grid refinement near the walls of the cylindrical corrugated spacer-filled channel. The coarsest mesh in the local grid refinement process was depicted due to better visibility. The region around the corrugation which goes under local refinement process is confined in a red square.

3.3. Numerical Settings and Configuration

All simulations presented in the following chapter were performed using the OpenFOAM version 4.1 software [44] on the IDUN cluster [54]. A summary of the numerical settings used in the current work are given in Table 1. The absolute residual for pressure, velocity, concentration, and electrical potential was set to 10^{-6}, while the relative residual for the parameters was set to 10^{-4}.

Table 1. Discretization schemes specified for the case studies.

Term	Scheme
Time	steadyState
Gradient	Gauss, linear
Divergence	Bounded, Gauss, linearUpwind
Laplacian	Gauss, linear, corrected

3.4. Factorial Design and Parametric Study

The influence of four quantitative parameters—inlet velocity, corrugation density, corrugation height, and temperature—on the resistivity and net peak power density were investigated. In addition to these four quantitative parameters, the effect of corrugation shape (cylindrical versus triangular) was considered to be a qualitative parameter. To determine the relative influence of each parameter on the power density, a parametric study was performed using a factorial design, as described by Montgomery [45]. The various factors and their corresponding levels are given in Table 2.

Table 2. Factors and levels used for the 2^4 design for cylindrical and triangular corrugated channels.

Factor	Name	High Level (+)	Low Level (−)
Inlet velocity	A	0.0258 m/s	0.0045 m/s
Temperature	B	55 °C	25 °C
Corrugation Density and L_{sp}	C	20 and 600 μm	16 and 800 μm
Corrugation Height	D	100 μm	50 μm

The corresponding values of the parameters for each geometry are given in Table 3. Notice that the Re numbers change, based on both the inlet velocity and the temperature, due to the change in viscosity. The pressure drop for Re numbers larger than 10 was so high that it resulted in a negative net peak power density in a unit cell and, therefore, the Re number in this study was limited to less than 10.

Table 3. Characteristic parameters of the studied geometries in factorial design and the input parameters. (The values of the current densities are dependent of the available area of the membranes for different topologies).

Parameter	Symbol	Value
Corrugation diameter	D_c	0.1 or 0.2 (mm)
Length of the channel	L	12.6 (mm)
Height of the channel	H	0.2 (mm)
Number of corrugations	N	16 or 20 (dimensionless)
Height of the corrugation	H_c	0.05 or 0.1 (mm)
Length of inlet and outlet section	L_i, L_o	0.25 and 0.85 (mm)
Distance of two successive corrugations center	L_{sp}	0.6 or 0.8 (mm)
Resistance of AEM and CEM	r_{AEM}, r_{CEM}	$1.0 \times 10^{-4} \Omega\ m^2$ [55]
Current densities	j	66, 68, 70 and 75 Am^{-2}

In the factorial design, each of the parameters (m parameters) were investigated at n levels, which gave us a set of $n \times m$ simulations, where the influence of each parameter, as well as their combined effect, could be determined. In this factorial design, the parameters were restricted to two levels, designated $+$ and $-$ (i.e., each parameter had a high and low level). Therefore, the results were restricted to a linear response for a given factor. There were 2^4 designs for cylindrical and 2^4 designs for triangular corrugation. Other fluid properties used for the current simulations are summarized in Table 4.

Table 4. Transport properties of the fluid at temperatures considered, reported diffusivities by Bastug and Kuyucak [48], and viscosities by Tseng et al. [56].

T (K)	$\mathcal{D}_-$ ($\frac{m^2}{s}$)	$\mathcal{D}_+$ ($\frac{m^2}{s}$)	ν ($\frac{m^2}{s}$)
298	2.03×10^{-9}	1.33×10^{-9}	9.05×10^{-7}
328	2.80×10^{-9}	2.10×10^{-9}	5.16×10^{-7}

4. Results and Discussion

Figure 5 shows the concentration contour for a dilute solution channel with cylindrical spacers. The corrugation height (radius) was 0.1 mm and the distance between two successive corrugation centers was 0.6 mm. The figure also shows the results for two different average inlet velocities (u = 4.5 and 25.8 $\frac{mm}{s}$) at two different temperatures (T = 25 °C and 55 °C).

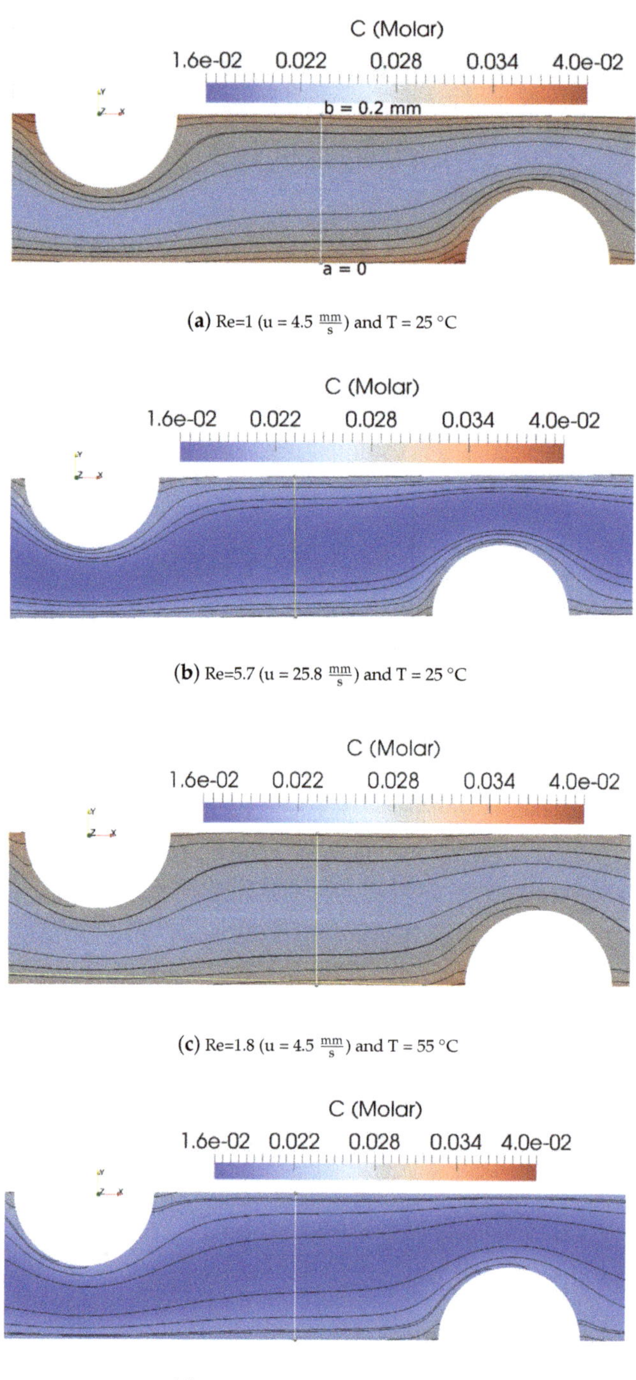

(**a**) Re=1 (u = 4.5 $\frac{mm}{s}$) and T = 25 °C

(**b**) Re=5.7 (u = 25.8 $\frac{mm}{s}$) and T = 25 °C

(**c**) Re=1.8 (u = 4.5 $\frac{mm}{s}$) and T = 55 °C

(**d**) Re=10 (u = 25.8 $\frac{mm}{s}$) and T = 55 °C

Figure 5. Concentration contour maps for dilute solution in a cylindrical corrugated channel at different Re numbers and temperatures.

Higher velocity and lower temperature resulted in less mixing of solutions and, therefore, lower average bulk concentration (shown by cold blue color in the concentration contour map in Figure 5b); thus, higher resistivities and lower power densities were expected. Enhanced mixing (higher average bulk concentration) was observed at lower velocity and higher temperature (shown by red and warmer blue colors of the concentration contour map in Figure 5c). The concentration profiles versus the height of the channel at the $a--b$ cross-section line is shown in Figure 5a, with a distance of X = 10.8 mm from the inlet of the dilute channel. Data sets from this line, for all geometries, are gathered and compared to each other in Figure 6. This was done for two different inlet average velocities (u = 4.5 and 25.8 $\frac{mm}{s}$) and two different temperatures (T = 25 °C and 55 °C). Again, this figure confirms that at higher velocities and lower temperatures, the average concentration became lower. For all cases, the current density along the wall of the channel was considered constant (i.e., the current density for the peak power density), while the concentration along the walls was not constant, due to the imposed boundary conditions. Furthermore, the conductivity profiles for four cases versus the height of the channel (the $a--b$ cross-section in Figure 5a) are compared in Figure 7. The solution conductivities can be calculated using Equation (19), in which conductivity is a function of the concentration of the solution.

$$\sigma = \frac{F^2 C}{RT}\left(\mathcal{D}_+ + \mathcal{D}_-\right). \tag{19}$$

The higher conductivity of the dilute channel agreed with the lower resistivity of the channel and, thus, a higher power density could be achieved. Figure 7 shows that the channel with lower flow velocity (u = 4.5 $\frac{mm}{s}$) and higher temperature (T = 55 °C) had enhanced mixing, with the highest calculated power density among these four cases.

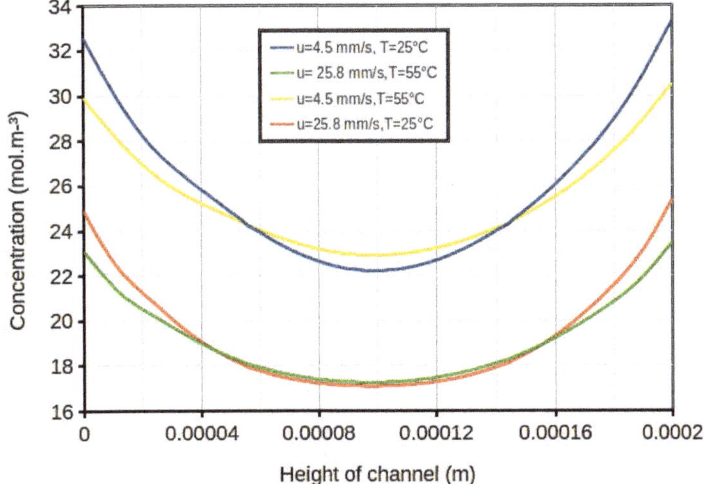

Figure 6. Concentration profiles versus the height of the channel (the $a--b$ cross-section in Figure 5a) at X = 0.0108 m from the inlet of the dilute channel at two different inlet average velocities (u = 4.5 and 25.8 $\frac{mm}{s}$) and T = 25 °C and T = 55 °C, resulting in four different Re numbers: Re = 1, 1.8, 5.7, and 10.

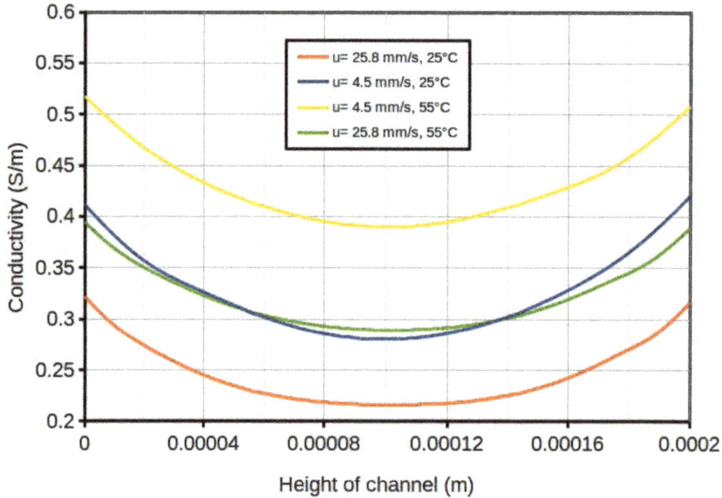

Figure 7. Conductivity versus the height of the channel (the $a - -b$ cross-section in Figure 5a) at X = 0.0108 m from the inlet of the dilute channel at two different inlet average velocities (u = 4.5 and 25.8 $\frac{mm}{s}$) and T = 25 °C and T = 55 °C, resulting in four different Re numbers: Re =1, 1.8, 5.7, and 10.

4.1. Parametric Study

The performance of reverse electrodialysis is influenced by several parameters. Their single or combined impacts were investigated, using a parametric study and a factorial design. Results for the area resistance and power density are summarized for cylindrical corrugation in Table 5, and for triangular corrugation in Table 6.

The simulated net power densities found in this numerical study were comparable to the maximum power densities for RED reported by the authors in another publication [2], which were calculated by using conceptual analytical models with similar channel dimensions, as well as similar temperature and concentration ranges. In addition, the power densities obtained in this study at 25 °C were close to the calculated power densities reported by Long et al. [40], at similar temperature and isothermal conditions, by applying numerical modeling for the investigation of nanofluidic salinity gradient energy harvesting [40]. The simulated net peak power densities were also in the range of the net power densities reported by Vermaas et al. [14], who calculated the theoretical RED net power density for different spacer-filled channels with channel thicknesses between 1–200 µm, and with residence time (defined as the length of the channel divided by the inlet flow velocity) between 0.5–200 s, in addition to changing the channel length and the resistivity of the AEM and the CEM. The residence time in the current work was within 0.2 to 2 s for the high- and low-level cases, respectively. As the resistivity of the channel decreased, the net power density for a RED unit cell increased for all system configurations. This observation was valid both for cylindrical and triangular corrugations, and was due to reduced lower-ohmic and non-ohmic losses. Increasing the temperature had a positive effect on the net peak power density, due to higher open-circuit potential, enhanced diffusivity, and improved mixing of concentrated and dilute solutions, as well as a lower pressure drop due to lower fluid viscosity at elevated temperature. Similar observations were reported experimentally by Luo et al. [37], Benneker et al. [57], and Daniilidis et al. [58]. Increasing the flow velocity had an adverse effect on the net power density, as a result of decreased mass transfer and increased pressure losses.

Vermaas et al. [55] also reported that the RED net power density was reduced for flows with Re numbers larger than 1 in channels with different thicknesses. The corrugation density and corrugation height had both positive and negative effects on the net peak power density. The corrugation height had an adverse effect on net power density, as pressure loss and consumed energy increase with higher corrugation height. This occurs even if the resistivity is lightly reduced, due to the increased corrugation height. In summary, one can conclude that the optimum parameters among the studied cases (i.e., for maximizing the net power density) was when the temperature was 55 °C, the flow velocity was 4.5 $\frac{mm}{s}$, the corrugation density was 20, and the corrugation height was 0.05 mm (for both the cylindrical and the triangular corrugation); see Tables 5 and 6.

Table 5. Summary of factors, area resistance of the dilute solution compartment, and net peak power densities of the unit cell in the 2D model of a **cylindrical** corrugated channel: A, velocity; B, temperature; C, corrugation density; D, corrugation height. Case 1 is the base case.

	Factor				Response	
Name	A	B	C	D	Area Resistance ($\Omega.cm^2$)	Net Peak Power Density (W/m^2)
Case 1	−	−	−	−	7.15	6.18
Case 2	+	−	−	−	8.56	5.43
Case 3	−	+	−	−	5.47	8.86
Case 4	+	+	−	−	6.51	7.96
Case 5	−	−	+	−	7.02	6.26
Case 6	+	−	+	−	8.40	5.50
Case 7	−	+	+	−	5.37	8.95
Case 8	+	+	+	−	6.39	8.05
Case 9	−	−	−	+	7.91	5.80
Case 10	+	−	−	+	9.45	5.02
Case 11	−	+	−	+	6.05	8.35
Case 12	+	+	−	+	7.19	7.48
Case 13	−	−	+	+	7.88	5.82
Case 14	+	−	+	+	9.41	5.03
Case 15	−	+	+	+	6.04	8.36
Case 16	+	+	+	+	7.16	7.46

Table 6. Summary of factors, area resistance of the dilute solution compartment, and net peak power densities of the unit cell in the 2D model of a **triangular** corrugated channel: A, velocity; B, temperature; C, corrugation density; D, corrugation height. Case 1 is the base case.

	Factor				Response	
Name	A	B	C	D	Area Resistance ($\Omega.cm^2$)	Net Peak Power Density (W/m^2)
Case 1	−	−	−	−	7.08	6.23
Case 2	+	−	−	−	8.47	5.47
Case 3	−	+	−	−	5.41	8.91
Case 4	+	+	−	−	6.44	8.01
Case 5	−	−	+	−	6.92	6.27
Case 6	+	−	+	−	8.28	5.45
Case 7	−	+	+	−	5.29	9.02
Case 8	+	+	+	−	6.29	8.12
Case 9	−	−	−	+	7.52	5.99
Case 10	+	−	−	+	8.99	5.22
Case 11	−	+	−	+	5.76	8.60
Case 12	+	+	−	+	6.84	7.72
Case 13	−	−	+	+	7.37	6.07
Case 14	+	−	+	+	8.80	5.29
Case 15	−	+	+	+	5.64	8.70
Case 16	+	+	+	+	6.69	7.80

The triangular spacer corrugation configuration had slightly better performance, compared to the cylindrical one, which was in agreement with the previous studies reported by Ahmad et al. [20]

and Jalili et al. [31]. The estimated effect of each factor is shown in Tables 7 and 8. The tables reveal that temperature was the most dominant factor, followed by inlet velocity, corrugation density, and corrugation height, respectively.

Table 7. Sign and percent contribution of area resistance and power density for each of the factors in the **cylindrical** corrugated channel shown in Table 5.

Factor	A	B	AB	C	AC	BC	ABC	D
Sign Area resistance	+	−	−	−	−	+	+	+
%	26.6	62.5	< 1	< 1	< 1	< 1	< 1	9.90
Factor	AD	BD	ABD	CD	ACD	BCD	ABCD	
Sign Area resistance	+	−	−	+	+	−	+	
%	< 1	< 1	< 1	< 1	< 1	< 1	< 1	
Factor	A	B	AB	C	AC	BC	ABC	D
Sign Power density	−	+	−	+	−	+	−	−
%	9.29	87.5	< 1	< 1	< 1	< 1	< 1	3.10
Factor	AD	BD	ABD	CD	ACD	BCD	ABCD	
Sign Power density	−	−	+	−	−	−	−	
%	< 1	< 1	< 1	< 1	< 1	< 1	< 1	

Table 8. Sign and percent contribution of area resistance and power density for each of the factors in the **triangular** corrugated channel shown in Table 6.

Factor	A	B	AB	C	AC	BC	ABC	D
Sign Area resistance	+	−	−	−	−	+	+	+
%	28.3	67	< 1	< 1	< 1	< 1	< 1	3.46
Factor	AD	BD	ABD	CD	ACD	BCD	ABCD	
Sign Area resistance	+	−	−	+	+	+	+	
%	< 1	< 1	< 1	< 1	< 1	< 1	< 1	
Factor	A	B	AB	C	AC	BC	ABC	D
Sign Power density	−	+	−	+	−	+	+	−
%	9	90	< 1	< 1	< 1	< 1	< 1	< 1
Factor	AD	BD	ABD	CD	ACD	BCD	ABCD	
Sign Power density	+	−	+	+	+	−	−	
%	< 1	< 1	< 1	< 1	< 1	< 1	< 1	

It is worth mentioning that the simulated power densities in this study were larger than the experimentally measured power densities, such as those reported by Zhu et al. [59]. The current mathematical model was developed for incompressible, steady-state, isothermal, and laminar flow with only the presence of monovalent ions. Therefore, the results of the CFD model might not be representative when the flow regime is turbulent, the system is in unsteady state, or if multivalent ions exist. In addition, this CFD model is proposed for a unit cell; thus, it does not represent a full RED stack. The influences of anion and cation exchange membranes or water osmosis of the membranes are ignored in this study. Other sources of energy losses, such as pumping losses through the collector and distributor of the stack, are also neglected, as well as the practical issues relating to 3D flow distribution.

4.2. Concentration Polarization

The area resistances reported in Tables 5 and 6 were calculated based on the electrical potential drop across the channel height for the whole channel, thereby accounting for both ohmic and non-ohmic contributions. By comparing the corresponding conductivities in Tables 5 and 6 with the conductivities in Figure 7, in which only ohmic contributions are considered, we can obtain the non-ohmic contribution (i.e., the share of polarization in the system), as shown in Table 9. In fact, the resistivity calculated by Equation (10) is the area-weighted total resistivity which depends on the area-weighted electrical potential loss, and is obtained directly from solving Equations (1)–(7), provided the boundary conditions. Equation (19) provides the average conductivity of the channel solution based on the average concentration. The reverse of the average conductivity is the average ohmic resistivity. The difference between the total and the average ohmic resistivity, gives the average non-ohmic resistivity.

Table 9. The contribution of ohmic and non-ohmic resistance for Cases 13, 14, 15, and 16 of Table 5 (i.e., with cylindrical corrugation).

Resistivity (Ω.m)	Case 13	Case 14	Case 15	Case 16
Total	3.94	4.71	3.02	3.58
Ohmic	3.08	4.09	2.36	3.13
Non-ohmic	0.86	0.61	0.66	0.45

By comparing the resistivities in Table 9, three observations can be made: First, the share of non-ohmic losses (i.e., concentration polarization effects) was significantly lower than ohmic losses. Second, by increasing the flow velocity at a constant temperature or reducing the temperature at a constant inlet velocity, the ohmic losses increase. Third, increasing the flow velocity and temperature results in the reduction of the non-ohmic losses share of the total resistivity; that is, increasing the Re number (by enhancing the temperature or increasing the inlet flow velocity) will assist in reducing the concentration polarization effect in RED systems. These are consistent with the experimental observations reported by Vermaas et al. [55].

5. Conclusions

The effect of flow velocities, temperature, and spacer topology on the resistivity and net peak power density of a reverse electrodialysis (RED) unit cell were explained, based on CFD modeling which enabled the simulation of flow, pressure drop, concentration, electrical potential, and power density. Our parametric study revealed that while increasing the temperature and corrugation density had positive effects on the net produced power density, increasing the flow velocity and corrugation height had adverse effects. Among the studied parameters, temperature was the most dominating factor, followed by inlet velocity, corrugation density, and corrugation height, respectively. Increasing the temperature benefited the system performance by decreasing the non-ohmic resistance and the corresponding energy losses. Increasing the temperature also benefited the system performance by decreasing ohmic resistances. Moreover, elevating the temperature led to a system with a better performance increase than varying the flow velocity. The increase of temperature can be realized by use of low-grade waste heat, as discussed in [1] for instance.

Author Contributions: Z.J.'s contribution is developing the idea of the research work, building the models and running the simulations, post-processing and analyzing of the results, and writing the draft of the manuscript, including the literature review, results, and discussions. K.E.E. and O.S.B. contributed to proposing the first idea of the research work and furthermore developing the idea, as well as supervising and revising the manuscript. All authors participated in discussing the results. All authors have read and agreed to the published version of the manuscript.

Funding: This research was funded by ENERSENSE (Energy and Sensor Systems) group (project number 68024511) at Norwegian University of Science and Technology (NTNU).

Acknowledgments: Financial support from ENERSENSE (Energy and Sensor Systems) group at Norwegian University of Science and Technology (NTNU) is greatly acknowledged. NTNU IDUN/EPIC computing cluster was used for performing simulations, which is also highly appreciated. Zohreh Jalili would like to thank Vahid Alipour Tabrizy for all help and support throughout this work, without which this research would not have been made possible.

Conflicts of Interest: The authors declare no conflict of interest.

Abbreviations

The following abbreviations are used in this manuscript:

RED	Reverse electrodialysis
CFD	Computational fluid dynamics
SGE	Salinity gradient energy
MD	Membrane distillation
r	Resistivity of the stack
A	Membrane area
Q	Volumetric flow rate
H	Height of the channel
L	Length of the channel
u	Average velocity in the channel
P	Power density
F	Faraday constant
I	Electric current
E_{OCP}	Open-circuit potential
ΔP	Pressure difference between inlet and outlet
$\mathcal{D}_i$	Diffusivity
C	Concentration
$\vec{j}$	Current density
ϕ	electrostatic potential
σ	conductivity

References

1. Krakhella, K.W.; Bock, R.; Burheim, O.S.; Seland, F.; Einarsrud, K.E. Heat to H2: Using waste heat for hydrogen production through reverse electrodialysis. *Energies* **2019**, *12*, 3428.
2. Jalili, Z.; Krakhella, K.; Einarsrud, K.; Burheim, B. Energy generation and storage by salinity gradient power: A model-based assessment. *J. Energy Storage* **2019**, *24*, 100755.
3. Raka, Y.; Karoliussen, H.; Lien, K.; Burheim, B. Opportunities and challenges for thermally driven hydrogen production using reverse electrodialysis system. *Int. J. Hydrogen Energy* **2020**, *45*, 1212–1225.
4. Yip N.Y.; Brogioli D.; Hamelers H.V.M.; Nijmeijer K.; Salinity gradients for sustainable energy: Primer, progress, and prospects. *Environ. Sci. Technol.* **2016**, *50*, 12072–12094.
5. Ramon G. Z.; Feinberg B. J.; Hoek E. M.; Membrane-based production of salinity-gradient power. *Environ. Sci. Technol.* **2011**, *4*, 4423–4434.
6. Yip N. Y.; Tiraferri A.; Phillip W. A.; Schiffman J. D.; Hoover L. A.; Kim Y. C.; Elimelech M. Thin-film composite pressure retarded osmosis membranes for sustainable power generation from salinity gradients. *Environ. Sci. Technol.* **2011**, *45*, 4360–4369.
7. Rica R. A.; Ziano R.; Salerno D.; Mantegazza F.; van Roij R.; Brogioli D. *Entropy* **2013**, *15*, 1388–1407.
8. Vermaas D. A.; Veerman J.; Yip N. Y.; Elimelech M.; Saakes M.; Nijmeijer K. High efficiency in energy generation from salinity gradients with reverse electrodialysis. *ACS Sustain. Chem. Eng.* **2013**, *1*, 1295–1302.

9. Sales, B.B.; Saakes, M.; Post, J.W.; Buisman, C.J.N.; Biesheuvel, P.M.; Hamelers, H.V.M. Direct power production from a water salinity difference in a membrane-modified supercapacitor flow cell. *Environ. Sci. Technol.* **2010**, *44*, 5661–5665.

10. Burheim, O.S.; Sel, F.; Pharoah, J.G.; Kjelstrup, S. Improved electrode systems for reverse electro-dialysis and electro-dialysis. *Desalination* **2012**, *285*, 147–152.

11. Burheim, O.S. *Engineering Energy Storage*, 1st ed.; Academic Press: USA, 2017; ISBN 9780128141007.

12. Pattle, R.E. Production of electric power by mixing fresh and salt water in the hydroelectric pile. *Nature* **1954**, *174*, 660.

13. Kingsbury, R.S.; Chu, K.; Coronell, O. Energy storage by reversible electrodialysis: The concentration battery. *J. Memb. Sci.* **2015**, *495*, 502–516.

14. Vermaas, D.A.; Güler, E.; Saakes, M.; Nijmeijer, K. Theoretical power density from salinity gradients using reverse electrodialysis. *Energy Procedia* **2012**, *20*, 170–184.

15. Veerman, J. *Reverse Electrodialysis: Design and Optimization by Modeling and Experimentation*; University of Groningen: Groningen, The Netherlands, 2010; Volume 225.

16. Gurreri, L.; Tamburini, A.; Cipollina, A.; Micale, G.; Ciofalo, M.. CFD prediction of concentration polarization phenomena in spacer-filled channels for reverse electrodialysis. *J. Memb. Sci.* **2014**, *468*, 133–148.

17. Gurreri, L.; Ciofalo, M.; Cipollina, A.; Tamburini, A.; Van Baak, W.; Micale, G. CFD modelling of profiled-membrane channels for reverse electrodialysis. *Desalin. Water Treat.* **2015**, *55*, 3404–3423.

18. Vermaas, D.A.; Saakes, M.; Nijmeijer, K. Power generation using profiled membranes in reverse electrodialysis. *J. Memb. Sci.* **2011**, *385*, 234–242.

19. Schwinge, J.; Wiley, D.E.; Fletcher, D.F. Simulation of the flow around spacer filaments between narrow channel walls. 1. Hydrodynamics. *Ind. Eng. Chem. Res.* **2002**, *41*, 4879–4888.

20. ; Ahmad, A.L.; Lau, K.K.; Bakar, M.A. Impact of different spacer filament geometries on concentration polarization control in narrow membrane channel. *J. Memb. Sci.* **2005**, *262*, 138–152.

21. Zhu, X.; He, W.; Logan, B.E. Influence of solution concentration and salt types on the performance of reverse electrodialysis cells. *J. Memb. Sci.* **2015**, *486*, 215–221.

22. Długołęcki, P.; Dąbrowska, J.; Nijmeijer, K.; Wessling, M. Ion conductive spacers for increased power generation in reverse electrodialysis. *J. Memb. Sci.* **2010**, *347*, 101–107.

23. Tadimeti, J.G.D.; Kurian, V.; Ch, ra, A.; Chattopadhyay, S. Corrugated membrane surfaces for effective ion transport in electrodialysis. *J. Memb. Sci.* **2016**, *499*, 418–428.

24. Vermaas, D.A.; Saakes, M.; Nijmeijer, K. Enhanced mixing in the diffusive boundary layer for energy generationin reverse electrodialysis.*J. Memb. Sci.* **2014**, *453*, 312–319.

25. Pawlowski, S.; Crespo, J.G.; Velizarov, S. Profiled ion exchange membranes: A comprehensible review. *Int. J. Mol. Sci.* **2019**, *20*, 165.

26. Gurreri, L.; Tamburini, A.; Cipollina, A.: Micale, G. Coupling CFD with a one-dimensional model to predict the performance of reverse electrodialysis stacks. *Desalin. Water Treat.* **2012**, *48*, 390–403.

27. Pawlowski, S.; Geraldes, V.; Crespo, J. G.; Velizarov, S. Computational fluid dynamics (CFD) assisted analysis of profiled membranes performance in reverse electrodialysis. *J. Memb. Sci.* **2017**, *502*, 179–190.

28. Pawlowski, S.; Rijnaarts, T.; Saakes, M.; Nijmeijer, K.; Crespo, J. G.; Velizarov, S. Improved fluid mixing and power density in reverse electrodialysis stacks with chevron-profiled membranes. *J. Memb. Sci.* **2017**, *531*, 111–121.

29. La Cerva, M.; Liberto, D. M.; Gurreri, L.; Tamburini, A.: Cipollina, A.; Micale, G.; Ciofalo, M, Coupling CFD with a one-dimensional model to predict the performance of reverse electrodialysis stacks. *J. Memb. Sci.* **2017**, *541*, 595–610.

30. Mehdizadeh S.; Yasukawa M.; Takakazu A.; Kakihana Y.; Higa M. Effect of spacer geometry on membrane and solution compartment resistances in reverse electrodialysis. *J. Memb. Sci.* **2019**, *572*, 271–280.

31. Jalili, Z.; Pharoah, J.; Stokke, Burheim, O.; Einarsrud, K. Temperature and velocity effects on mass and momentum transport in spacer-filled channels for reverse electrodialysis: A numerical study. *Energies* **2018**, *11*, 2028.

32. Jalili Z.; Burheim O.S.; Einarsrud K. E.; New insights into computational fluid dynamic modeling of the resistivity and overpotential in reverse electrodialysis. *ECS Trans.* **2018**, *85*, 129–144.

33. Ortiz-Imedio, R.; Gomez-Coma, L.; Fallanza, M.; Alfredo, O.; Ibanez, R.; Ortiz, I. Comparative performance of Salinity Gradient Power-Reverse Electrodialysis under different operating conditions. *Desalination* **2019**, *457*, 8–21.

34. Dong, F.; Jin, D.; Xu, S.; Xu, L.; Wu, X.; Wang, P.; Leng, Q.; Xi, R. Numerical simulation of flow and mass transfer in profiled membrane channels for reverse electrodialysis. *Chem. Eng. Res. Des.* **2020**, *157*, 77–91.

35. Long, R.; Li, B.; Liu, Z.; Liu, W. Performance analysis of reverse electrodialysis stacks: Channel geometry and flow rate optimization. *Energy* **2018** , *158*, 427–436.

36. Long, R.; Li, B.; Liu, Z.; Liu, W. Reverse electrodialysis: modelling and performance analysis based on multi-objective optimization. *Energy* **2018** , *151*, 1–10.

37. Luo, X.; Cao, X.; Mo, Y.; Xiao, K.; Zhang, X.; Liang, P.; Huang, X. Power generation by coupling reverse electrodialysis and ammonium bicarbonate: Implication for recovery of waste heat. *Electrochem. Commun* **2012**, *19*, 25–28.

38. Micari, M.; Cipollinaa, A.; Giacalonea, F.; Kosmadakise, G.; Papapetroua, M.; Zaragozab, G.; Micalea, G.; Tamburinia, A. Towards the first proof of the concept of a Reverse ElectroDialysis-Membrane Distillation Heat Engine. *Desalinationn* **2019**, *453*, 77–88.

39. Giacalone, F.; Vassallo, F.; Scargiali, F.; Tamburini, A.; Cipollina, A.; Micale, G. The first operating thermolytic reverse electrodialysis heat engine. *J. Memb. Sci.* **2020** , *595*, 117522.

40. Long, R.; Kuang, Z.; Liu, Z.; Liu, W. Ionic thermal up-diffusion in nanofluidic salinity-gradient energy harvesting. *Natl. Sci. Rev.* **2019** , *6*, 1266–1273.

41. Long, R.; Kuang, Z.; Liu, Z.; Liu, W. Effects of heat transfer and the membrane thermal conductivity on the thermally nanofluidic salinity gradient energy conversion. *Nano Energy* **2020** , *67*, 104284.

42. Sonin, A.A.; Probstein, R.F. A hydrodynamic theory of desalination by electrodialysis.*Desalination* **1968**, *5*, 293–329.

43. Lacey R.E. Energy by reverse electrodialysis. *Ocean Eng.* **1980**, *7*, 1–47.

44. OpenFOAM. The OpenFOAM Foundation. Availabel online: http://www.openfoam.org (accessed on 1 September 2018).

45. Montgomery, D.C. *Design and Analysis of Experiments*; John Wiley and Sons: Hoboken, NJ, USA, 2019; ISBN 978-1-119-49244-3.

46. Newman, J.; Thomas-Alyea, K.E. *Electrochemical Systems*; John Wiley and Sons: Hoboken, NJ, USA, 2012, ISBN 978-0-471-47756-3.

47. Kirby, B.J. *Micro-and Nanoscale Fluid Mechanics: Transport in Microfluidic Devices*; Cambridge university press: Cambridge, UK, 2010, ISBN 9780521119030.

48. Baştuğ, T.; Kuyucak, S. Temperature dependence of the transport coefficients of ions from molecular dynamics simulations. *Chem. Phys. Lett.* **2005**, *408*.

49. OpenFOAMwiki. Available online: https://openfoamwiki.net/index.php/How_to_add_temperature_to_icoFoam, (accessed on 26 July 2020).

50. Burheim, O.S.; Pharoah, J.G.; Vermaas, D.; Sales, B.B.; Nijmeijer, K.; Hamelers, H.V. Reverse Electrodialysis. *Encyclopedia of Membrane Science and Technology* 2013, pp. 1–20.

51. Zourmand, Z.; Faridirad, F.; Kasiri, N.; Mohammadi, T. Mass transfer modeling of desalination through an electrodialysis cell. *Desalination* **2015**, *359*, 41–51.

52. Da Costa, A.R.; Fane, A.G.; Wiley, D.E. Spacer characterization and pressure drop modelling in spacer-filled channels for ultrafiltration. *J. Memb. Sci.* **1994**, *87*, 79–98.

53. Haaksman, V.; Siddiqui, A.; Schellenberg, C.; Kidwell, J.; Vrouwenvelder, J.S.; Picioreanu, C. Characterization of feed channel spacer performance using geometries obtained by X-ray computed tomography. *J. Memb. Sci.* **2017**, *522*, 124–139.

54. Själander, M.; Jahre, M.; Tufte, G.; Reissmann, N. EPIC: An Energy-Efficient, High-Performance GPGPU Computing Research Infrastructure. *arXiv* **2019**, arXiv:1912.05848.

55. Vermaas, D.A.; Saakes, M.; Nijmeijer, K. Doubled power density from salinity gradients at reduced intermembrane distance. *Environ. Sci. Technol.* **2011**, *45*, 7089–7095.

56. Tseng, S.; Li, Y.M.; Lin, C.Y.; Hsu, J.P. Salinity gradient power: influences of temperature and nanopore size. *Nanoscale* **2016**, *8*, 2350–2357.

57. Benneker, A.M.; Rijnaarts, T.; Lammertink, R.G.; Wood, J.A. Effect of temperature gradients in (reverse) electrodialysis in the Ohmic regime. *J. Memb. Sci.* **2018**, *548*, 421–428.

58. Daniilidis, A.; Vermaas, D.A.; Herber, R.; Nijmeijer, K. Upscale potential and financial feasibility of a reverse electrodialysis power plant. *Renew. Energy* **2014**, *64*, 123–131.

59. Zhu, X.; He, W.; Logan, B.E. Reducing pumping energy by using different flow rates of high and low concentration solutions in reverse electrodialysis cells. *J. Memb. Sci.* **2015**, *486*, 215–221.

 *membranes*

Article

A Comprehensive Membrane Process for Preparing Lithium Carbonate from High Mg/Li Brine

Wenhua Xu, Dongfu Liu, Lihua He * and Zhongwei Zhao *

School of Metallurgy and Environment, Central South University, Changsha 410083, China;
xuwenhua@csu.edu.cn (W.X.); liudongfu@csu.edu.cn (D.L.)
* Correspondence: helihua@csu.edu.cn (L.H.); zhaozw@csu.edu.cn (Z.Z.)

Received: 21 October 2020; Accepted: 18 November 2020; Published: 26 November 2020

Abstract: The preparation of Li_2CO_3 from brine with a high mass ratio of Mg/Li is a worldwide technology problem. Membrane separation is considered as a green and efficient method. In this paper, a comprehensive Li_2CO_3 preparation process, which involves electrochemical intercalation-deintercalation, nanofiltration, reverse osmosis, evaporation, and precipitation, was constructed. Concretely, the electrochemical intercalation-deintercalation method shows excellent separation performance of lithium and magnesium, and the mass ratio of Mg/Li decreased from the initial 58.5 in the brine to 0.93 in the obtained lithium-containing anolyte. Subsequently, the purification and concentration are performed based on nanofiltration and reverse osmosis technologies, which remove mass magnesium and enrich lithium, respectively. After further evaporation and purification, industrial-grade Li_2CO_3 can be prepared directly. The direct recovery of lithium from the high Mg/Li brine to the production of Li_2CO_3 can reach 68.7%, considering that most of the solutions are cycled in the system, the total recovery of lithium will be greater than 85%. In general, this new integrated lithium extraction system provides a new perspective for preparing lithium carbonate from high Mg/Li brine.

Keywords: membrane process; Li_2CO_3; electrochemical intercalation deintercalation; high Mg/Li brine

1. Introduction

The fast development of electric vehicles, storage devices, and hand-held electronic devices has dramatically increased the demands for lithium [1–4]. Lithium carbonate is an important raw material for preparing lithium-ion battery cathode materials [5]. In recent years, global lithium (Li) demand has reached 180,000 tons of lithium carbonate equivalent in 2015, with forecasts as high as 1.6 M tons by 2030 [6,7].

Nowadays, lithium resources mainly exist in solid ore (such as spodumene and lepidolite) and brine, and over 70% of exploitable lithium in the world existed in the brine [8,9]. Compared with the lithium extraction from these two kinds of resources, lithium extraction from brine is more effective, simpler, and cheaper [8]. Most lithium resources in continental brines are found in a small region in South America, often referred to as the "Lithium Triangle" [9,10]. A notable feature of brines in the "Lithium Triangle" region is the low mass ratio of Mg/Li. In contrast, the grade of brine in other regions is much worse. In China, the major lithium-containing brines are located in the Qinghai–Tibet plateau [8,11], and most of the lithium-containing brines in this area are mostly magnesium sulfate subtype [12]. A typical feature of magnesium sulfate subtype brines is the mass ratio of Mg/Li, which has a long span (from tens to hundreds, even more than 1000) [13]. Therefore, how to effectively realize the separation of magnesium and lithium is the key to produce Li_2CO_3 from high Mg/Li brines.

Multifarious methods such as solvent extraction [14], membrane separation [15–17], adsorption [18,19], and electrochemical intercalation-deintercalation (EID) method [17,20–23] have been developed for lithium extraction from high Mg/Li brine. Solvent extraction is an efficient separation technology; both of the separation factors (SF_{Li-Na}, SF_{Li-Mg}) can reach hundreds or even more than one thousand [24,25]. However, the extraction reagent has a slight solubility in aqueous solution [26], which is not suitable for treating brine directly. The ion-sieve absorption method is considered to be an effective approach to extract lithium from the high Mg/Li ratio brines thanks to its low cost, high selectivity, and nontoxicity [27]. However, the ion-sieve absorption method faces the following problems: (1) it is difficult to prepare the high absorption capacity absorbent; and (2) there is a significant loss of capacity in the desorption process when acids or oxidants are used as desorption agents. The above problems seriously restrict its large-scale industrial application [20].

Nanofiltration (NF), as an important membrane separation technology, has been successfully applied for separating lithium and magnesium from a high Mg/Li brine because of its selective rejection of divalent ions and monovalent ions based on Donnan exclusion [28,29]. However, it also suffers from the following problems: (1) This technology can only treat brine with very low sodium and potassium content, and it usually takes 1–2 years to obtain this kind of brine [30–32]. (2) The salinity in the type of brine after potassium removal is too high to meet the operation condition for this purpose, which needs to be diluted with water (the amount of water used for dilution is usually several times than the brine). This process not only needs to consume a large amount of fresh water, but also increases the amount of water to be treated.

In our previous work, we have proved that the EID method shows an excellent lithium extraction properties from the high mass ratio of brine [20,21,33]; the mass ratio of brine can be decreased from the initial 58.5 in the brine to 0.93 in the obtained anolyte. Although the mass ratio of Mg/Li in the anolyte is much lower than the original brine, the lithium concentration in the anolyte is only 1–2 $g \cdot L^{-1}$, which is far from the lithium concentration required to precipitate lithium carbonate. For this reason, we need to concentrate the anolyte and remove the residual impurities (e.g., Mg^{2+}, Ca^{2+}, and SO_4^{2-}) in it. Theoretically, all kinds of concentration methods (like reverse osmosis, electrodialysis, evaporation, and so on) [34,35] and impurity removal methods (like nanofiltration, solvent extraction, and so on) [24,28,29] can be used to treat the obtained anolyte. Notably, the total salt concentration of the obtained anolyte is between 20 and 30 $g \cdot L^{-1}$, which is an ideal range for NF and reverse osmosis (RO) treatment. Therefore, we proposed an integrated lithium carbonate preparation process combining EID, NF, RO, evaporation, and precipitation processes to prepare Li_2CO_3 from a high Mg/Li brine. The aim of the main processes are as follows: (1) the EID method is used to maximize the separation of magnesium and lithium from the brine to obtain a low Mg/Li anolyte; (2) removing the multivalent ions (e.g., Mg^{2+}, Ca^{2+}, and SO_4^{2-}) from the obtained anolyte via the NF method; (3) concentrating the permeate flow produced by NF with the RO method; (4) further increasing the lithium concentration by evaporation; and (5) precipitating Li_2CO_3 by adding Na_2CO_3.

2. Materials and Methods

2.1. Membranes

The membrane used in the EID method is a heterogeneous anionic membrane (MA-3475), which was purchased from Beijing Anke Membrane Separation Technology & Engineering Co., LTD. (Beijing, China agent). The heterogeneous anionic membrane selectively allows the anions to pass through and reject the cations. The NF (NF2) and RO (RO5) membranes used for the experiment are disc tube membranes, which were made by RisingSun Membrane Technology Co., Ltd., (Beijing, China). Specifically, the membrane areas of the NF membrane and RO membrane are both 2.2 m^2, and the operation pH are in the range of 3–11. The permeate flux and desalination rate of NF were 42 $L \cdot m^{-2} \cdot h^{-1}$ and 98%, respectively, which were obtained at 25 °C, operating pressure of 0.7 MPa, and test salt concentration of $MgSO_4$ of 2 $g \cdot L^{-1}$. Further, the permeate flux and desalination rate of RO were 42 $L \cdot m^{-2} \cdot h^{-1}$ and 99.5%, respectively, which were obtained at 25 °C, operating pressure of 1.55 MPa, and test salt concentration of NaCl of 2 $g \cdot L^{-1}$.

2.2. Experimental Illustration

2.2.1. Methods

LiFePO$_4$/FePO$_4$ electrodes' preparation: LiFePO$_4$ electrode was prepared as follows: (1) weighing LiFePO$_4$, polyvinylidene fluoride (PVDF), and acetylene black (C) in a mass ratio of 8:1:1; (2) dissolving PVDF into N-methylpyrrolidone (NMP) and then adding C and LiFePO$_4$ in order; (3) coating the above-mixed slurry on a carbon fiber sheet; and (4) drying the prepared carbon fiber sheet in a vacuum oven at 95 °C for 12 h. The FePO$_4$ electrode was obtained by deintercalating lithium from the LiFePO$_4$ electrode. Concretely, an electrolytic cell is divided into an anode chamber and cathode chamber by anion membrane, LiFePO$_4$ electrode (anode) and nickel foam (cathode) were placed into the anode and cathode chamber, respectively. Both of the chambers were filled with 5 g·L^{-1} NaCl solution and the pH value of the catholyte was controlled to 2–3 using HCl. The voltage used in electrolysis is 1.0 V, and the electrolysis ends until the current density is less than 0.05 mA·cm^{-2}.

EID method for lithium extraction: The device for the EID method is shown in our previous work [20]. The device of the EID system was divided into two chambers by the anion membrane, where LiFePO$_4$ and FePO$_4$ is used as anode and cathode, respectively. The anode and cathode chambers are filled with supporting electrolyte and brine, respectively. The entire working process is shown below: (1) lithium deintercalated from LiFePO$_4$ to the supporting electrolyte (LiFePO$_4$ – e = Li$^+$ + FePO$_4$); (2) lithium existed in the brine intercalated into FePO$_4$ (Li$^+$ + FePO$_4$ + e = LiFePO$_4$); and (3) the Cl$^-$ in the brine diffused into the anode chamber through the anionic membrane to maintain the electroneutrality of the anolyte and brine, and LiCl was obtain in the anolyte.

The brine used for the lithium extraction was from West Taijnar Salt Lake (Golmud, China) with the Mg/Li ratio of 58.8 (Table 1), and the lithium extraction process was carried out via instrument LANHE-CT2001A (Wuhan, China). The effective size of the electrodes was 17 × 20 cm^2, and the electrodes of LiFePO$_4$ and FePO$_4$ worked as anode and cathode, respectively. The electrode coating density was about 85 mg (LiFePO$_4$)·cm^{-2}. The electrolytic cell was comprised of two chambers, which were separated by an anion exchange membrane (MA-3475, Beijing Anke Membrane Separation Technology Engineering Co. LTD, Beijing, China agent). The anode chamber was filled with 1.5 L of 5 g·L^{-1} NaCl as supporting electrolyte, and the cathode chamber was filled with 1.5 L brine. The entire electrolysis was performed with a constant current of 0.6 A until the voltage reached 0.35 V, and then worked at a constant voltage until the current dropped to 0.1 A to end the electrolysis process.

Table 1. The components of the West Taijnar Salt Lake brine (g·L^{-1}).

Element	Li$^+$	Na$^+$	K$^+$	Mg^{2+}	Ca^{2+}	SO$_4^{2-}$	Cl$^-$
West Taijnar	2.05	0.81	0.48	120.56	0.04	31.01	360.9

NF for purification: The obtained anolyte was purified by NF, the volume of the feed used in the nanofiltration was 200 L, and the composition of the feed used was configured according to the composition of the anolyte obtained by the EID system. The NF process was carried out at a constant pump power of 2 KW until the operation pressure reached 8 MPa. The total volume of the collected permeate solution was 180 L.

RO for concentration: The collected permeate solution after the NF treatment was concentrated by RO process, and only 175 L feed liquor was used in the process. The whole RO process was performed at the room temperature and ended until the volume of permeate reached 105 L. The RO process was also carried out at a constant pump power of 2 KW.

Evaporation and concentration: The evaporation process was carried out by an electric furnace, and the initial volume of the solution used for the evaporation was 5 L.

Precipitation of Li$_2$CO$_3$: The lithium-containing solution after evaporation process was precipitated by Na$_2$CO$_3$ (280 g·L^{-1}) at 95 °C. The addition of sodium carbonate is 1.05 times the dosage of theoretical amount used in the lithium precipitation reaction. When all of the Na_2CO_3 wass added into the LiCl solution, the solution was stirred for 1 h to mature the lithium carbonate, and then the Li_2CO_3 was filtered out. The obtained lithium carbonate was washed twice with deionized water and dried to obtain the Li_2CO_3 product.

In general, the comprehensive membrane process is shown in Figure 1.

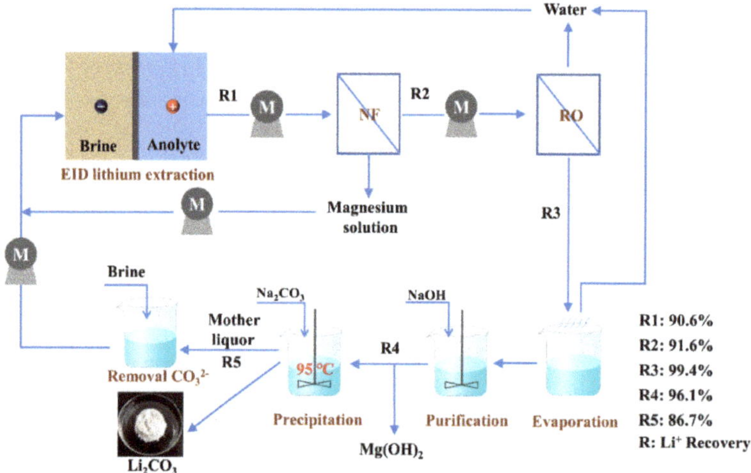

Figure 1. Schematic diagram of the comprehensive membrane process. EID, electrochemical intercalation-deintercalation; NF, nanofiltration; RO, reverse osmosis.

2.2.2. Analytical Methods

The concentration of Li^+, Na^+, K^+ and Mg^{2+}, and Ca^{2+} in the solutions was measured by inductively coupled plasma-optical emission spectrometry (ICP-OES, Thermo Scientific iCAP-7200, Shanghai, China agent), and the concentration of SO_4^{2-} was measured by ion chromatography (ICS-5000/DIONEX, Thermofisher Scientific, Shanghai, China agent). The X-ray Diffraction (XRD) patterns were measured via a BRUKER D8 ADVANCE using Cu-Kα radiation (λ = 1.54056 Å). The morphology of Li_2CO_3 was detected by a scanning electron microscope (SEM, JEOL JSM-6490LV, JEOL (BEIJING) CO., LTD., Beijing, China agent).

2.2.3. Calculation

The separation factor (SF) of lithium and magnesium was calculated as Equation (1):

$$\text{SF} = \frac{C_{Li}/C_{Mg}}{C'_{Li}/C'_{Mg}} \tag{1}$$

where SF is the separation factor of Li^+ and Mg^{2+}, C_{Li} is the concentration of lithium in the obtained solution (g·L^{-1}), C_{Mg} is the concentration of magnesium in the obtained solution (g·L^{-1}), C'_{Li} is the concentration of lithium in the feed (g·L^{-1}), and C'_{Mg} is the concentration of magnesium retained in the feed (g·L^{-1}).

The recovery of lithium (R_E) for the electrolytic intercalation-deintercalation system was calculated as Equation (2):

$$R_E = \frac{C_0V_0 - \int_0^t C_tV_t}{C_0V_0} \times 100\% \tag{2}$$

where R_E is the recovery of lithium in the brine, C_0 is the initial concentration of lithium in the brine $(g \cdot L^{-1})$, V_0 is the initial volume of the brine (L), t is the sampling time (h), C_t is the concentration of lithium in brine at t $(g \cdot L^{-1})$, and V_t is the volume of brine at t (L).

The retention ratio (*R*) refers to the permeability of ions, which is the main index to evaluate the separation performance. The corresponding calculation process is shown in Equation (3).

$$R = \frac{C_F V_F - C_P V_P}{C_F V_F} \times 100\% \tag{3}$$

where R represent the retention ratio and C_F and C_P are the concentrations of ions of the feed and permeate solution $(g \cdot L^{-1})$, respectively. V_F and V_P are the volume of the feed and permeate solution (L).

2.2.4. Membrane Cleaning

The membranes need to be washed when the transmembrane pressure difference is greater than 0.35 MPa. For the membrane scaling caused by inorganic salts, 1% (wt) ethylenediamine tetraacetic acid disodium salt (EDTA) + citric acid solution (citric acid is used to adjust the pH of the solution to 3–4) is generally used for cleaning at room temperature for about 1 h.

3. Results and Discussion

3.1. Lithium Extraction From the Brine

The primary contents of the West Taijinar used for the lithium extraction are shown in Table 1, and the experimental results are exhibited in Figure 2. From Figure 2a, it can be seen that the concentration of lithium reached 2.1 $g \cdot L^{-1}$ at the end of the second cycle, and the concentration of lithium in the brine decreased from the initial 2.05 $g \cdot L^{-1}$ to 0.18 $g \cdot L^{-1}$, while the total recovery of lithium reached 90.6% at the end of the second cycle. In the same way, the decline rate of lithium in the second cycle is slightly lower than that in the first cycle, which is mainly owing to the continuous decline of lithium concentration in the brine.

Figure 2b shows the voltage and current curves in the first two cycles. It can be seen that the first cycle took 13.5 h, while the second cycle only lasted 10.5 h. In addition, the constant current process in the first cycle lasts longer than in the second cycle. Correspondingly, the voltage growth rate in the first cycle is also slower. The above results are attributed to the fact that the lithium concentration in the second cycle is lower than that in the first cycle, which leads to more serious polarization of lithium extraction in the second cycle.

Figure 2c shows the cyclic voltammetry (CV) curves of LiFePO$_4$ in the brine; it can be seen that there are a couple of obvious peaks for the deintercalation/intercalation of lithium located at 0.337 V (vs. saturated calomel electrode (SCE)) and 0.178 V (vs. SCE), which correspond to the deintercalation of lithium from LiFePO$_4$ (LiFePO$_4$ $-$ e $=$ Li$^+$ $+$ FePO$_4$) and the intercalation of lithium to FePO$_4$ (FePO$_4$ $+$ Li$^+$ $+$ e $=$ LiFePO$_4$), respectively. There also exists a weak reduction peak at -0.443 V (vs. SCE), which corresponds to the intercalation of magnesium (FePO$_4$ $+$ 0.5 Mg^{2+} $+$ e $=$ Mg$_{0.5}$FePO$_4$). Obviously, magnesium is more difficult to insert into FePO$_4$ than lithium, which means that FePO$_4$ can selectively extract lithium from a high Mg/Li brine via potential control. In addition, the inset illustration in Figure 2c shows that the mass ratio of Mg/Li in the obtained anolyte is only 0.93, which is far lower than 58.5 in the brine. The above results show that the new EID system has excellent separation performance for lithium and magnesium.

Figure 2d shows the charge/discharge curves of LiFePO$_4$ in the West Taijinar brine. It can be seen that the charging and discharging curves of the 20 cycles are relatively stable, which means that LiFePO$_4$ can operate stably in the brine.

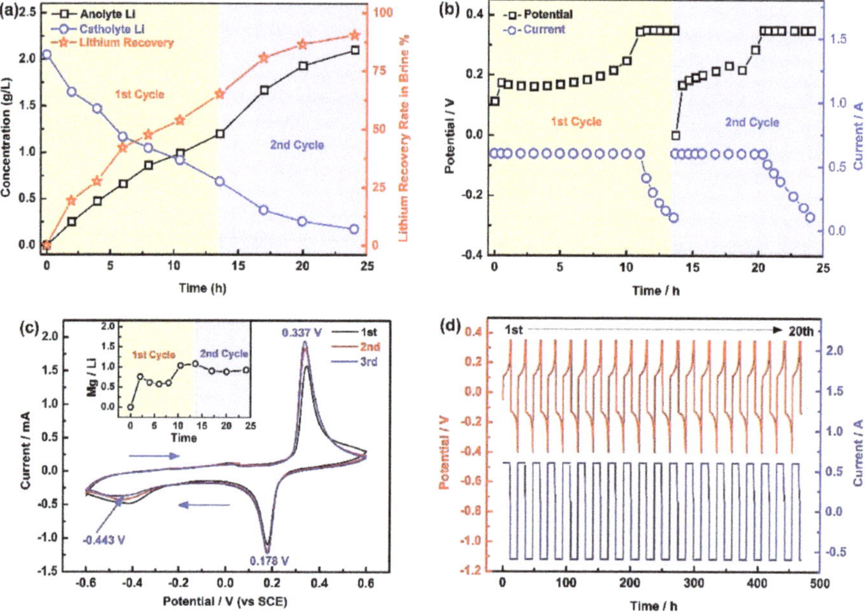

Figure 2. The EID system for lithium extraction. (**a**) Li$^+$ concentration and Li$^+$ recovery rate in the first two cycles; (**b**) current and voltage changes in two cycles; (**c**) the cyclic voltammetry (CV) curves of brine and the illustration shows the Mg/Li in the obtained anolyte; (**d**) charge and discharge cycle performance of the brine. SCE, saturated calomel electrode.

Furthermore, the analysis results of the main ions in the produced anolyte are shown in Table 2. From Table 2, it can be seen that the main ions in the anolyte are Li$^+$, Na$^+$, and Mg^{2+}. Compared with the Mg^{2+} concentration in the brine, the penetration of magnesium into the anolyte is negligible. The rejection rates of the impurities such as K$^+$, Mg^{2+}, and SO$_4^{2-}$ are 92.2%, 98.5%, and 99.2%, respectively. The retention of cations by the anion membrane is mainly due to the charge repulsion of the fixed cationic groups of the membrane itself to the cations in the solution [36,37]. The interception of divalent sulfate is mainly due to the fact that the ionic radius of sulfate is larger than that of chloride ions, and the concentration of chloride ions is much greater than that of sulfate, which makes the content of sulfate permeable through the membrane very low in the process of lithium extraction. In general, the concentration of the impurities in the obtained anolyte is very low, which is facilitation for the subsequent purification process.

Table 2. The concentration of the main ions in the obtained anolyte (g·L^{-1}).

Components	Li$^+$	Na$^+$	K$^+$	Mg^{2+}	Ca^{2+}	SO$_4^{2-}$	Cl$^-$
Concentration	2.1	1.9 *	0.04	1.95	0.004	0.26	19.1
Recovery & Rejection %	90.6	-	92.2	98.5	-	99.2	-

* The initial concentration of Na$^+$ added in the form of NaCl is 2.0 g·L^{-1}.

Therefore, the EID system shows excellent separation properties of lithium and magnesium. It is an efficient, environmentally friendly, and stable process without using acid, alkali, or any toxic reagents, nor does it produce any solid waste. The brine after the lithium extraction can be directly discharged back to the salt fields, without affecting the environment.

3.2. NF and RO Processes

In order to precipitate lithium carbonate, the lithium-riched anolyte needs to be deeply purified and concentrated. In this paper, NF and RO were used for deep purifying of the divalent ions and concentrating of the penetrating fluid, respectively. Both NF and RO were carried out only once and the corresponding results of the NF and RO processes are shown in below.

3.2.1. NF Process

The corresponding experimental results are shown in Figure 3 and Table 3. As shown in Figure 3a, the initial operation pressure of the nanofiltration was 2.0 MPa. The water flux decreases slowly from the initial 52 $L \cdot m^{-2} \cdot h^{-1}$ to 50.5 $L \cdot m^{-2} \cdot h^{-1}$ at first, and then rapidly declines from 50.5 $L \cdot m^{-2} \cdot h^{-1}$ to 31.6 $L \cdot m^{-2} \cdot h^{-1}$. Inversely, the operation pressure increases at first and then rapidly reaches 8 MPa. There are three main reasons for this phenomenon: (1) the increase of the osmotic pressure in the retentate solution; (2) the precipitation of the salts on the surface of the NF membrane; and (3) the compaction of the NF membrane.

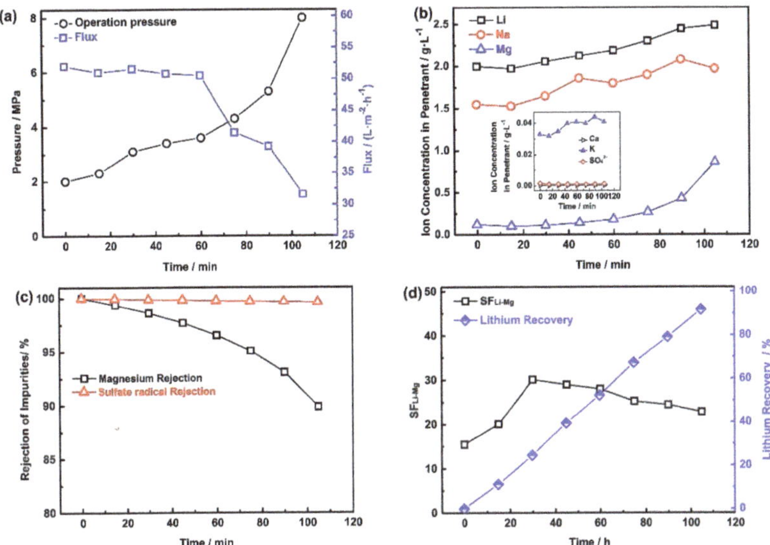

Figure 3. The NF process for purification. (**a**) The relationship of the operation pressure and the flux of the membrane; (**b**) concentration of Li^+, Na^+, K^+, Mg^{2+}, Ca^{2+}, and SO_4^{2-} in the permeate flow; (**c**) the rejection rate of Mg^{2+} and SO_4^{2-}; (**d**) the recovery of lithium and the separation factor of lithium and magnesium (SFLi-Mg).

Table 3. The main analytical results in the collected retentate during the nanofiltration (NF) process.

Time	Conductivity	T	Flow	Concentration of Ions/$g \cdot L^{-1}$						
min	$ms \cdot cm^{-1}$	°C	$L \cdot min^{-1}$	Li^+	Na^+	K^+	Mg^{2+}	Ca^{2+}	Cl^-	SO_4^{2-}
0	37.5	24.7	16	2.10	1.91	0.04	1.95	0.004	19.1	0.26
30	40.9	25.3	16	2.26	1.82	0.038	3.64	0.0053	24.5	0.35
60	44.9	26.2	16	2.32	1.91	0.037	5.34	0.0085	30.2	0.56
90	52.0	27.0	16	2.28	1.69	0.036	9.78	0.018	42.1	1.21
105	108.2	27.6	16	2.31	1.96	0.041	18.23	0.02	66.8	2.37

The concentration of the ions in the permeate flow and collected retentate during the NF process are shown in Table 3 and Figure 3b, respectively. From Table 3, it can be seen that the conductivity increased slowly at the beginning of the initial stage (increased from 37.5 mS·cm^{-1} to 52.0 mS·cm^{-1}). Subsequently, a significant increase followed after 90 min from the start of the NF, and the conductivity reached 108.2 mS·cm^{-1}. The temperature rose slowly throughout the experiments (from 24.7 °C to 27.5 °C), and the rise in water temperature comes from two aspects: (1) the mechanical friction of the high pressure pump produce a great deal of heat; and (2) the friction of the fluid and the pipe, which also generates heat. The flow rate of the entire NF process was kept at 16 L·min^{-1}. There was no significant change in the concentration of monovalent ions such as Li$^+$, Na$^+$, and K$^+$, while divalent ions such as Mg^{2+}, Ca^{2+}, and SO$_4{}^{2-}$ are abundantly enriched in the retentate solution. Moreover, the concentration of Mg^{2+}, Ca^{2+}, and SO$_4{}^{2-}$ at the end of the NF process reached 18.23 g·L^{-1}, 0.02 g·L^{-1}, and 2.41 g·L^{-1}, respectively. It can be found that Mg^{2+}, Ca^{2+}, and SO$_4{}^{2-}$ were concentrated 9.3 times, 5 times, and 9.1 times, respectively. The concentrated times of Ca^{2+} were lower than those of Mg^{2+} and SO$_4{}^{2-}$. Notably, the main anion in the collected retentate is Cl$^-$, which was rejected to maintain the electrical neutrality of the collected retentate.

As shown in Figure 3b, the concentration of Li$^+$, Na$^+$, Mg^{2+}, and K$^+$ in the permeate flow increased obviously, while the concentration of Ca^{2+}, and SO$_4{}^{2-}$ is very low and can almost be ignored (Ca^{2+} and SO$_4{}^{2-}$ are 3.1 × 10^{-4} g·L^{-1} and 1.07 × 10^{-3} g·L^{-1} at the end of the NF process). Specifically, the concentration of Li$^+$ and Na$^+$ increased from 2 g·L^{-1} to 2.54 g·L^{-1} and 1.55 g·L^{-1} to 1.97 g·L^{-1}, respectively. Moreover, the concentration of K$^+$ also increased slowly from 0.033 g·L^{-1} to 0.041 g·L^{-1}. In contrast, the concentration of Mg^{2+} increased sharply from 0.12 g·L^{-1} to 0.86 g·L^{-1}. Combining the concentration of ions (Li$^+$, Na$^+$, and Mg^{2+}) in the collected retentate, it can be found that there is basically no interception of monovalent ions, while the interception rate of multivalent ions is very high. The reason for the higher rejection of Mg^{2+} can be explained using Donnan exclusion. The concentration of counter ions (ions with charge opposite to the fixed charge in the membrane) in the membrane is higher than that in the bulk solution, while the concentration of homonymous ions in the membrane is lower than that in the bulk solution. The Donnan difference prevents the diffusion of homonymic ions from the bulk solution into the membrane. In order to maintain electrical neutrality, the counter ions are also trapped by the membrane. The coulomb repulsion of the multivalent ions is greater than that of the monovalent ions, which explains why the rejection of Mg^{2+} is higher than that of Li$^+$ and Na$^+$.

The rejection rate of the divalent ions is shown in Figure 3c. It can be seen that the rejection rates of SO$_4{}^{2-}$ are higher than 99%, while the retention rates of magnesium gradually drop to 89.9%. Combining the data presented in Figure 3b, it can be found that the concentration of Ca^{2+} and SO$_4{}^{2-}$ in the permeate flow can almost be ignored, which means that sulfate and calcium ions can hardly pass through the nanofiltration membrane. In order to determine whether there is precipitation in the NF process, the solubility of all chlorides and sulfates in the solution at 20 °C is listed, as shown in Table 4.

Table 4. The solubility of all the chloride and sulfate exist in the collected retentate.

Compound	LiCl	NaCl	KCl	MgCl$_2$	CaCl$_2$ *	Li$_2$SO$_4$	Na$_2$SO$_4$	CaSO$_4$ *	MgSO$_4$	K$_2$SO$_4$
solubility/g	83.5	35.9	34.2	54.6	74.5	34.8	19.5	0.255	33.7	11.1

* The solubility of calcium chloride and calcium sulfate refers to the solubility of their hydrated salts; they are CaCl$_2$·6H$_2$O and CaSO$_4$·2H$_2$O, respectively.

According to the results provided by Tables 3 and 4, all of the soluble salts that exist in the collected retentate are not saturated. Notably, there is 0.004 g·L^{-1} Ca^{2+} and 0.26 g·L^{-1} SO$_4{}^{2-}$ in the beginning of the NF, which results in 0.04 g·L^{-1} Ca^{2+} and 2.6 g·L^{-1} SO$_4{}^{2-}$ at an assumed retention of 100%, and the concentration of Ca^{2+} and SO$_4{}^{2-}$ has not reached the K_{sp} of CaSO$_4$·2H$_2$O (the solubility of CaSO$_4$·2H$_2$O is 0.255 g at 20 °C, which means the K_{sp} of CaSO$_4$·2H$_2$O is 2.2 × 10^{-4}) [38]. Because of the retention of divalent ions by the NF membrane and the influence of the electric double layer, a large amount of divalent ions will be enriched on the surface of the NF membrane. When the sulfate and calcium in the bulk retentate solution have not reached the conditions for CaSO$_4$·H$_2$O precipitation, there is already $CaSO_4 \cdot H_2O$ precipitation on the surface of the nanofiltration membrane. That is the reason the concentration of Ca^{2+} in the bulk collected retentate is only 0.02 $g \cdot L^{-1}$, as the feed solution has concentrated 10 times. Further, because the total amount of Ca^{2+} is much lower than that of SO_4^{2-}, this results in the lower concentrated times of Ca^{2+} than SO_4^{2-}. In order to reduce the membrane scaling caused by calcium sulfate precipitation, it is better to wash the membranes after the NF operation. By contrast, Mg^{2+} can only be continuously accumulated in the collected retentate without precipitation, resulting in a higher concentration of Mg^{2+} in the permeate flow. In other words, the more Mg^{2+} that enters the permeate flow, the lower the retention rate of Mg^{2+}.

The separation factor of lithium and magnesium (SF_{Li-Mg}) and lithium recovery are shown in Figure 3d. It can be seen that the SF_{Li-Mg} rose from 15.4 to 30.1 in the first 30 min, and then gradually decrease from 30.1 to 22.8 in the next 75 min. The increasing concentration of Mg^{2+} in the collected retentate is unhelpful for the separation of lithium and magnesium. In addition, the lithium recovery increased almost linearly, and reached 91.6% at the end of the NF process. Noteworthily, the total salinity in the retentate liquid is too high, and the residual lithium cannot be directly recycled by NF, but this retentate liquid can be returned to the EID system to separate lithium and magnesium, which can reduce the waste of lithium.

The final compositions of the permeate flow produced by NF are shown in Table 5. From Table 5, it can be seen that the major cationic ions in the permeate flow are Li^+, Na^+, and Mg^{2+}, and the main anionic ion is Cl^-. The concentration of K^+ is only 0.03 $g \cdot L^{-1}$, and other impurities such as Ca^{2+} and SO_4^{2-} can almost be ignored.

Table 5. The compositions of the permeate flow produced by NF ($g \cdot L^{-1}$).

Elements	Li^+	Na^+	K^+	Mg^{2+}	Ca^{2+}	Cl^-	SO_4^{2-}	Lithium Recovery %
Concentration	2.2	1.7	0.03	0.21	3.1×10^{-4}	14.39	0.0013	91.6

3.2.2. RO Process

The permeate flow produced by the NF process was treated by the RO process, and the main results are shown in Figure 4. Figure 4a has shown that the operation pressure increased from the initial 3 MPa to 5.5 MPa during the RO process, while the flux of the water decreased from 49 $L \cdot m^{-2} \cdot h^{-1}$ to 21.8 $L \cdot m^{-2} \cdot h^{-1}$. Figure 4b shows that the concentration of ions such as Li^+, Na^+, Mg^{2+}, and K^+ in the collected retentate increased almost linearly. Concretely, Li^+ has increased from 2.2 $g \cdot L^{-1}$ to 5.4 $g \cdot L^{-1}$ and Mg^{2+} increased from 0.21 $g \cdot L^{-1}$ to 0.525 $g \cdot L^{-1}$. Figure 4c shows that the concentration of Li^+, Na^+, and Mg^{2+} in the permeate flow increased significantly with the concentration process, but the maximum concentration of lithium is still lower than 0.04 $g \cdot L^{-1}$, and the lithium loss is almost negligible.

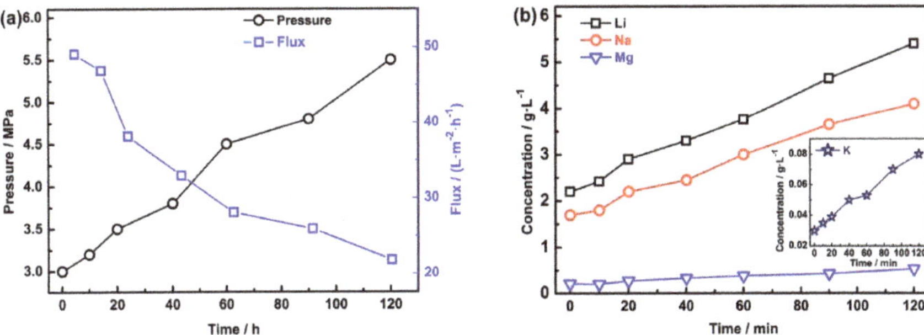

Figure 4. *Cont.*

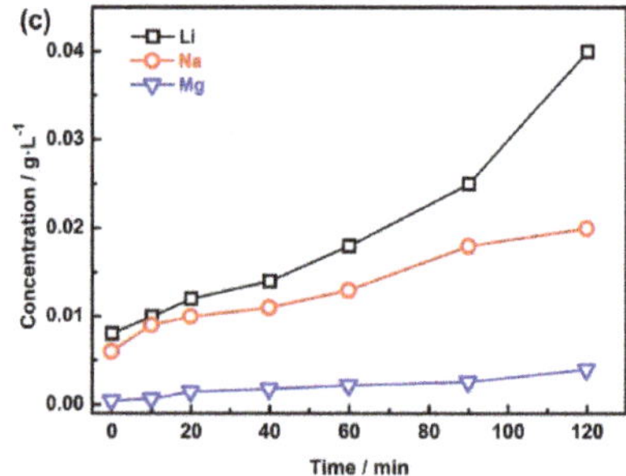

Figure 4. The main results in the RO process. (**a**) The relationship of the operation pressure and the flux of the membrane; (**b**) concentration of Li^+, Na^+, K^+, and Mg^{2+} in the collected retentate; (**c**) concentration of Li^+, Na^+, K^+, and Mg^{2+} in the permeate flow.

The final composition of the permeate flow and the collected retentate produced by RO is shown in Table 6. As shown in Table 6, the concentration of ions in the permeate flow is very low, the loss of the lithium in the RO permeate flow almost can be ignored, and the recovery of lithium can reach 99.4%. Moreover, the permeate flow with such a low salinity content can be used to prepare the supporting electrolyte for the EID system.

Table 6. The final compositions of the permeate flow and collected retentate.

Elements	Li^+	Na^+	K^+	Mg^{2+}	Ca^{2+}	$SO_4{}^{2-}$	Lithium Recovery %
Permeate flow	0.021	0.011	4×10^{-4}	0.002	/	/	/
Collected retentate	5.4	4.4	0.08	0.525	6.3×10^{-4}	0.003	99.4

3.3. Precipitation of Li₂CO₃

The concentrations of Li^+ and Mg^{2+} after RO are 5.4 $g \cdot L^{-1}$ and 0.525 $g \cdot L^{-1}$, respectively. This solution cannot be used directly for the precipitation of Li_2CO_3, and generally requires evaporation and impurity removal. Subsequently, we use an electric furnace to evaporate 5 L of solution to 1.2 L, and add NaOH to adjust the pH of the solution to 12.5 for further removal of magnesium (Mg^{2+} precipitates in the form of $Mg(OH)_2$ when the solution is alkaline). The composition of the solution after magnesium removal is shown in Table 7.

Table 7. The composition of the solution after magnesium removal ($g \cdot L^{-1}$).

Elements	Li^+	Na^+	K^+	Mg^{2+}	Ca^{2+}	$SO_4{}^{2-}$	Lithium Recovery %
Concentration	21.6	23.9	0.34	0.002	2.9×10^{-4}	0.018	96.1

As shown in Table 7, the concentration of Li^+ is enriched to 21.6 $g \cdot L^{-1}$; the mass ratio of Na/Li is slightly greater than 1; and other ions such as K^+, Mg^{2+}, Ca^{2+}, and $SO_4{}^{2-}$ are very low. The recovery of lithium in this process can reach 96.1%; such a low lithium loss is attributed to the effective removal of magnesium by the NF, which greatly reduces the generation of $Mg(OH)_2$ and improves the recovery rate of lithium. In the actual production process, the water generated by evaporation can also be returned to the EID system to prepare the supporting electrolyte.

The solution with 21.6 $g \cdot L^{-1}$ lithium was used for the precipitation of Li_2CO_3 with 280 $g \cdot L^{-1}$ Na_2CO_3. Moreover, the concentration of the mother liquor is shown in Table 8. From Table 8, it can be seen that the main ions in the mother liquor are Na^+ and Li^+. Noteworthily, only 86.7% lithium was precipitated by Na_2CO_3, and the concentration of lithium in the mother liquor is still 1.8 $g \cdot L^{-1}$. In the same way, the mother liquor contains a small amount of excess carbonate, which can be neutralized by part of the brine with high Mg^{2+} ions, and then the mother liquor is returned to the EID system to recover the residual lithium.

Table 8. The main parameters of the mother liquor ($g \cdot L^{-1}$).

Element	Li^+	Na^+	K^+	Lithium Recovery %
Concentration	1.8	90	0.2	86.7

The phase and morphology analysis of the obtained solid is shown in Figure 5. From Figure 5a, it can be seen that the XRD pattern of the obtained powder is indexed to Li_2CO_3 (JCPDS card 22-1141). Morphology analysis by SEM, as shown in Figure 5b, indicated that the particles were columnar and rod, mostly clusters, and have a relatively flat surface. The chemical composition of the prepared Li_2CO_3 is shown in Table 9, and the composition of the obtained Li_2CO_3 meets the national standard (Li_2CO_3-0, GB/T 11075-2013).

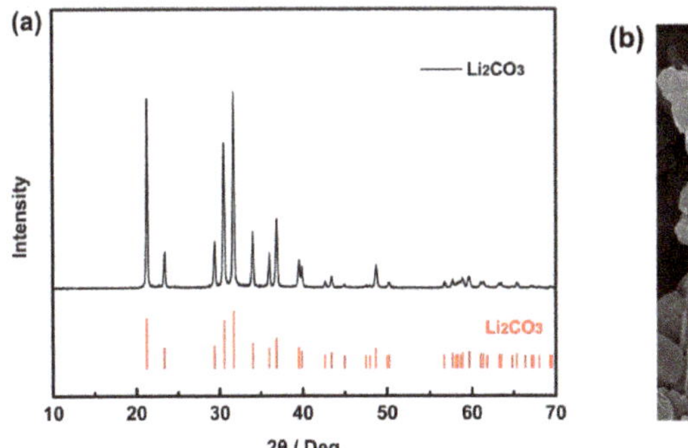

Figure 5. (**a**) XRD and (**b**) scanning electron microscope (SEM) of the obtained solid by precipitation.

Table 9. The chemical composition of the prepared Li_2CO_3.

Constituents	Li_2CO_3	Na	Mg	Fe	Ca	SO_4^{2-}	Cl^-
Content (%)	99.6	0.026	0.005	0.0013	0.011	0.007	0.012

In general, the direct recovery of lithium from the high Mg/Li brine to the production of Li_2CO_3 can reach 68.7%, which was calculated by the product of the recovery of lithium in each process; considering that most of the solutions are cycled in the system (except the lithium loss by the precipitation of $Mg(OH)_2$), the total recovery of lithium will be greater than 85%.

3.4. Comparison of Methods for Lithium Extraction from High Mg/Li Brine

Table 10 shows the comparison of methods for lithium extraction from high Mg/Li brine.

Table 10. Comparison of the Li^+ recovery between this study and conventional methods.

Methods	Li^+ Concentration in Brine/$g \cdot L^{-1}$	Mg/Li in Brine	Li^+ Recovery Rate %	References
Solvent extraction	2.088	44.06	90.93 *	[39]
Ion sieve	0.259	95	82.1 *	[40]
Electrodialysis	0.148	60	72.1 *	[41]
This study	2.05	58.5	>85	This study

* The asterisk only indicates the recovery rate of the separation of magnesium and lithium from brine.

From Table 10, it can be seen that the total Li^+ recovery rate in this paper is superior to that of ion sieve method and electrolysis method, but slightly lower than that of solvent extraction method. However, the extractant used in the solvent extraction method has a slight dissolution in the brine, which will cause greater environmental pollution. Noteworthily, this comprehensive membrane process has environmental protection significance.

4. Conclusions

In this paper, we constructed an integrated membrane process combining the EID system and NF and RO processes to prepare Li_2CO_3 from a high mass ratio of Mg/Li brine. This method successfully realizes the separation of lithium and magnesium in brine with a high Mg/Li ratio, which relies on the anion membrane to retain cations and the selective characteristics of $LiFePO_4$ to adsorb lithium. Most of the bivalent ions in the prepared lithium-riched solution were removed by nanofiltration membrane. After concentration, purification, and precipitation, we prepared industrial-grade Li_2CO_3. Noteworthily, the removal of magnesium by nanofiltration can reduce the amount of alkali and reduce the entrainment loss of lithium caused by the massive production of magnesium hydroxide. In general, this process can efficiently realize the selective separation of magnesium and lithium without pollution to the environment and provide a new perspective for extracting lithium from salt lakes.

Author Contributions: Conceptualization and methodology (Z.Z., L.H., W.X.), formal analysis (W.X., L.H., D.L.); original draft preparation (W.X.); writing—review and editing, (W.X., L.H., Z.Z.); supervision (Z.Z.); project administration (Z.Z., L.H.); funding acquisition (Z.Z., L.H.). All authors have read and agreed to the published version of the manuscript.

Funding: This study is supported by the Major Program of National Natural Science Foundation of China (51934010), National Science Foundation of Hunan province (Grant No. 2019JJ40377), and Innovation-Driven Project of Central South University (No. 2020CX026).

Acknowledgments: The XRD and SEM data were obtained by using the equipment provided by Changsha Research Institute of Mining and Metallurgy Co., Ltd.

Conflicts of Interest: The authors declare no conflict of interest.

References

1. Armand, M.; Tarascon, J.-M. Building better batteries. *Nature* **2008**, *451*, 652–657. [CrossRef]
2. An, J.W.; Kang, D.J.; Tran, K.T.; Kim, M.J.; Lim, T.; Tran, T. Recovery of lithium from Uyuni salar brine. *Hydrometallurgy* **2012**, *117*, 64–70. [CrossRef]
3. Goodenough, J.B.; Kim, Y. Challenges for Rechargeable Li Batteries. *Chem. Mater.* **2010**, *22*, 587–603. [CrossRef]
4. Grosjean, C.; Miranda, P.H.; Perrin, M.; Poggi, P. Assessment of world lithium resources and consequences of their geographic distribution on the expected development of the electric vehicle industry. *Renew. Sustain. Energy Rev.* **2012**, *16*, 1735–1744. [CrossRef]
5. Hamzaoui, A.; M'Nif, A.; Hammi, H.; Rokbani, R. Contribution to the lithium recovery from brine. *Desalination* **2003**, *158*, 221–224. [CrossRef]
6. Naumov, A.V.; Naumova, M.A. Modern state of the world lithium market. *Russ. J. Non-Ferr. Met.* **2010**, *51*, 324–330. [CrossRef]
7. Martin, G.; Rentsch, L.; Höck, M.; Bertau, M. Lithium market research—Global supply, future demand and price development. *Energy Storage Mater.* **2017**, *6*, 171–179. [CrossRef]
8. Kesler, S.E.; Gruber, P.W.; Medina, P.A.; Keoleian, G.A.; Everson, M.P.; Wallington, T.J. Global lithium resources: Relative importance of pegmatite, brine and other deposits. *Ore Geol. Rev.* **2012**, *48*, 55–69. [CrossRef]
9. Flexer, V.; Baspineiro, C.F.; Galli, C.I. Lithium recovery from brines: A vital raw material for green energies with a potential environmental impact in its mining and processing. *Sci. Total. Environ.* **2018**, *639*, 1188–1204. [CrossRef]
10. Yaksic, A.; Tilton, J.E. Using the cumulative availability curve to assess the threat of mineral depletion: The case of lithium. *Resour. Policy* **2009**, *34*, 185–194. [CrossRef]

11. Song, J.F.; Nghiem, L.D.; Li, X.-M.; He, T. Lithium extraction from Chinese salt-lake brines: Opportunities, challenges, and future outlook. *Environ. Sci. Water Res. Technol.* **2017**, *3*, 593–597. [CrossRef]

12. Zheng, M.; Liu, X. Hydrochemistry of Salt Lakes of the Qinghai-Tibet Plateau, China. *Aquat. Geochem.* **2009**, *15*, 293–320. [CrossRef]

13. Wang, H.; Zhong, Y.; Du, B.; Zhao, Y.; Wang, M. Recovery of both magnesium and lithium from high Mg/Li ratio brines using a novel process. *Hydrometallurgy* **2018**, *175*, 102–108. [CrossRef]

14. Shi, C.; Jing, Y.; Xiao, J.; Wang, X.; Yao, Y.; Jia, Y. Solvent extraction of lithium from aqueous solution using non-fluorinated functionalized ionic liquids as extraction agents. *Sep. Purif. Technol.* **2017**, *172*, 473–479. [CrossRef]

15. Zhao, Y.; Wang, H.; Li, Y.; Wang, M.; Xiang, X. An integrated membrane process for preparation of lithium hydroxide from high Mg/Li ratio salt lake brine. *Desalination* **2020**, *493*, 114620. [CrossRef]

16. Somrani, A.; Hamzaoui, A.; Pontie, M. Study on lithium separation from salt lake brines by nanofiltration (NF) and low pressure reverse osmosis (LPRO). *Desalination* **2013**, *317*, 184–192. [CrossRef]

17. Liu, X.; Chen, X.; He, L.; Zhao, Z. Study on extraction of lithium from salt lake brine by membrane electrolysis. *Desalination* **2015**, *376*, 35–40. [CrossRef]

18. Gao, A.; Sun, Z.; Li, S.; Hou, X.; Li, H.; Wu, Q.; Xi, X. The mechanism of manganese dissolution on Li1.6Mn1.6O4 ion sieves with HCl. *Dalton Trans.* **2018**, *47*, 3864–3871. [CrossRef]

19. Yang, F.; Chen, S.C.; Shi, C.T.; Xue, F.; Zhang, X.X.; Ju, S.G.; Xing, W.H. A Facile Synthesis of Hexagonal Spinel lambda-MnO2 Ion-Sieves for Highly Selective Li+ Adsorption. *Processes* **2018**, *6*, 59. [CrossRef]

20. He, L.; Xu, W.; Song, Y.; Luo, Y.; Liu, X.; Zhao, Z. New Insights into the Application of Lithium-Ion Battery Materials: Selective Extraction of Lithium from Brines via a Rocking-Chair Lithium-Ion Battery System. *Glob. Chall.* **2018**, *2*, 1700079. [CrossRef]

21. Zhao, Z.W.; Si, X.F.; Liu, X.H.; He, L.H.; Liang, X.X. Li extraction from high Mg/Li ratio brine with LiFePO4/FePO4 as electrode materials. *Hydrometallurgy* **2013**, *133*, 75–83. [CrossRef]

22. Zhao, Z.W.; Si, X.F.; Liang, X.X.; Liu, X.H.; He, L.H. Electrochemical behavior of Li$^+$, Mg^{2+}, Na$^+$ and K$^+$ in LiFePO$_4$/FePO$_4$ structures. *Trans. Nonferr. Met. Soc. China* **2013**, *23*, 1157–1164. [CrossRef]

23. Liu, X.H.; Chen, X.Y.; Zhao, Z.W.; Liang, X.X. Effect of Na$^+$ on Li extraction from brine using LiFePO$_4$/FePO$_4$ electrodes. *Hydrometallurgy* **2014**, *146*, 24–28. [CrossRef]

24. Li, Z.; Binnemans, K. Selective removal of magnesium from lithium-rich brine for lithium purification by synergic solvent extraction using β-diketones and Cyanex 923. *AIChE J.* **2020**, *66*, e16246. [CrossRef]

25. Wang, J.; Yang, S.; Bai, R.; Chen, Y.; Zhang, S. Lithium Recovery from the Mother Liquor Obtained in the Process of Li$_2$CO$_3$ Production. *Ind. Eng. Chem. Res.* **2019**, *58*, 1363–1372. [CrossRef]

26. Seeley, F.; Baldwin, W. Extraction of lithium from neutral salt solutions with fluorinated β-diketones. *J. Inorg. Nucl. Chem.* **1976**, *38*, 1049–1052. [CrossRef]

27. Ooi, K.; Miyai, Y.; Katoh, S.; Maeda, H.; Abe, M. Lithium-ion Insertion/Extraction Reaction with λ-MnO$_2$ in the Aqueous Phase. *Chem. Lett.* **1988**, *17*, 989–992. [CrossRef]

28. Chen, Y.; Liu, F.; Wang, Y.; Lin, H.; Han, L. A tight nanofiltration membrane with multi-charged nanofilms for high rejection to concentrated salts. *J. Membr. Sci.* **2017**, *537*, 407–415. [CrossRef]

29. Bai, X.; Zhang, Y.; Wang, H.; Zhang, H.; Liu, J. Study on the modification of positively charged composite nanofiltration membrane by TiO$_2$ nanoparticles. *Desalination* **2013**, *313*, 57–65. [CrossRef]

30. Lithium, N. The 3Q Project. 2019. Available online: https://www.neolithium.ca/project.php (accessed on 18 October 2020).

31. SQM. Sustainability of Lithium Production in Chile. 2020. Available online: https://www.sqm.com (accessed on 18 October 2020).

32. Kim, S.; Lee, J.; Kang, J.S.; Jo, K.; Kim, S.; Sung, Y.E.; Yoon, J. Lithium recovery from brine using a lambda-MnO2/activated carbon hybrid supercapacitor system. *Chemosphere* **2015**, *125*, 50–56. [CrossRef]

33. Zhao, M.Y.; Ji, Z.Y.; Zhang, Y.G.; Guo, Z.Y.; Zhao, Y.Y.; Liu, J.; Yuan, J.S. Study on lithium extraction from brines based on LiMn2O4/Li1-xMn$_2$O$_4$ by electrochemical method. *Electrochim. Acta* **2017**, *252*, 350–361. [CrossRef]

34. Wagh, P.; Islam, S.Z.; Deshmane, V.G.; Gangavarapu, P.; Poplawsky, J.; Yang, G.; Sacci, R.; Evans, S.F.; Mahajan, S.; Paranthaman, M.P.; et al. Fabrication and Characterization of Composite Membranes for the Concentration of Lithium Containing Solutions Using Forward Osmosis. *Adv. Sustain. Syst.* **2020**. [CrossRef]

35. Misyura, S.Y. Evaporation of aqueous solutions of LiBr and LiCl salts. *Int. Commun. Heat Mass Transf.* **2020**, *117*, 104727. [CrossRef]
36. Al-Rashdi, B.; Johnson, D.; Hilal, N. Removal of heavy metal ions by nanofiltration. *Desalination* **2013**, *315*, 2–17. [CrossRef]
37. Wu, D.; Yu, S.; Lawless, D.; Feng, X. Thin film composite nanofiltration membranes fabricated from polymeric amine polyethylenimine imbedded with monomeric amine piperazine for enhanced salt separations. *React. Funct. Polym.* **2015**, *86*, 168–183. [CrossRef]
38. Speight, J.G. *Land's Handbook of Chemistry*, 16th ed.; McGraw-Hill: New York, NY, USA, 2005.
39. Shi, C.; Duan, D.; Jia, Y.; Jing, Y. A highly efficient solvent system containing ionic liquid in tributyl phosphate for lithium ion extraction. *J. Mol. Liq.* **2014**, *200*, 191–195. [CrossRef]
40. Lai, X.; Yuan, Y.; Chen, Z.; Peng, J.; Sun, H.; Zhong, H. Adsorption–Desorption Properties of Granular EP/HMO Composite and Its Application in Lithium Recovery from Brine. *Ind. Eng. Chem. Res.* **2020**, *59*, 7913–7925. [CrossRef]
41. Feng, W.X. The Research on Separation of Magnesium and Lithium from Brine by Electrodialysis. Ph.D. Thesis, Hebei University of Technology, Tianjin, China, 2016.

Publisher's Note: MDPI stays neutral with regard to jurisdictional claims in published maps and institutional affiliations.

 *membranes*

Perspective

The Acid–Base Flow Battery: Sustainable Energy Storage via Reversible Water Dissociation with Bipolar Membranes

Ragne Pärnamäe [1], Luigi Gurreri [2,*], Jan Post [1], Willem Johannes van Egmond [3], Andrea Culcasi [2], Michel Saakes [1], Jiajun Cen [3,4], Emil Goosen [3], Alessandro Tamburini [2], David A. Vermaas [3,5] and Michele Tedesco [1,*]

[1] Wetsus, European Centre of Excellence for Sustainable Water Technology, 8911 MA Leeuwarden, The Netherlands; ragne.parnamae@wetsus.nl (R.P.); jan.post@wetsus.nl (J.P.); michel.saakes@wetsus.nl (M.S.)

[2] Dipartimento di Ingegneria, Università degli Studi di Palermo, viale delle Scienze Ed. 6, 90128 Palermo, Italy; andrea.culcasi@unipa.it (A.C.); alessandro.tamburini@unipa.it (A.T.)

[3] AquaBattery B.V., Lijnbaan 3C, 2352 CK Leiderdorp, The Netherlands; janwillem.vanegmond@aquabattery.nl (W.J.v.E.); jiajun.cen@aquabattery.nl (J.C.); emil.goosen@aquabattery.nl (E.G.); david.vermaas@aquabattery.nl (D.A.V.)

[4] Imperial College London, Department of Chemical Engineering, South Kensington Campus, London SW7 2AZ, UK

[5] Department of Chemical Engineering, Delft University of Technology, Van der Maasweg 9, 2629 HZ Delft, The Netherlands

* Correspondence: luigi.gurreri@unipa.it (L.G.); michele.tedesco@wetsus.nl (M.T.)

Received: 12 November 2020; Accepted: 8 December 2020; Published: 10 December 2020

Abstract: The increasing share of renewables in electric grids nowadays causes a growing daily and seasonal mismatch between electricity generation and demand. In this regard, novel energy storage systems need to be developed, to allow large-scale storage of the excess electricity during low-demand time, and its distribution during peak demand time. Acid–base flow battery (ABFB) is a novel and environmentally friendly technology based on the reversible water dissociation by bipolar membranes, and it stores electricity in the form of chemical energy in acid and base solutions. The technology has already been demonstrated at the laboratory scale, and the experimental testing of the first 1 kW pilot plant is currently ongoing. This work aims to describe the current development and the perspectives of the ABFB technology. In particular, we discuss the main technical challenges related to the development of battery components (membranes, electrolyte solutions, and stack design), as well as simulated scenarios, to demonstrate the technology at the kW–MW scale. Finally, we present an economic analysis for a first 100 kW commercial unit and suggest future directions for further technology scale-up and commercial deployment.

Keywords: flow battery; energy storage; bipolar membrane; reverse electrodialysis; bipolar membrane electrodialysis; water dissociation

---

1. Introduction

The awareness of climate change and its alarming impact has resulted in the recognition of urgent need for decarbonization to stop further change. As coal-fired power plants alone account for almost a third of global CO_2 emissions [1], the energy sector is under increased attention for its potential to remarkably reduce the emissions. To realize that potential, several climate mitigation strategies must be deployed at scale, such as carbon capture, utilization, and storage (CCUS), and increasing the share of nuclear power and renewables as energy source. However, the intermittent nature of renewables, such as solar and wind, presents a new challenge for electric grids, where the equality of power generation and consumption needs to be ensured with rapid adjustments. Hence, new technologies are needed to guarantee rapid adjustments and stabilization in modern grids with increasing share of renewables. In this regard, energy storage systems provide an excellent option for system stabilization. By storing energy while supply is larger than demand (and discharging energy back to the grid when the opposite occurs), energy storage systems can improve the flexibility and reliability of the grid. Moreover, since renewables often have distinct seasonal variations, there is especially a need for long-term (i.e., seasonal) energy storage.

Although there are a number of technologies available for energy storage (Figure 1), only few of them are commercially deployed. Today, pumped hydro energy storage (PHS) is the most mature long-duration electricity storage system, and the only one commercially available at a large scale [1–3]. PHS systems store energy by moving water to a reservoir at elevated heights during times of low demand, and releasing it through a turbine into a lower reservoir during peak demand. PHS holds today the largest share among storage methods with over 120 GW installed electricity storage capacity for pure PHS plants (not receiving natural inflows, "closed-loop" plants) and almost 1.2 TW storage capacity for mixed PHS plants (both stored water and natural inflow used for generating electricity, "pump-back" plants) [4]. Due to its ability to store energy for up to months [5], and the emerging need to firm the seasonal fluctuations of renewables in grids, its installed capacity is anticipated to increase much more. However, PHS has major geographical constraints, as it needs large amounts of water, and elevated heights at site. Therefore, PHS is not suitable for flat and dry regions, as the construction of PHS plants would be clearly uneconomical in such locations. Suitable installation sites for PHS plants are mountainous regions with rivers (which are often protected natural areas). This raises some ecological and social concerns that need to be overcome when opting for PHS.

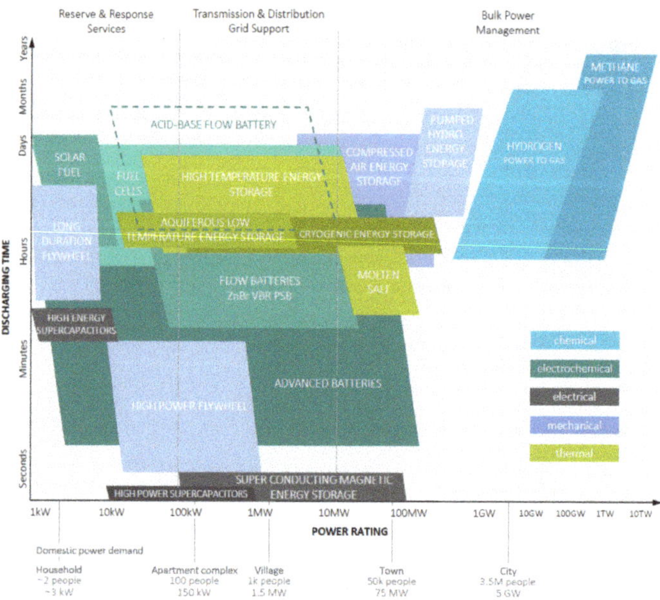

Figure 1. Energy storage systems and their conceptual comparison in terms of discharging time and power range. The figure is simplified, to give a qualitative comparison, and is not intended to be exhaustive; many of the storage systems can have broader operational ranges than shown. The domestic power demand scale is based on peak electric load demand of 3 kW per average EU household (~2 people). Adapted from Reference [6].

Electrochemical energy storage has received increasing attention as an alternative storage system [7–11], and several battery technologies have been making rapid advances in the past years. However, although batteries are commercial on a smaller scale, they are not yet widespread on larger scale nor in connection with electric grids. Larger-scale applications require specifically designed batteries. For example, lithium ion batteries are economically viable only for short-duration energy storage (<10 h discharge), where the value of the energy that they generate is higher than their own cost [12]; thus, they are unsuitable for the long-duration storage needed for renewables.

Development, Principles, and State-of-the-Art of the Acid–Base Flow Battery

The acid–base flow battery (ABFB) can be considered as a modification of the concentration gradient flow battery [13], which relies on two opposite processes, i.e., electrodialysis (ED) and reverse electrodialysis (RED). Electrodialysis [14,15] exploits electric energy to desalinate a feed stream (typically brackish water). In conventional ED an electric field is applied over a membrane stack consisting of alternating anion- and cation-exchange membranes (AEMs and CEMs) that, by selective ion transport, separate the feed solution into a concentrate and a diluate stream, thus creating salinity gradients over the membranes. Over the past years, the opposite process, i.e., reverse electrodialysis (RED) [16,17], has also been widely investigated: In RED, two streams at a different salt concentration (i.e., concentrate and diluate) are fed to an analogous stack (with alternating AEMs and CEMs), so that the fluxes of cations and anions are driven by the concentration difference. As a result of the diffusive drive force inside the stack and the selective transport through the membranes, an ionic current can be harvested as electric current at the electrodes. Thus, in RED salinity gradients are used to produce electricity [18]. Interestingly, by coupling ED and RED processes in the same device, it is possible to create an energy storage system (known as 'concentration gradient flow battery', CGFB), to store electric energy in salinity gradients [19]. During the CGFB charging step (i.e., ED mode), electric energy is stored in generated salinity gradients. During the battery discharging step (RED mode), the previously generated salinity gradients are used to produce electricity. The CGFB technology has been demonstrated on laboratory scale [19] and on pilot scale [20]. A 1 kW/10 kWh pilot used to supply energy to a nearby student housing is operational since 2018 and located in Delft (The Netherlands).

In an ABFB, bipolar membranes are added auxiliary to CEMs and AEMs to generate a pH gradient in addition to the salinity gradient [21]. A bipolar membrane (BPM) is a composite membrane consisting of oppositely charged ion-exchange layers. In contrast to CEM or AEM that allow selective ion transport, the BPM (ideally) allows no transport of ions across it. Instead, the BPM is used to produce ions by dissociating water at the junction of its two layers [22,23]. Notably, unlike conventional water splitting occurring on electrode surfaces that produces gas (H_2 and O_2), the BPM-assisted water dissociation only generates ions (H^+ and OH^-), and occurs at a lower voltage (i.e., 0.83 V across a BPM separating 1 M HCl and 1 M NaOH instead of 1.23 V needed for water splitting in conventional electrolysis). No gas production inside the ABFB is also advantageous in terms of safety, compared to other battery systems based on electrolysis [24].

The principle of ABFB is shown in Figure 2. During charging (Figure 2a), an electric field is applied over the stack and inside the BPMs water is dissociated into protons and hydroxyl ions. The produced ions leave the BPM junction through a respective BPM layer—protons through the cation-exchange layer (AEL), and hydroxide ions through the anion-exchange layer (AEL), meaning that H^+ and OH^- ions leave the BPM on opposite sides. An additional salt (e.g., NaCl) is added to a third compartment in the repeating unit of the battery. As a result of the "salt ion" transport across monopolar membranes (i.e., Cl^- through AEMs and Na^+ through CEMs), an acidic solution is obtained in one compartment (adjacent to the CEL of the BPM), and an alkaline solution in the other compartment (adjacent to the AEL of the BPM). Thus, two concentration gradients are produced over the BPM—(I) a pH gradient due to an acidic solution produced on one side, and an alkaline solution on the other side of the BPM, and (II) a salinity gradient due to the different composition of acidic and alkaline solutions. During battery discharge (Figure 2b), the electric field over the stack is opposite and the electric current flows through an external load, leading to the neutralization of acid and base solutions: H^+ and OH^- ions flow into the BPM junction, where they recombine into water. Therefore, the ABFB charging step is bipolar membrane electrodialysis (BMED), and the discharging step bipolar membrane reverse electrodialysis (BMRED). Introducing BPMs (and, consequently, an additional pH gradient) in the battery increases the energy density of the battery significantly, i.e., by more than three times compared to the concentration gradient flow battery [25].

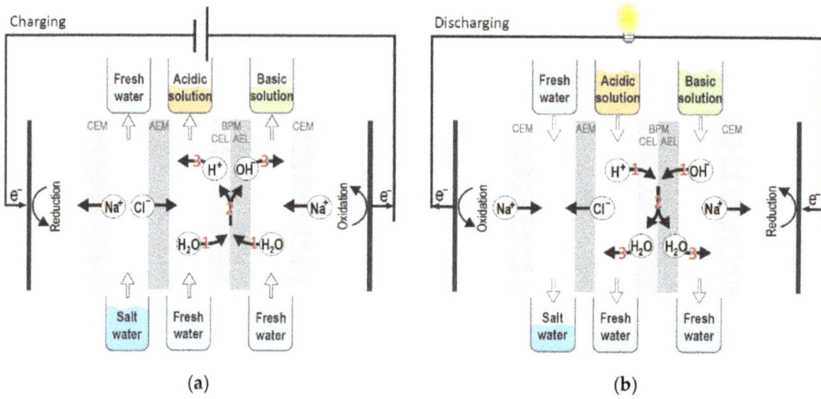

Figure 2. Schematic principle of the acid–base battery during (**a**) charging mode (bipolar membrane electrodialysis, BMED) and (**b**) discharging mode (bipolar membrane reverse electrodialysis, BMRED). Adapted from Reference [25].

The ABFB technology is still at its early stage of development, with a very limited amount of works reported in the literature. The earliest study on this technology is from 1983 by Emrén and Holmström, who reported extremely low energy efficiency (0.1%), caused by poor permselectivity of the membranes and a high resistance of the (early stage) BPMs used at that time [26]. Pretz and Staude used the ABFB concept for a fuel cell application [27]. They operated the cell with acid–base concentrations up to 1 M HCl-NaOH, but observed irreversible water accumulation in the junction of the BPM, which led to delamination of the BPM. Zholkovskij et al. tested a similar battery to Emrén and Holmström (recirculating the salt solution while keeping acid and base compartments stagnant) [28]. They charged the battery only up to 0.03 M acid and base solutions, which also explains the low battery performance metrics (see Table 1). Kim et al. introduced Fe^{2+}/Fe^{3+} redox couple in the electrode compartments of the ABFB to avoid electrolysis and subsequent gas formation at the electrodes [29]. However, because the electrode compartments were separated from the rest of the cell with CEMs, the battery performance suffered from iron ion migration from the electrode towards the base compartment, thus causing precipitation of iron salts. Van Egmond et al. demonstrated stable ABFB operation (at 150 A/m^2 current density during charge, and 15 A/m^2 during discharge) over a wide pH range (pH = 0–14), and analyzed the contribution of different energy loss sources [25]. They estimated the total energy lost by co-ion transport to be the biggest factor, contributing 39–65% of the total losses. Xia et al. investigated the ABFB on both single-cell [30] and stack (5–20 cell units) level [31], and concluded that the single cell performance can be extrapolated to the stack performance. However, additional energy losses by parasitic currents (also known as shortcut currents [32], shunt currents [31,33], or leakage currents [34]) through the manifolds need to be taken into account in the stack [31]. This aspect was highlighted by Culcasi et al., who modeled ABFB systems predicting a loss in round-trip efficiency in the range of 25–35% due to parasitic currents [35]. More recently, Zaffora et al. investigated the ABFB under different conditions of acid–base concentration (focusing on the discharge phase, similarly to Pretz and Staude [27]), and reported a maximum power density of 17 W/m^2, and energy density of 10 kWh/m^3 (1 M HCl-NaOH, at 100 A/m^2 current density during discharge) [36].

Table 1. Overview of experimental conditions and performance parameters of previous works on acid–base flow batteries (ABFBs) (chronological order).

Authors, Year (Source)	Battery Composition	Membranes	Charge/Discharge Conditions	Performance *
Emrén and Holmström, 1983 [26]	7-triplet stack, copper electrodes. 0.85 M NaCl (all compartments)	Self-made BPM (modified polysulphones), CEM/AEM: not specified. Membrane active area: 7 cm^2	Charge: 1.4–56 A/m^2 (for 2 h) Discharge: constant load	Voltage 1.8 V at 1.4 A/m^2; EE 0.1%
Pretz and Staude, 1998 [27]	20-triplet stack, platinized Ti electrodes. Electrolyte composition: 0.1, 0.5, or 1.0 M NaCl-HCl-NaOH	BPM: Stantech, or self-made by casting or gluing. CEM/AEM: self-made from polysulphones, Thomapor MC3470/MA3475, Tokuyama CMS/ACS	Only discharge: 0–50 A/m^2	Power density: 3.63 W/m^2 (10.9 W/m^2 triplet) with 0.5 M HCl-NaOH; EE 22%
Zholkovskij et al., 1998 [28]	1-triplet stack + extra salt comp., platinized Ti electrodes. Batchwise operation (no flow). 0.03 M HCl-NaOH (acid/base comp.)	BPM: Stantech; CEM/AEM: Selemion CMV/AMV. Membrane active area: 28 cm^2	Charge: 3.57 A/m^2 for 50 min, 35.7 A/m^2 for 5 min, or 357 A/m^2 for 0.5 min. Discharge: 0.4 mA/m^2 for 30 h (slow discharge), 1.7 A/m^2 for 10 min (fast discharge)	At slow discharge (30 h): specific capacity: 0.3 Ah/kg; energy density: 0.1 Wh/kg; max power density: 0.005 W/kg product. At fast discharge (10 min): specific capacity: 0.15 Ah/kg; energy density: 3×10^{-2} Wh/kg; max power density: 0.5 W/kg product. Efficiency: 45–61%
Kim et al., 2016 [29]	1-triplet stack, carbon-felt electrodes. Electrolyte composition: 0.1 M NaCl, 0.1, 0.2, … , 0.6, or 0.7 M HCl-NaOH. Electrode rinse solution: 0.01 M $FeSO_4/Fe_2(SO_4)_3$, 0.01 M Na_2SO_4.	BPM: Tokuyama BP-1; CEM/AEM: Tokuyama CMS, CMX/AM-1	13 cycles with 0.5 M HCl-NaOH system at 2.9 A/m^2. Cycle voltage range: 1.25–0.40 V.	Max power density 2.9 W/m^2 (11.6 W/m^2 per single-cell stack) with 0.6 M HCl-NaOH. Cyclability tests with 0.5 M HCl-NaOH: CE 98.3%, VE 76.0%, EE 77.3% from 2nd to 9th cycle on average. Average charge capacity 1.12 Ah/L, average discharge capacity 1.11 Ah/L. Rapid decline in charge/discharge capacities after 9th cycle.
Van Egmond et al., 2018 [25]	1-triplet stack, Ir/Ru-coated Ti electrodes. Electrolyte composition: 0.214 M NaCl (salt comp.), 1 M HCl + 0.5 M NaCl (acid comp.), 1 M NaOH + 0.5 M NaCl (base comp.). Electrode rinse solution: 0.5 M Na_2SO_4.	BPM: Fumasep FBM. CEM: Nafion N117. AEM: Fumasep FAB-PK-30 (proton blocking). Membrane active area: 100 cm^2.	9 charge/discharge cycles. Voltage range: 0–0.83 V. Charge: 50–150 A/m^2; discharge: 5–15 A/m^2. Discharge time: 7 h at 5 A/m^2, and 5 h at 15 A/m^2	Power density up to 3.7 W/m^2. Energy density 2.9 Wh/L
Xia et al., 2018 [30]	1-triplet stack + extra salt comp., platinized Ti mesh electrodes. Electrolyte composition: 0.5 M NaCl (salt comp), 0.25, 0.5, 0.75, or 1.0 M HCl-NaOH (acid/base comp.). Electrode rinse solution: 0.25 M Na_2SO_4.	BPM: Fumasep FBM; CEM/AEM: Fumasep FKB/FAB. Membrane active area: 25 cm^2.	20 cycles with 0.75 M HCl-NaOH system at 400 A/m^2 charge/discharge; 20 min charge, 20 min discharge per cycle	OCV 0.775 V; mean voltage over the BPM: 0.87 V at charge, 0.63 V at discharge (BPM VE 72%)
Xia et al., 2020 [31]	20-triplet stack. Electrolyte composition: 0.5 M NaCl (salt comp), 0.5 or 1.0 M HCl-NaOH (acid/base comp). Electrode rinse solution: 0.25 M Na_2SO_4	BPM: Fumasep FBM; CEM/AEM: Fumasep FKB/FAB. Membrane active area: 100 cm^2.	90 A/m^2 charge and discharge, both 5 min.	~15 W/m^2 excluding electrode losses, for 20- triplet stack with 1 M HCl-NaOH at 100 A/m^2.
Zaffora et al., 2020 [36]	5–38-triplet stack. Ti/mixed-metal oxide electrodes. Electrolyte composition: 0.25 M NaCl (salt comp.), 0.2, 0.6 or 1.0 M HCl-NaOH (acid/base comp). Electrode rinse solution: 0.25 M Na_2SO_4 or 0.5 M $FaCl_2/FeCl_3$.	BPM: Fumasep FBM; CEM/AEM: Fumasep FKB/FAB. Membrane active area: 100 cm^2.	Only discharge with single pass, up to 100 A/m^2.	~17 W/m^2 for 10-triplet stack with 1 M HCl-NaOH at 100 A/m^2. Estimated energy density of 10.3 kWh/m^3 acid for complete discharge.

* Power density values are normalized per m^2 of total membrane area (to be multiplied by a factor 3 to obtain a power density per m^2 of BPM, or triplet). BPM, bipolar membrane; CEM, cation-exchange membrane; AEM, anion-exchange membrane.

The aim of this work is to describe the current development and technological challenges of the ABFB technology as a novel energy storage system. In particular, we focus on the main aspects related to the development of battery components (membranes, electrolyte solutions, stack design), and on modelled scenarios to demonstrate the technology at kW-scale. Finally, we present a preliminary techno-economic analysis of the technology, and suggest future direction for large-scale implementation.

2. Battery Components and Design

2.1. Bipolar and Monopolar Membranes

The core element of the ABFB is the bipolar membrane (BPM), which is responsible for the reversible water dissociation, and therefore, for the pH gradient in the battery. A BPM is an ion-exchange membrane consisting of two layers: a cation-exchange layer (CEL) and an anion-exchange layer (AEL). Contrary to conventional ion-exchange membranes (CEMs and AEMs), the function of a BPM is not to selectively transport ions from one side to the opposite one, as no ions can cross both layers of the membrane. In fact, ion transport across the BPM is unwanted. Instead, the function of a BPM is to dissociate water to protons and hydroxide ions at the junction (J) of its two layers (Figure 3).

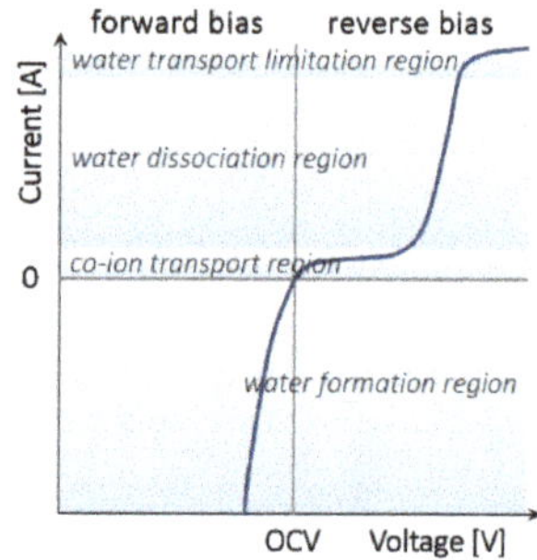

Figure 3. Typical current–voltage curve for a bipolar membrane. For the acid–base flow battery, forward bias corresponds to the battery discharge mode, and reverse bias to battery charge mode.

If a high enough voltage is applied over the bipolar membrane, the water diffused into the membrane is dissociated into H^+ and OH^- ions at the bipolar junction that carry the current demanded by the applied voltage. Commonly, a catalyst is introduced into the bipolar junction in order to promote water dissociation at lower voltage, thus lowering the energy requirements of the process [37–41].

The unique function of the bipolar membrane grants it some distinctive properties. While high water permeability is unwanted for ion-exchange membranes intended for separation processes (including the cation-and anion-exchange membranes in the ABFB which should have low water permeability to avoid diluting the acid and base compartments), it is a desired property for BPMs. When water is dissociated into ions at the BPM junction (battery charging mode, i.e., reverse bias in BPM literature), the junction must be constantly replenished with a diffusive water flux to ensure the membrane can withstand the current [42]. Oppositely, when H^+ and OH^- ions are recombined at the junction (battery discharging mode, i.e., forward bias in BPM literature), the formed water must be able to diffuse out of the junction fast enough to avoid water accumulation at the junction and the consequent delamination of the bipolar membrane layers. Unlike all other bipolar membrane assisted processes [23], the acid–base flow battery uses the BPMs under both reverse bias (i.e., water dissociation) and forward bias (water recombination). Current commercial membranes are designed only for dissociating water, while operating BPMs under water recombination (which occurs during the discharging process in the ABFB), has been so far overlooked by membrane scientists and manufacturers. Hence, when using commercial BPMs in ABFB the discharge current densities of the battery are today limited by membrane delamination [25,30], and are thus relatively low. In particular, van Egmond et al. reported successful stable performance (up to 9 cycles) for a 1 M HCl-1 M NaOH battery operated at 15 A/m^2 discharge current density [25], while Xia et al. noted water accumulation in the BPM junction when operating a 0.75 M HCl–0.75 M NaOH battery at discharge current densities above 200 A/m^2 [30].

As with any electromembrane process, selectivity of the membranes is an important parameter also for the ABFB application. Both bipolar and the monopolar membranes should have high permselectivity to reduce co-ion leakage [43]. Co-ion leakage of "salt" ions (Na$^+$, Cl$^-$) causes the formation of neutral salt in the acid/base compartments, while co-ion leakage of water ions (H$^+$ and OH$^-$), i.e., the ion crossover through the entire BPM, causes water recombination in the outer solution (i.e., outside the BPM). In any case, the co-ion fluxes through (monopolar or bipolar) membranes lead to self-discharge of the battery. In addition, high selectivity of the monopolar membranes is crucial to avoid leakage of the electrode rinse solution into the acid and base compartments, as monopolar membranes are used for separating the electrode compartments from the rest of the cell, and the electrode rinse solution has a different composition from the rest of the battery [29]. Finally, due to the presence of highly acidic and alkaline solutions, all membranes in the ABFB need to be chemically stable in a wide pH range (pH = 0–14).

2.2. Battery Chemistry: Optimizing Electrolytes for Acid–Base Flow Batteries

The chemistry of acid–base flow batteries is based on the added electrolyte-the produced acid will consist of a proton from dissociation of water and the anion from the electrolyte, and the produced base of a hydroxide ion and the electrolyte cation. Thus, the choice of the electrolyte directly decides the composition of the acid and base produced for storing energy. In principle, any salt that is highly soluble in water, cheap, abundant, and that gives highly conductive solutions, could be potentially used as electrolyte in ABFBs. The main constraint for such salt is that it must be soluble not only in neutral conditions (aqueous solution), but also in acidic and alkaline conditions. In case of co-ion leakage into acid or base compartment an insoluble salt would otherwise precipitate and cause scaling on the membranes (CEM or BPM). For the same reason, the solubility requirement also applies for the acid and base produced from the salt, as well as for the electrode rinse solution. Thus, multivalent ions that lead to precipitation of hydroxides (e.g., Ca^{2+}, Mg^{2+}, and Fe^{3+}), are not preferred as electrolytes in ABFBs [29,44–46].

The solubility limit of electrolytes in water is also directly connected to the battery storage capacity: A higher concentration of the acid and base solutions corresponds to a larger amount of energy stored in the battery. For example, considering the solubility limits of HCl and NaOH in water (12 and 19 M at 25 °C, respectively), the ABFB could theoretically be charged up to 12 M HCl-NaOH. Due to the solubility limit of NaCl in water being 6 M, the volume of NaCl solution should in such case be twice the volume of acid and base solutions in the battery. The Gibbs free energy for neutralization reaction of H$^+$ with OH$^-$ at room temperature is equal to the following:

$$\Delta G = -R \cdot T \cdot \ln K_{eq} = -1.99 \cdot 10^{-3} \cdot 298.15 \cdot \ln \frac{(1.0 \cdot 10^{-14})}{55.5} \cong -25 \text{ Wh/mol}_{water}, \tag{1}$$

where R is the gas constant, T is temperature, and K_{eq} is the water dissociation constant. Accordingly, the theoretical storage capacity for 1 m^3 of both HCl and NaOH solutions at 12 M concentration is ~300 kWh, which is remarkably high for a flow battery. However, uncontrolled mixing of such concentrated acid and base can be explosive, which raises new safety concerns when operating the battery. Furthermore, reaching such high acid–base concentrations in the battery is unpractical today, as for current commercial bipolar membranes 1 M acid–base concentration (theoretical energy density of ~25 kWh/m^3) is the maximum practical value without sacrificing permselectivity [23]. Likewise, commercial monopolar membranes would suffer from severe co-ion leakage at such high concentrations. In other words, the bottleneck for increasing the ion concentration in ABFBs lies in the selectivity of monopolar/bipolar membranes. In particular, if new membranes with improved selectivity in highly concentrated solutions will be available, the ABFB capacity could be increased remarkably. Notably, even with a 2 M acid–base concentration (for a NaCl-HCl-NaOH system), the ABFB power density would be comparable with vanadium redox flow batteries [47]. Overall, the chemistry of the ABFB electrolyte is still largely unexplored, as all the reported studies so far focus only on the NaCl-HCl-NaOH system, with NaCl concentration in the range of 0.1 to 1 M (see Table 1).

2.3. Stack Design

The design of an acid–base flow battery resembles the typical design of a 3-compartment BMED stack [23], where a series of CEM, BPM, and AEM is used to create the repeating unit, or "triplet" (Figure 2). Each membrane is separated from its adjacent membranes with net spacers to create the compartments for salt, acid, and base solutions, and promote mixing. A gasket, either integrated with the spacer or not, is placed between two membranes to make the cell leak-proof. The gasket materials should be able to withstand highly concentrated acid and base solutions (i.e., using fluoroelastomers such as PVDF, FKM, FFKM, etc.). Long-term battery operation is possible only if the cell is acid and base resistant, has no internal leakages, and no extensive co-ion transport.

To increase power generation, multiple triplets are piled in one stack. An extra monopolar membrane is then added to close the membrane pile, and the whole series is placed between electrodes to form an ABFB stack. The electrode compartments can be rinsed with a different solution, for example Na_2SO_4 [30], to avoid the production of Cl_2 at the anode (i.e., the oxidation product when using only NaCl solutions at the electrodes). The electrodes do not necessarily need to be of metal, but can also be of more environmentally friendly carbon material [25], in which case an electrode rinse solution with a redox couple must be used. Using such electrode rinse solution has the advantage of opposite electrode reactions occurring at the anode and cathode, meaning that by recirculating the rinse solution no net change of the chemical composition occurs, and the thermodynamic voltage of the electrode reactions is zero. In contrast to redox flow batteries, where electrodes or bipolar plates are needed between each repeating unit, a single ABFB stack contains only two end electrodes.

The feed flow in the triplets of the stack can be either parallel or serial. In case of parallel flow, all the compartments are simultaneously fed directly from the external electrolyte solution vessels. In case of serial flow, the feed solutions from the external vessel are fed into the first cell unit of the stack and the next cell unit receives the solution from the previous cell unit. At a given total flow rate, parallel flow has the advantage of lower pressure drop compared to serial flow, where the solutions flow through the entire stack. However, parallel flow has less homogenous flow as the compartment resistances throughout the stack might vary. In addition, it causes higher parasitic current losses, which has been described by Xia et al. [31,48]. When an external voltage is applied over the stack, protons and hydroxide ions migrate through the manifolds in opposite direction from one side of the stack to the other. This means that at the center of the stack the sum of their parasitic ion fluxes is the highest. The parasitic H^+ and OH^- ion fluxes, and the compensating H^+ and OH^- ion fluxes in the opposite direction at the stack center lead to water recombination inside the central BPMs. In practice, this results in self-discharge of the battery. As such, self-discharge phenomenon is limited only to the stack and does not include the feed storage vessels, reducing the manifolds size (by instance reducing the diameter) is necessary to decrease the effect of parasitic currents and self-discharge of the battery. In addition, the stack should have an optimal number of cell units to reduce the manifolds length and therefore the risk of self-discharging. Thus, instead of stacking together hundreds of cells, it can be more practical to connect together multiple stacks. These stacks should be hydraulically connected in parallel, so all stacks would work at the same concentration gradient, but electrically in series (to avoid reverse polarity instances between stacks). Another option to reduce parasitic currents is using independent hydraulic circuits for small blocks of triplets.

The ABFB energy and power ratings are independent of each other, as is the case with any flow battery. In practice, this means that the volume and concentration of the electrolyte solutions define the storage capacity of the battery, while the active area of the stack determines its power rating. This is a clear advantage for the ABFB in terms of scalability, especially when using an abundant and cheap salt (such as NaCl) as electrolyte. Moreover, the energy cost per kWh decreases with increasing battery capacity [49].

3. Simulation of Upscaled Scenarios for Technology Demonstration at kW–MW Scale

To evaluate the feasibility of the ABFB technology at larger scale, we performed a sensitivity analysis with a large-scale multi-stage (battery stacks hydraulically in series) ABFB under different scenarios, especially focusing on the effect of the number of battery stacks in series. The ABFB was simulated by using the process model previously developed and validated against laboratory experimental data by Culcasi et al. [35]. This model requires electrochemical and transport properties of the membranes as input parameters, and can predict the behavior and performance of the ABFB for different geometrical configurations and operating conditions as output.

The modelling tool has distributed parameters and is based on a hierarchical simulation strategy (multi-scale approach [35]), which can be briefly described as follows. The lowest level of simulation is represented by a single channel, where the model computes the physical properties of the electrolyte solutions, and computational fluid dynamics (CFD) correlations are used to calculate concentration polarization phenomena and pressure losses. The middle-low level is given by the "triplet" model (i.e., the repeating unit of the ABFB), which describes the mass balance and transport of water and ions (Nernst–Planck–Donnan approach) through both monopolar and bipolar membranes. Moreover, the "triplet" model evaluates also electrical variables as the resistance of the repetitive unit and the electromotive force (Nernst equation for multi-electrolyte solutions [50]). These first two modelling levels present a computational domain discretized along the channel length (where 30 intervals were sufficient to reach numerical accuracy at the simulated battery size). The middle-high level simulates the hydraulic and the electrical behavior of the stack. Note that the design features adopted in the present simulations were purposely chosen to minimize pressure drops and parasitic currents ($<{\sim}5\%$ of the gross power). Finally, the highest-scale model is able to simulate the external hydraulic circuit, including the dynamic mass balance in all the vessels, as well as pressure drops in the external piping. In the present work, we have assumed only once-through operations with negligible pressure drops in the external circuit. For the sake of brevity, only the main modeling results and the definition of output variables are reported in this work, while a more detailed mathematical description of the model can be found in Reference [35].

The sensitivity analysis was performed by using the input parameters shown in Table 2. In particular, different multi-stage operations were simulated by varying the number of battery stacks from 4 up to 17 (i.e., batteries hydraulically connected in series, fed with single pass through all 4 to 17 stacks, or "stages"). In all simulations, we assumed a battery state of charge (SOC) of 0% at 0.05 M and 100% at 1.00 M HCl, thus fixing the concentration targets at the outlet of the last stage for both charge and discharge. The electric current was tuned accordingly to achieve the concentration targets in all the simulated scenarios. For the sake of simplicity, the same electric current was used in all the sequential stages. Steady-state simulations were performed, assuming a single charge–discharge cycle.

The following model outputs are defined to characterize the battery performance. The gross power density (*GPD*, in watt per m^2 of total membrane area) of the k-th stage is calculated as

$$GPD_k = \frac{I_{ext}U_{ext,k}}{3\,N\,b\,L},\tag{2}$$

where I_{ext} is the electric current in the external circuit (note that I_{ext} changes with the number of stages, but is the same for all sequential stages), $U_{ext,k}$ is the predicted voltage on the external load (during discharge phase) or the power supply (charge phase), N is the number of repeating units (i.e., 10 triplets) of each stack, and b and L are the width and length of the membrane active area. Note that the power density in Equation (2) is normalized for the total membrane area, i.e., taking into account three membranes (CEM, AEM, and BPM) for each triplet, and it must be multiplied by a factor 3, to obtain a power density per square meter of BPM or triplet (as used in other works in the literature).

The power output of the k-th stack (P_k) is equal to $P_k = I_{ext}U_{ext,k}$. The resulting gross power density of the multi-stage ABFB system is given by the average GPD over all the N_s sequential stages:

$$\overline{GPD} = \frac{\sum_{k=1}^{N_s} GPD_k}{N_s}. \tag{3}$$

The gross power (P) of an ABFB plant of generic size (i.e., of a given N_p number of stacks hydraulically in parallel) is equal to the following:

$$P = N_p \sum_{k=1}^{N_s} P_k. \tag{4}$$

The discharge gross energy density (GED_d) is calculated as follows:

$$GED_d = \frac{\sum_{k=1}^{N_s} P_{k,d}}{Q_a}, \tag{5}$$

in which $P_{k,d}$ is the discharge power of the k-th stage and Q_a is the volumetric flow rate of acid solution. The discharge energy efficiency (η_d) is calculated as follows:

$$\eta_d = \frac{GED_d}{GED_{th,d}}, \tag{6}$$

in which $GED_{th,d}$ is the theoretical energy density, equal to 24 kWh/m^3 (normalized on the volume of one feed solution) at 1 M HCl-NaOH [25]. Finally, the main figure of merits characterizing the charge–discharge cycle of the ABFB are the coulombic efficiency (CE), voltage efficiency (VE), and round-trip efficiency (RTE), defined as follows:

$$CE = \frac{I_{ext,d}}{I_{ext,c}}, \tag{7}$$

$$VE = \frac{\overline{U_{ext,d}}}{\overline{U_{ext,c}}}, \tag{8}$$

$$RTE = \frac{I_{ext,d} \times \overline{U_{ext,d}}}{I_{ext,c} \times \overline{U_{ext,c}}} = CE \times VE, \tag{9}$$

in which, $\overline{U_{ext,d}}$ and $\overline{U_{ext,c}}$ are the average values of external voltage during discharge and charge, respectively, and $I_{ext,d}$ and $I_{ext,c}$ are the corresponding external currents (equal for all stages).

For comparison, we have also simulated the ABFB pilot plant as described in Section 4 (i.e., four hydraulically parallel stacks, with 56 triplets per stack). The main stack features and model input correspond to those reported in Table 2. In addition, each stack of the pilot plant is divided into eight blocks with independent hydraulic circuit. Accordingly, one seven-triplet block was simulated with an additional resistance of 6.5 Ω cm^2 attributed to the supplementary components used for dividing a stack into multiple blocks. The ABFB pilot was simulated as a single stage system operating in dynamic mode with the solutions continuously recirculated in the tanks, which were assumed with perfect mixing. The volumes of solutions per block were considered 62.5 L for the acid and base solutions, and 312.5 L for the salt solution (i.e., 500 L of acid and base solutions per stack, and 2500 L of salt solution per stack). All the performance parameters were calculated as time averages with corresponding changes with respect to the above definitions. Both charge and discharge phases were simulated with fixed current density of 100 A/m^2. The CE was calculated as the ratio between the total charge transferred in the discharge phase and that in the charge phase.

Table 2. Overview of main input parameters used in the sensitivity analysis. Adapted from Reference [35].

Geometrical Parameters of the Stack				
	Units	**Value**		
Spacer length, L	cm	50		
Spacer width, b	cm	50		
Spacer thickness	μm	475		
Membrane Properties				
	Units	**AEM**	**CEM**	**BPM**
Thickness	μm	130	130	190
Areal resistance	$\Omega\,cm^2$	4.0	3.5	5.0
H^+ diffusivity [a]	m^2/s	2.0×10^{-11}	0.7×10^{-11}	-
Na^+ diffusivity	m^2/s	1.6×10^{-11}	0.5×10^{-11}	-
Cl^- diffusivity	m^2/s	1.7×10^{-11}	0.6×10^{-11}	-
OH^- diffusivity	m^2/s	1.9×10^{-11}	0.6×10^{-11}	-
Fixed charge density	mol/m^3	5000	5000	-
Feed Conditions in the First Stage				
Feed Composition	**Units**	**Charge phase (0% SOC)**	**Discharge (100% SOC)**	
HCl in acid compartment	mol/m^3	50	1000	
NaCl in acid compartment	mol/m^3	250		
HCl in salt solution compartment	mol/m^3	10		
NaCl in salt solution compartment	mol/m^3	1000		
NaOH in base compartment	mol/m^3	50		
NaCl in base compartment	mol/m^3	250		
Fluid flow velocity	cm/s	1.0	1.0	
Electrode Compartments and Triplets				
	Units	**Value**		
Blank resistance [b]	$\Omega\,cm^2$	12		
Number of triplets (repeating units) per stack, N	-	10		

[a] Ion diffusivities estimated from experimental measurements with NaCl solutions and a two-chamber diffusion cell, assuming ion diffusivities inversely proportional to hydrated radius (Stokes–Einstein equation); [b] Blank resistance obtained experimentally by using $FeCl_2/FeCl_3$ as electrode rinse solution. SOC, state of charge.

3.1. Performance of Upscaled Multi-Stage ABFB System as Function of the Number of Stages

The main simulation results of the multi-stage ABFB are reported in Figure 4, as well as the applied values of current density, highlighting the effect of the number of stages on the process performance. In particular, Figure 4a shows the current density (i.e., current divided by the membrane active area) that was required to achieve the same (inlet–outlet) concentration difference for the acid solution, as a function of the number of ABFB stages. As expected, the required current density decreases as the number of sequential stages N_s increases, due to the accompanying increase in total membrane area (at fixed feed flow rate). However, the total electric current is not constant. As N_s increases, the total electric current increases from ~250 to ~270 A in charge, and it decreases from ~235 to ~221 A in discharge, thus indicating a decreasing current efficiency in both phases. This is simply caused by the increasing total membrane area, and consequently, increasing total mass transported by undesired fluxes of co-ions and water.

The predicted values of voltage over all stacks are reported in Figure 4b. During the discharge phase, the external voltage decreases along the stages due to the decreasing driving force (i.e., pH and concentration difference). Likewise, the voltage increases along the stages during charge, as a result of the increasing concentration difference. As the number of stages in series increases (Figure 4b), the voltage profiles along the stages tend towards the open circuit conditions (both for discharge and charge), as a result of the decreasing current density in the stacks (Figure 4a). The average gross power density ($\overline{GPD}$) of the stacks series (Figure 4c) exhibits a decreasing trend with the number of stages (N_s), similarly to the electric current. During the charge phase both the electric current and the average voltage decreases with

N_s, thus causing a more pronounced reduction of $\overline{GPD}$ than during the discharge phase. The predicted $\overline{GPD}$ values are in the range of 18.5–98.2 W/m^2 for charge, and 11.7–30.6 W/m^2 for discharge.

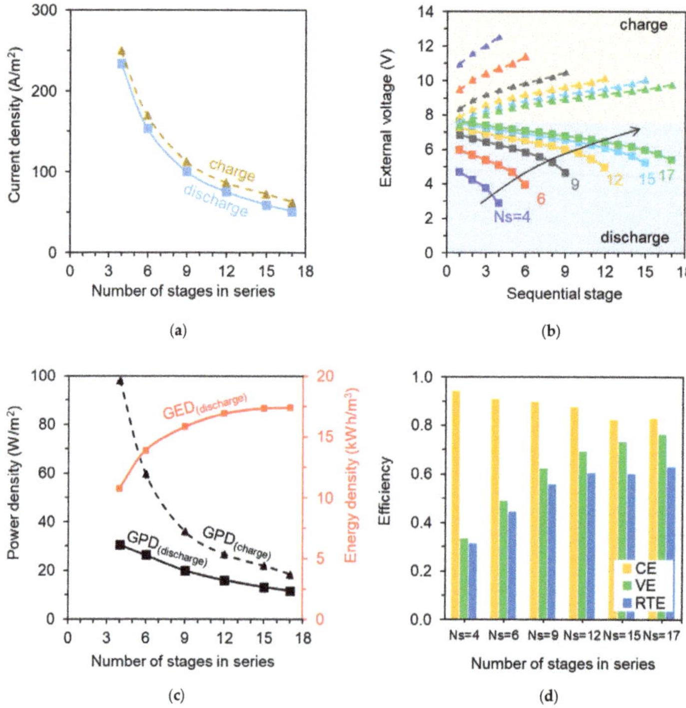

Figure 4. Model predictions of the effect of number of stages in series for an acid–base flow battery system: (**a**) charge and discharge current density (fixed equal for all stages); (**b**) profiles of charge/discharge external voltage at each sequential ABFB stage; (**c**) average charge/discharge gross power density (*GPD*) and (discharge) gross energy density (*GED*); (**d**) Coulombic efficiency (*CE*), voltage efficiency (*VE*), and round-trip efficiency (*RTE*). Each stage is simulated as an ABFB stack with a membrane active area of 0.5 × 0.5 m^2.

The discharge gross energy density (GED_d) is reported also in Figure 4c. The gross energy density is equal to 10.4 kWh/m$^3_{acid}$ for the four-stage system (discharge energy efficiency η_d of ~45%), and it increases with the number of stages due to the increasing cumulative power, reaching a plateau at GED_d = 17.4 kWh/m$^3_{acid}$ for the 17-stage system ($\eta_d \approx 72\%$). This indicates that the overall power ($\sum_{k=1}^{N_s} P_{k,d}$) increased at decreasing rate as a function of N_s, eventually reaching a maximum value. This is because the simulations at higher numbers of stages were based on lower current density values, thus distant from peak power conditions in discharge. The discharge electrical efficiency, defined as the power delivered to the external load divided by the total dissipated power (i.e., the sum of the internal and external power) was comprised between 47% (N_s = 4) and 79% (N_s = 17). Moreover, the less than proportional increase of the overall power with the number of stages justifies the reduction of $\overline{GPD}$ Figure 4d shows the efficiency of the process, in terms of coulombic efficiency (*CE*), voltage efficiency (*VE*), and round-trip efficiency (*RTE*). The *CE* decreases from 94% to 82% as N_s increases. Such high values of coulombic efficiency mean that the battery is characterized by high current efficiencies in both phases (charge/discharge). In particular, since parasitic currents in the manifolds are negligible in this case (due to small manifolds size and low number of triplets), the current efficiency is affected only by undesired fluxes (of co-ions and water) through the membranes. The voltage efficiency (*VE*) increases with the number of stages (from 33% up to 76%), as a result of the more homogeneous voltage distribution among sequential stages both during charge and discharge (Figure 4b). As overall result, the round-trip efficiency (*RTE*) trend is essentially determined by the increase of the *VE*, with *RTE* values in the range of 31–63% by increasing the number of stages.

The results of the pilot plant simulation are summarized in Table 3. The multi-stage ABFB and the pilot plant have different features, and thus a direct comparison of their performance parameters is not possible. However, it can be observed that the pilot plant (four stacks in parallel) performance is similar to the performance of a nine-stage ABFB (nine stacks in series), where similar values of current density (i.e., ~100 A/m^2) were applied.

Table 3. Main results predicted by the simulation of the pilot plant. The current density was fixed at 100 A/m^2 for both charge and discharge.

Quantity	Units	Value
Average external voltage in charge [1]	V	6.7
Average external voltage in discharge [1]	V	4.3
Average Gross Power Density in charge	W/m^2	32.0
Average Gross Power Density in discharge	W/m^2	20.6
Gross Energy Density in discharge	kWh/m$^3_{acid}$	18.0
Current Efficiency	-	86.8%
Voltage Efficiency	-	64.4%
Round Trip Efficiency	-	55.9%

[1] Average voltage over seven triplets.

Figure 5 shows the predicted output power (Equation (4)) as a function of the total membrane area (increased by increasing the number of parallel stages) for the cases of $N_s = 4$ and $N_s = 17$ sequential stages. With the current performance of commercial (monopolar/bipolar) membranes, a discharge power of 1 kW could be achieved by using a four-stage ABFB system with a total membrane area of about 30 m^2 (i.e., each stage equipped with 10 triplets and a membrane active area of 0.5 × 0.5 m^2). For a 1 MW power system, the required membrane area increased up to 30,000 m^2 (i.e., ~10,000 parallel triplets for each stage of the four serial stages). For the case of 17 stacks in series, the membrane area providing the same discharge power is increased by 2.6 times (6730 parallel triplets for 1 MW). Given the energy density of the system (Figure 4c), the corresponding flow rates to supply 1 MW power are 93 and 58 m^3/h (for each solution) for the four-stage and 17-stage ABFB systems, respectively. These modelling results highlight that, to reach ABFB applications on the kW–MW scale, further optimization studies should focus on plant design to reduce the membrane area and volume of solutions. The pilot plant system is also represented in Figure 5. With a fixed current density of 100 A/m^2, the plant consisting of four stacks with eight blocks of seven triplets each (total membrane area of 168 m^2) could provide a power output of 3.46 kW. Compared to the four-stage system, it requires ~60% more membrane area to deliver the same power.

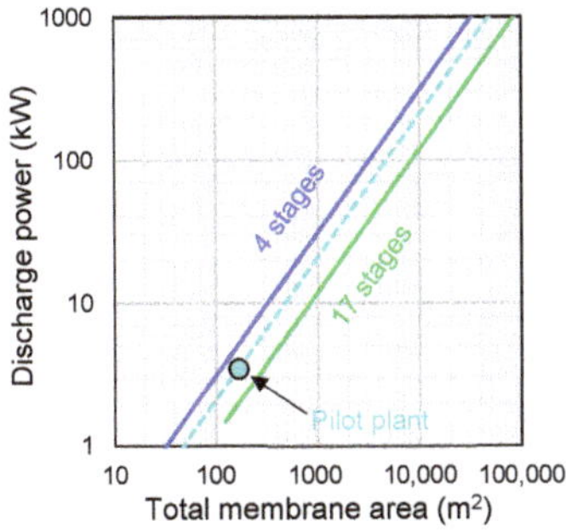

Figure 5. Discharge power as a function of the total membrane area for four-stage ($N_s = 4$) and 17-stage ($N_s = 17$) ABFB systems, and for the single stage pilot system (the highlighted point refers to the four-stack pilot plant). Each stage/stack is simulated with a membrane active area of 0.5 × 0.5 m^2.

4. Techno-Economic Assessment of First Pilot Plant and Technology Scale-Up

The development of ABFB technology is rapidly growing, and in 2020, a first pilot-scale demonstration plant with a target capacity of 1 kW/7 kWh was constructed by AquaBattery B.V. The pilot was recently installed in Pantelleria (a small Italian island in the Mediterranean Sea) and will be tested in the upcoming months as energy storage system at the local power plant, to provide seasonal storage during the high energy demand in summer months. In this section, we give a cost breakdown of the construction of this pilot plant, and we estimate a cost projection for a future 100× upscaled plant (i.e., a 100 kW/700 kWh full-scale unit), taking into account technology development and estimated prices in the next five years (2021–2025). Finally, we compare the cost of the ABFB with redox flow batteries.

The ABFB system, and hence the related costs, can be divided into three main subsystems: (i) a power subsystem, comprising all components related to the stack and the battery triplets (determining the power rating); (ii) an energy subsystem, comprising the volume of electrolyte solutions and associated components (determining the storage capacity); and (iii) the periphery, including all the auxiliary components that are not scale-dependent (e.g., battery management system). Location-dependent components needed for the battery integration into an existing built environment are not taken into consideration in this cost analysis.

The pilot plant consists of a four-stack (hydraulically parallel) ABFB system, where each stack contains 56 triplets with a membrane active area of 0.5×0.5 m^2 and spacer thickness of 475 μm, with a co-flow arrangement inside each stack. Therefore, each ABFB stack is designed to deliver a power output of 250 W, with an average power density of 6 W/m^2 membrane. Such low power density is due to the discharge current density of the pilot plant being limited to 30 A/m^2 during the preliminary testing phase for avoiding any risks of BPM delamination, and thus operating the stack far from the peak power condition (maximum of GPD-i curve). However, the plant is planned to operate closer to peak power at a later testing phase. The cost breakdown is summarized in Table 4. The total cost of membranes and spacers of this pilot plant was €63,000, which accounts for 74% of the total cost (€85,000) of the power unit. It should be noted that membrane and spacer costs are rather high (~€1000/m^2 of triplet), as the costs are related to a pilot-scale project, and are therefore affected by large R&D costs (e.g., due to the relatively small quantity of tailored non-commercial membranes that had to be produced). The energy storage capacity of the pilot (7 kWh), consisting of water storage tanks for the acid and base (2000 L each), salt (10,000 L tank for 4000 L solution), and electrode rinse solutions (25 L), cost €13,000. The periphery, containing the electrical cabinet and sensors among others, cost €39,000. Table 4 also shows the cost estimation for a First-of-a-Kind (FOAK) commercial unit.

Table 4. Capital expenditure (CAPEX) of current pilot-scale plant (1 kW/7 kWh), and cost estimation for First-of-a-Kind (FOAK) commercial unit (100 kW/700 kWh) in 2025.

CAPEX (Materials)	Demonstration Pilot (2020) 1 kW/7 kWh		FOAK Commercial Unit (2025) 100 kW/700 kWh	
Power subsystem (membrane, spacers, electrodes)	€85,000	€85,000/kW	€152,000	€1520/kW
Energy subsystem (storage tanks, electrolyte)	€13,000	€1900/kWh	€35,000	€50/kWh
Periphery (battery management systems, sensors)	€39,000	€39,000/unit	€22,000	€22,000/unit
System (total)	€137,000	€19,600/kWh	€328,000	€470/kWh

To estimate the costs for a 100 kW FOAK commercial unit, we consider a four-stack ABFB with each stack delivering a peak power output of 25 kW. This corresponds to upscaling the current demonstration pilot by a factor of 16 for the total membrane area and could be achieved by deploying four stacks with 224 triplets and membrane active area of 0.5×2.0 m^2. By upscaling the technology, the production cost of membranes and spacers will further decrease, leading to expected membrane/spacer costs in the range of €100/m^2 per triplet (i.e., for three membranes and spacers). Moreover, the development of bipolar membranes specifically tailored for the ABFB application (i.e., able to withstand high current densities under forward bias without delaminating) will allow to operate the ABFB at higher discharge current densities, and hence to increase the power density. Power density of 17 W/m^2 total membrane area has already been achieved at the lab scale [36], and simulations by our model predict power densities up to 30 W/m^2 (Section 3.1). Taking into consideration that the development of new membranes with lower resistance will decrease the internal resistance of the battery stack in the future, power densities in the range of 30–40 W/m^2 (100 W/m^2 triplet assumed for the cost calculations) could be realistically achieved. Accordingly, the power subsystem (only membranes) costs for the 100 kW commercial unit result in a total of €152,000 (Table 4). The substantial reduction in price compared to the pilot is a combination of the aforementioned factors: reduced membrane price, due to economies of scale and ongoing development of membrane manufacturing, and improved power density (6 to 30–40 W/m^2 in future). The costs of the energy subsystem (€35,000) include the use of low-cost water storage bags instead of water tanks for the feed solutions. The periphery costs account for additional €22,000.

Interestingly, according to Table 4, the power subsystem costs and energy subsystem costs for the FOAK commercial unit are €1520/kW and €50/kWh, respectively. These power unit costs are comparable with large-scale vanadium redox flow batteries mentioned in literature, i.e., in the range of ~€1000/kW for the power subsystem [51,52]. The energy unit costs are, however, significantly lower than for the competing vanadium-based flow battery technologies, which are in the range of €250–400/kWh [51–53]. Compared to other flow batteries, the ABFB is especially attractive for long-term storage, due to the relatively low cost of the energy subsystem.

Since the cost levels are dependent on both material and technology development, a crucial aspect that determines the costs of flow batteries is the cost of the active materials involved in the storage capacity (i.e., the redox couples in the case of redox flow batteries). Figure 6 presents an overview of the cost of active materials for several redox flow battery (RFB) technologies and the ABFB, and it shows that the resulting energy storage costs related to the storage medium for ABFB are far lower than for other flow batteries. In other words, since energy is stored in (abundantly available) salt solutions, the ABFB has the potential to be a truly sustainable and cost-effective battery technology for stationary energy storage at a large scale.

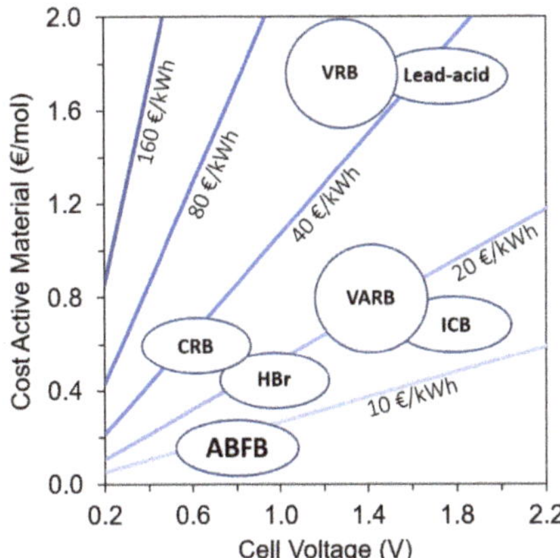

Figure 6. Energy costs (€/kWh) related to the active materials used for storage in the ABFB and various redox flow batteries. CRB, all-copper redox flow battery; HBr, hydrogen bromine redox flow battery; ICB, iron chromium redox flow battery; VARB, vanadium/air redox flow battery; VRB, vanadium redox flow battery.

5. Outlook and Perspectives

Acid–base flow batteries represent a promising technology to provide safe and sustainable storage in many applications. While the energy density of ABFB is comparable with PHS [25], it is relatively low compared to other batteries, and thus the potential applications of ABFBs are on a somewhat smaller scale (Figure 1). This, however, can be an advantage, as ABFBs can be brought closer to the consumer: Distributed batteries at the household level (i.e., "behind-the-meter" batteries) offer a larger amount of services and thus contribute to the electrical system the most. While an "in-front-of-the-meter" battery supports the grid, a behind-the-meter battery can additionally be used for customer services (e.g., for managing electric bills or, more importantly, for backup power). Although the ABFB technology is also suitable as an in-front-of-the-meter battery for secluded estates in regions with weak electric grids (for example small islands) and for solar or wind farms, its main application could be as a behind-the-meter battery for neighboring house groups or apartment complexes in areas with large share of renewables. The non-toxic and safe battery chemistry (which is based on NaCl and water) further justifies the suitability of ABFBs for household level, by eliminating many safety concerns relevant to other battery systems. In addition, as the battery chemistry is based on abundant salts, the costs for the active material of the battery are exceptionally low. To reduce the costs even further, ABFBs could in principle operate with natural saline waters (e.g., brackish water and seawater) or industrial waste waters as electrolyte solutions (though the use of natural feed waters would require additional pretreatment costs to avoid membrane scaling and fouling). Using industrial waste streams for energy storage via ABFBs can additionally reduce brine disposal costs and contribute to the development of zero liquid discharge processes in the future.

One of the main technical challenges of the ABFB technology is to improve the BPM performance, especially to increase its stability and selectivity under forward bias (i.e., water formation) conditions, which would allow for fast discharge of the battery. In fact, fast discharge of the ABFB is today unfeasible due to the limited performance of the available bipolar membranes; however, this is expected to change in the near future. The rapidly increasing attention towards BPMs over the last two decades has already led to major improvements in the BPM properties [23]. For instance, electrospun bipolar membranes with 3D junction have shown to be stable at unprecedentedly high values of current density [54], and could be therefore suitable for ABFB applications. A new class of BPMs might emerge in the future, with focus on high performance during both forward and reverse bias, and therefore with optimized properties for energy storage applications. This might lead to new insights also on the composition of the BPM junction, as there is no evidence that a good catalyst for water dissociation (i.e., reverse bias) can also catalyze the opposite process (water formation), since the behavior of BPMs under forward bias is essentially unexplored in the literature. Ultimately, bipolar membranes have become a quickly expanding market in recent years, with several new manufacturers (e.g., Xergy [55], Weifang Senya Chemical [56], and others [23]) on the market. Thus, it is justified to expect that the upscaling of ABFB technology (which has the bipolar membrane as its core element) will soon benefit from the R&D advances in the field.

Improvements in the properties of monopolar membranes will also benefit the process. Achieving low resistance, co-ion leakage, and water permeability at the same time is very challenging. However, research activities on the development of novel high-performing membranes are currently intense and promising. Moreover, the optimization of stack design and operating conditions will be crucial for the process competitiveness. In this regard, several flow layouts can be adopted, e.g., single or multi-stage/stack, batch (recirculation) or sequential [57]. Efforts on the component development and process optimization will lead to better performances, e.g., higher values of power density, RTE, and number of cycles. Finally, the abatement of the membrane cost is crucial for the techno-economic feasibility of ABFB systems and their actual implementation at a large scale.

6. Conclusions

The aim of this work is to present the state-of-the-art and latest developments of acid–base flow batteries (ABFBs) as a promising technology to provide seasonal energy storage by means of water dissociation with bipolar membranes. While still at the early stage of development, the ABFB technology is gaining attention and has been recently demonstrated at the pilot scale, for seasonal storage. To evaluate the feasibility of the ABFB technology at a larger scale, different scenarios of multi-stage (from 4 to 17) operation were simulated by fixing the same concentration target at the last stage. The results showed average values of discharge power density decreasing from 30.6 to 11.7 W/m^2 membrane, while the energy density increased from 10.7 to 17.4 kWh/m^3 acid. This means a total membrane area of ~30 to 86 m^2/kW discharge power and a volume of each electrolyte solution of 0.09–0.06 m^3/kWh. The round-trip efficiency increased from 31% to 63%. Improved membranes and optimized systems can lead to enhanced performances, thus reducing the electrolyte volumes and the membrane area. The simulation of the pilot plant showed results in line with the multi-stage systems, being potentially able to deliver a discharge power density of ~21 W/m^2 total membrane area.

As the performance of the ABFB is tightly connected to its core component, expected improvements of bipolar membranes in the near future will also directly improve the battery. In particular, tailored bipolar membranes that are able to withstand high current densities under forward bias (battery discharge mode) are needed to enable fast discharge of the ABFB. Despite all of the significant advancements on several battery technologies during the past decade, there is still a need for novel and sustainable energy storage systems for long-duration storage. In this regard, thanks to the safe and cost-effective battery chemistry, the acid–base flow battery can play a role towards the development of environmentally safe and sustainable energy storage systems.

Author Contributions: Conceptualization, R.P., M.T., and A.T.; methodology, L.G. and A.C.; software, A.C.; validation, L.G. and A.T.; formal analysis, A.C. and L.G.; investigation, A.C., J.P., and J.C.; resources, A.T.; data curation, L.G.; writing—original draft preparation, R.P., A.C., L.G., and J.P.; writing—review and editing, R.P., M.T., L.G., W.J.v.E., J.P., M.S., J.C., A.C., and A.T.; visualization, R.P., A.C., L.G., and W.J.v.E.; supervision, M.T., R.P., A.T., and L.G.; project administration, M.T., A.T., and D.A.V.; funding acquisition, J.P., A.T., D.A.V., and E.G. All authors have read and agreed to the published version of the manuscript.

Funding: This work was performed in the framework of the BAoBaB project (*Blue Acid/Base Battery: Storage and recovery of renewable electrical energy by reversible salt–water dissociation*). The BAoBaB project received funding from the European Union's Horizon 2020 Research and Innovation program, under Grant Agreement no. 731187 (www.baobabproject.eu).

Conflicts of Interest: W.J.v.E., J.C., D.A.V., and E.G. are affiliated with AquaBattery B.V. (The Netherlands), a company aiming to commercialize the ABFB technology. The other authors declare no conflict of interest.

References

1. International Energy Agency. Global Energy Review 2019: The Latest trends in Energy and Emissions in 2019. 2020. Available online: https://www.iea.org/reports/global-energy-review-2019 (accessed on 9 December 2020).
2. International Hydropower Association. Pumped Storage Tracking Tool. 2019. Available online: https://www.hydropower.org/hydropower-pumped-storage-tool (accessed on 30 August 2020).
3. International Renewable Energy Agency (IRENA). Renewable Energy Statistics 2020. 2020. Available online: https://www.irena.org/-/media/Files/IRENA/Agency/Publication/2020/Jul/IRENA_Renewable_Energy_Statistics_2020.pdf (accessed on 9 December 2020).
4. International Renewable Energy Agency (IRENA). Renewable Energy Capacity Statistics 2020. 2020. Available online: https://www.irena.org/-/media/Files/IRENA/Agency/Publication/2020/Mar/IRENA_RE_Capacity_Statistics_2020.pdf (accessed on 9 December 2020).
5. Hunt, J.D.; Byers, E.; Wada, Y.; Parkinson, S.; Gernaat, D.E.H.J.; Langan, S.; van Vuuren, D.P.; Riahi, K. Global resource potential of seasonal pumped hydropower storage for energy and water storage. *Nat. Commun.* **2020**, *11*, 947. [CrossRef] [PubMed]
6. Chen, H.; Cong, T.N.; Yang, W.; Tan, C.; Li, Y.; Ding, Y. Progress in electrical energy storage system: A critical review. *Prog. Nat. Sci.* **2009**, *19*, 291–312. [CrossRef]

7. Guo, Y.; Yang, Q.; Wang, D.; Li, H.; Huang, Z.; Li, X.; Zhao, Y.; Dong, B.; Zhi, C. A rechargeable Al–N 2 battery for energy storage and highly efficient N 2 fixation. *Energy Environ. Sci.* **2020**, *13*, 2888–2895. [CrossRef]

8. Comello, S.; Reichelstein, S. The emergence of cost effective battery storage. *Nat. Commun.* **2019**, *10*, 1–9. [CrossRef]

9. Boruah, B.D.; Mathieson, A.; Wen, B.; Feldmann, S.; Dose, W.M.; De Volder, M. Photo-rechargeable zinc-ion batteries. *Energy Environ. Sci.* **2020**, *13*, 2414–2421. [CrossRef]

10. International Renewable Energy Agency (IRENA). Electricity Storage and Renewables: Costs and Markets to 2030. 2017. Available online: https://www.irena.org/-/media/Files/IRENA/Agency/Publication/2017/Oct/IRENA_Electricity_Storage_Costs_2017.pdf (accessed on 9 December 2020).

11. Waldman, J.; Sharma, S.; Afshari, S.; Fekete, B. Solar-power replacement as a solution for hydropower foregone in US dam removals. *Nat. Sustain.* **2019**, *2*, 872–878. [CrossRef]

12. Albertus, P.; Manser, J.S.; Litzelman, S. Long-Duration Electricity Storage Applications, Economics, and Technologies. *Joule* **2020**, *4*, 21–32. [CrossRef]

13. Van Egmond, W.J.; Saakes, M.; Porada, S.; Meuwissen, T.; Buisman, C.J.N.; Hamelers, H.V.M. The concentration gradient flow battery as electricity storage system: Technology potential and energy dissipation. *J. Power Source* **2016**, *325*, 129–139. [CrossRef]

14. Bazinet, L.; Geoffroy, T.R. Electrodialytic Processes: Market Overview, Membrane Phenomena, Recent Developments and Sustainable Strategies. *Membranes* **2020**, *10*, 221. [CrossRef]

15. Gurreri, L.; Tamburini, A.; Cipollina, A.; Micale, G. Electrodialysis Applications in Wastewater Treatment for Environmental Protection and Resources Recovery: A Systematic Review on Progress and Perspectives. *Membranes* **2020**, *10*, 146. [CrossRef]

16. Jang, J.; Kang, Y.; Han, J.-H.; Jang, K.; Kim, C.-M.; Kim, I.S. Developments and future prospects of reverse electrodialysis for salinity gradient power generation: Influence of ion exchange membranes and electrodes. *Desalination* **2020**, *491*, 114540. [CrossRef]

17. Pawlowski, S.; Huertas, R.M.; Galinha, C.F.; Crespo, J.G.; Velizarov, S. On operation of reverse electrodialysis (RED) and membrane capacitive deionisation (MCDI) with natural saline streams: A critical review. *Desalination* **2020**, *476*, 114183. [CrossRef]

18. Pattle, R.E. Production of Electric Power by mixing Fresh and Salt Water in the Hydroelectric Pile. *Nature* **1954**, *174*, 660. [CrossRef]

19. Kingsbury, R.S.; Chu, K.; Coronell, O. Energy storage by reversible electrodialysis: The concentration battery. *J. Memb. Sci.* **2015**, *495*, 502–516. [CrossRef]

20. Aquabattery, B.V. Blue Battery Pilot Projects, Pilot I TheGreenVillage. 2018. Available online: http://aquabattery.nl/bluebatterypilotprojects/ (accessed on 29 October 2020).

21. Walther, J.F. Process for Production of Electrical Energy from the Neutralization of Acid and Base in a Bipolar Membrane Cell, US4311771A. 1980. Available online: https://patents.google.com/patent/US4311771A/en (accessed on 9 December 2020).

22. Frilette, V.J. Preparation and Characterization of Bipolar Ion Exchange Membranes. *J. Phys. Chem.* **1956**, *60*, 435–439. [CrossRef]

23. Pärnamäe, R.; Mareev, S.; Nikonenko, V.; Melnikov, S.; Sheldeshov, N.; Zabolotskii, V.; Hamelers, H.V.M.; Tedesco, M. Bipolar membranes: A review on principles, latest developments, and applications. *J. Memb. Sci.* **2020**, *617*, 118538. [CrossRef]

24. Sáez, A.; Montiel, V.; Aldaz, A. An Acid-Base Electrochemical Flow Battery as energy storage system. *Int. J. Hydrogen Energy* **2016**, *41*, 17801–17806. [CrossRef]

25. van Egmond, W.J.; Saakes, M.; Noor, I.; Porada, S.; Buisman, C.J.N.; Hamelers, H.V.M. Performance of an environmentally benign acid base flow battery at high energy density. *Int. J. Energy Res.* **2018**, *42*, 1524–1535. [CrossRef]

26. Emrén, A.T.; Holmström, V.J.M. Energy storage in a fuel cell with bipolar membranes burning acid and hydroxide. *Energy* **1983**, *8*, 277–282. [CrossRef]

27. Pretz, J.; Staude, E. Reverse electrodialysis (RED) with bipolar membranes, an energy storage system. *Ber. Bunsenges. Phys. Chem.* **1998**, *102*, 676–685. [CrossRef]

28. Zholkovskij, E.K.; Müller, M.C.; Staude, E. The storage battery with bipolar membranes. *J. Memb. Sci.* **1998**, *141*, 231–243. [CrossRef]

29. Kim, J.H.; Lee, J.H.; Maurya, S.; Shin, S.H.; Lee, J.Y.; Chang, I.S.; Moon, S.H. Proof-of-concept experiments of an acid-base junction flow battery by reverse bipolar electrodialysis for an energy conversion system. *Electrochem. Commun.* **2016**, *72*, 157–161. [CrossRef]

30. Xia, J.; Eigenberger, G.; Strathmann, H.; Nieken, U. Flow battery based on reverse electrodialysis with bipolar membranes: Single cell experiments. *J. Memb. Sci.* **2018**, *565*, 157–168. [CrossRef]

31. Xia, J.; Eigenberger, G.; Strathmann, H.; Nieken, U. Acid-Base Flow Battery, Based on Reverse Electrodialysis with Bi-Polar Membranes: Stack Experiments. *Processes* **2020**, *8*, 99. [CrossRef]

32. Veerman, J.; Post, J.W.; Saakes, M.; Metz, S.J.; Harmsen, G.J. Reducing power losses caused by ionic shortcut currents in reverse electrodialysis stacks by a validated model. *J. Memb. Sci.* **2008**, *310*, 418–430. [CrossRef]

33. Tang, A.; McCann, J.; Bao, J.; Skyllas-Kazacos, M. Investigation of the effect of shunt current on battery efficiency and stack temperature in vanadium redox flow battery. *J. Power Source* **2013**, *242*, 349–356. [CrossRef]

34. Culcasi, A.; Gurreri, L.; Zaffora, A.; Cosenza, A.; Tamburini, A.; Cipollina, A.; Micale, G. Ionic shortcut currents via manifolds in reverse electrodialysis stacks. *Desalination* **2020**, *485*, 114450. [CrossRef]

35. Culcasi, A.; Gurreri, L.; Zaffora, A.; Cosenza, A.; Tamburini, A.; Micale, G. On the modelling of an Acid/Base Flow Battery: An innovative electrical energy storage device based on pH and salinity gradients. *Appl. Energy* **2020**, *277*, 115576. [CrossRef]

36. Zaffora, A.; Culcasi, A.; Gurreri, L.; Cosenza, A.; Tamburini, A.; Santamaria, M.; Micale, G. Energy Harvesting by Waste Acid/Base Neutralization via Bipolar Membrane Reverse Electrodialysis. *Energies* **2020**, *13*, 5510. [CrossRef]

37. Kishino, M.; Yuzuki, K.; Fukuta, K. Bipolar Membrane, EP3444295A1. 2017. Available online: https://patents.google.com/patent/EP3444295A1/en (accessed on 9 December 2020).

38. Zabolotsky, V.; Utin, S.; Bespalov, A.; Strelkov, V. Modification of asymmetric bipolar membranes by functionalized hyperbranched polymers and their investigation during pH correction of diluted electrolytes solutions by electrodialysis. *J. Memb. Sci.* **2015**, *494*, 188–195. [CrossRef]

39. Simons, R. Preparation of a high performance bipolar membrane. *J. Memb. Sci.* **1993**, *78*, 13–23. [CrossRef]

40. McDonald, M.B.; Freund, M.S.; Hammond, P.T. Catalytic, Conductive Bipolar Membrane Interfaces through Layer-by-Layer Deposition for the Design of Membrane-Integrated Artificial Photosynthesis Systems. *ChemSusChem* **2017**, *10*, 4599–4609. [CrossRef] [PubMed]

41. Kang, M.S.; Choi, Y.J.; Lee, H.J.; Moon, S.H. Effects of inorganic substances on water splitting in ion-exchange membranes: I. Electrochemical characteristics of ion-exchange membranes coated with iron hydroxide/oxide and silica sol. *J. Colloid Interface Sci.* **2004**, *273*, 523–532. [CrossRef] [PubMed]

42. Strathmann, H.; Krol, J.J.; Rapp, H.J.; Eigenberger, G. Limiting current density and water dissociation in bipolar membranes. *J. Memb. Sci.* **1997**, *125*, 123–142. [CrossRef]

43. Blommaert, M.A.; Verdonk, J.A.H.; Blommaert, H.C.B.; Smith, W.A.; Vermaas, D.A. Reduced Ion Crossover in Bipolar Membrane Electrolysis via Increased Current Density, Molecular Size, and Valence. *ACS Appl. Energy Mater.* **2020**, *3*, 6. [CrossRef]

44. Mikhaylin, S.; Bazinet, L. Fouling on ion-exchange membranes: Classi fi cation, characterization and strategies of prevention and control. *Adv. Colloid Interface Sci.* **2016**, *229*, 34–56. [CrossRef]

45. Haddad, M.; Mikhaylin, S.; Bazinet, L.; Savadogo, O.; Paris, J. Electrochemical acidification of Kraft black liquor by electrodialysis with bipolar membrane: Ion exchange membrane fouling identification and mechanisms. *J. Colloid Interface Sci.* **2017**, *488*, 39–47. [CrossRef]

46. Wang, Y.; Huang, C.; Xu, T. Which is more competitive for production of organic acids, ion-exchange or electrodialysis with bipolar membranes? *J. Memb. Sci.* **2011**, *374*, 150–156. [CrossRef]

47. Kim, S. Vanadium Redox Flow Batteries: Electrochemical Engineering. In *Energy Storage Devices*; Demirkan, M.T., Attia, A., Eds.; IntechOpen Limited: London, UK, 2019.

48. Rubinstein, I.; Pretz, J.; Staude, E. Open circuit voltage in a reverse electrodialysis cell. *Phys. Chem. Chem. Phys.* **2001**, *3*, 1666–1667. [CrossRef]

49. Trahey, L.; Brushett, F.R.; Balsara, N.P.; Ceder, G.; Cheng, L.; Chiang, Y.-M.; Hahn, N.T.; Ingram, B.J.; Minteer, S.D.; Moore, J.S.; et al. Energy storage emerging: A perspective from the Joint Center for Energy Storage Research. *Proc. Natl. Acad. Sci. USA* **2020**, *117*, 12550–12557. [CrossRef]

50. Tanaka, Y. Chapter 7 Donnan Dialysis. In *Ion Exchange Membranes: Fundamentals and Applications*; Elsevier: Amsterdam, The Netherlands, 2007; pp. 495–503.

51. Minke, C.; Kunz, U.; Turek, T. Techno-economic assessment of novel vanadium redox flow batteries with large-area cells. *J. Power Source* **2017**, *361*, 105–114. [CrossRef]

52. Viswanathan, V.; Crawford, A.; Stephenson, D.; Kim, S.; Wang, W.; Li, B.; Coffey, G.; Thomsen, E.; Graff, G.; Balducci, P.; et al. Cost and performance model for redox flow batteries. *J. Power Source* **2014**, *247*, 1040–1051. [CrossRef]

53. Noack, J.; Wietschel, L.; Roznyatovskaya, N.; Pinkwart, K.; Tübke, J. Techno-economic modeling and analysis of redox flow battery systems. *Energies* **2016**, *9*, 627. [CrossRef]

54. Shen, C.; Wycisk, R.; Pintauro, P.N. High performance electrospun bipolar membrane with a 3D junction. *Energy Environ. Sci.* **2017**, *10*, 1435–1442. [CrossRef]

55. Xergy Inc. XION™ Composite Bipolar Membranes. 2020. Available online: https://www.xergyincstore.com/product-category/xion-composite-bipolar-membranes/ (accessed on 19 October 2020).

56. Weifang Senya Chemical, Ltd. The Bipolar (Anion/Cation) Exchange Membrane, (n.d.). Available online: http://ion-exchangemembrane.com/BipolarMembrane.html (accessed on 19 October 2020).

57. Liu, B.; Zheng, M.; Sun, J.; Yu, Z. No-mixing design of vanadium redox flow battery for enhanced effective energy capacity. *J. Energy Storage* **2019**, *23*, 278–291. [CrossRef]

Publisher's Note: MDPI stays neutral with regard to jurisdictional claims in published maps and institutional affiliations.

MDPI
St. Alban-Anlage 66
4052 Basel
Switzerland
Tel. +41 61 683 77 34
Fax +41 61 302 89 18
www.mdpi.com

Membranes Editorial Office
E-mail: membranes@mdpi.com
www.mdpi.com/journal/membranes